# The Rafter Book

## *A Modern Handbook for Roof Framers*

David Mahin McIntire

Joseph Publishing ■ Maui

*The Rafter Book* is available at your local bookstore, or may be ordered direct. An order form has been provided on the last page of the book for your convenience. $15.95 includes handling and book-rate postage. Add $2.00 for Priority Mail postage. Send check or money order to:

The Rafter Book
P.O. Box 1142
Haiku, HI 96708-1142

Dealer inquiries are invited. Discounts are available for schools and apprenticeship programs.

 Comments and suggestions are appreciated and will be acknowledged. Write to the address above.

**Library of Congress Catalog Card 95-79996**
**ISBN 0-9647784-0-8**

3rd Printing, 2000

*Printed in the United States of America*
*Book design by Sara Patton, Maui, HI*

# Contents

# *Foreword*

This book is written with the modern roof framing carpenter in mind. In addition to full length rafter tables for hip, valley, and common rafters for roof pitches from 1/4 in 12 to 25 in 12, much information is provided that simply is not available in other books. Cuts are given in degrees, for use with compound miter or radial arm saws, **and** with pairs of numbers for layout with framing square. Settings are given **in degrees for motor-driven saws (Skilsaws) for all cuts,** including:

1. Jack Rafters
2. Bevel (backing) cuts for Hip and Valley Rafter Chamfers
3. Hip and Valley Rafter tail cuts for roofs with square cut common rafters
4. Fascia cuts, including Hip and Valley miters for roofs with square tail cut common rafters
5. Cuts for Frieze Blocks against Hip and Valley Rafters
6. Hip and valley Roof Sheathing cuts
7. Special information for hip and valley cuts using plywood roof sheathing.

# 1. Definitions

**Span:** The width of the building, or section of building, to be spanned by the roof.

**Run:** One-half the span. The horizontal distance covered from the ridge to the outside of the building.

**Rise:** The vertical distance that the roof rises from the plate to the ridge.

**Pitch or Slope**: Rise in relation to run. Generally (and in this book) given in inches of rise to twelve inches of run. The pitch in Illustrations 1 and 2 is **6 in 12**, or 26$^1/_2$°.

**Overhang or Tail**: That portion of a rafter that extends outside the building.

**Projection**: The horizontal distance covered by the overhang. The run of the overhang.

**Seat Cut, or Bird's Mouth**: A notch or cutout near or at the end of a rafter, consisting of a plumb and a level cut, which allows the rafter to fit over a bearing wall.

**Building Line**: The outside edge of the framing, against which the notch of the Seat Cut fits.

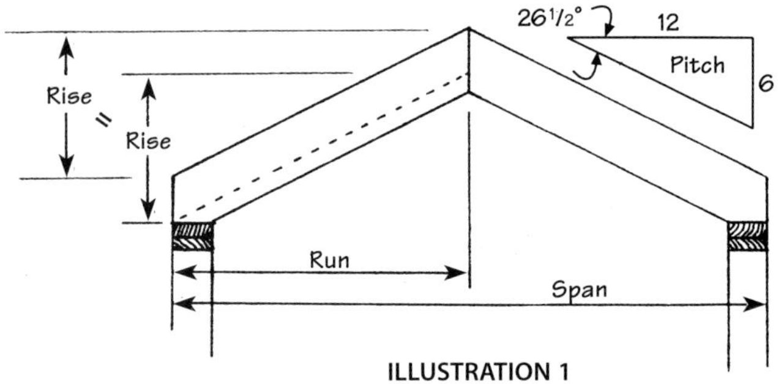


ILLUSTRATION 1

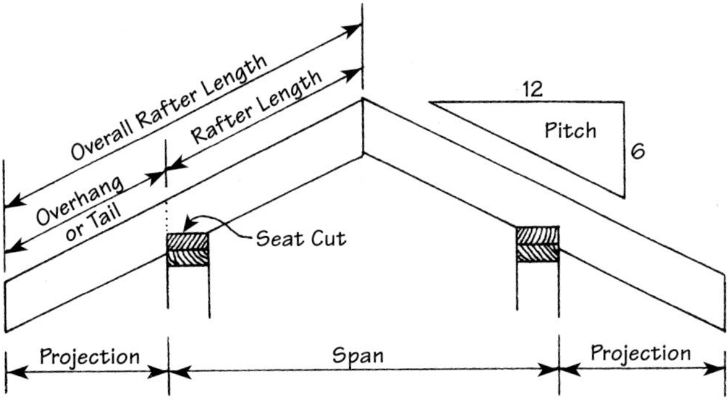


ILLUSTRATION 2

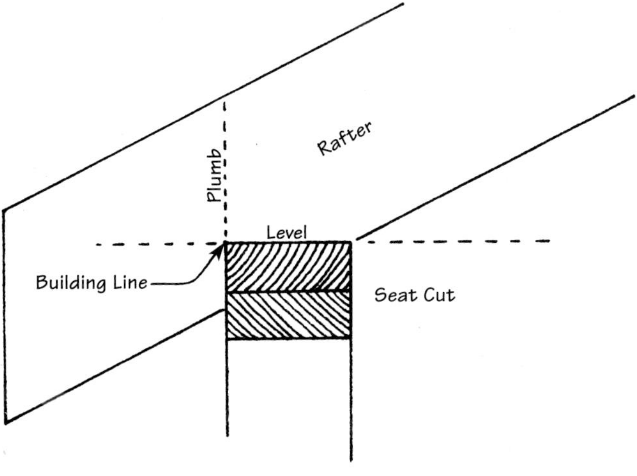


ILLUSTRATION 3

## 2. Laying Out Cuts With a Framing Square

### a. Plumb and Level Cuts on Common and Jack Rafters

In Illustration 4 below, an 8 in 12 roof pitch is taken as an example. A framing square laid as indicated on a board to be cut for a Common or Jack Rafter defines two cuts: the Plumb (vertical) Cut and the Level (horizontal) Cut. The Plumb Cut would be called in this book an **8 and 12 cut,** and the Level Cut a **12 and 8 cut. The number that goes first is the cut.** Note that on a roof steeper than 12 in 12, as the 18 in 12 roof pitch shown in Illustration 5, the first number is greater for the Plumb Cut, and less for the Level Cut.

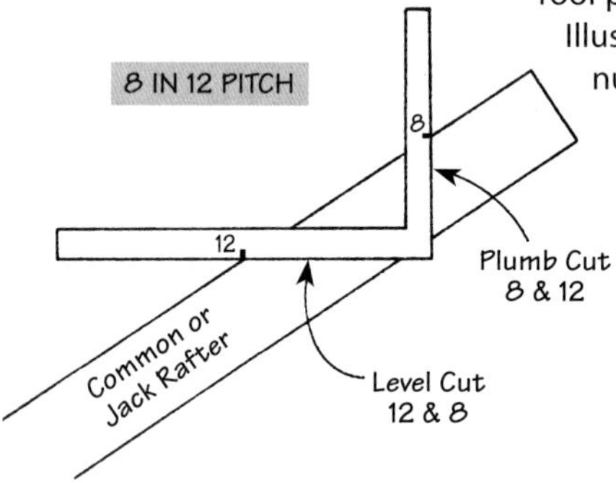


ILLUSTRATION 4

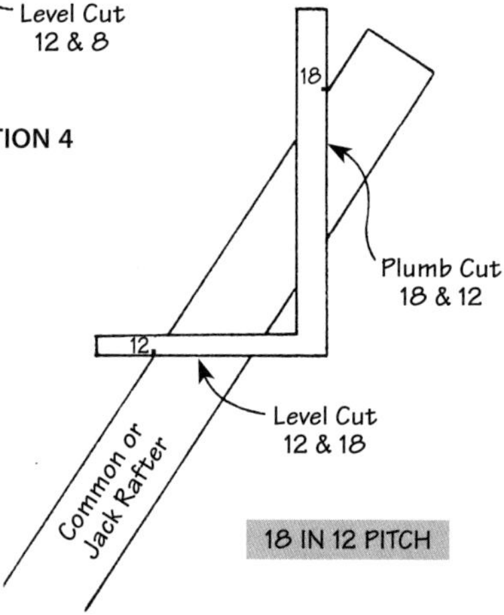


ILLUSTRATION 5

## b. Plumb and Level Cuts on Hip and Valley Rafters

With Hip and Valley Rafters, Plumb and Level Cuts are laid out **with the first number of the pitch, and 17**. The Plumb Cut on a Hip or Valley Rafter on an 8 in 12 pitch roof, for example, would be **8 and 17**, and the Level Cut **17 and 8**. (The reason for this is that the diagonal of a 12 × 12 square is 16.97, or approximately 17.)

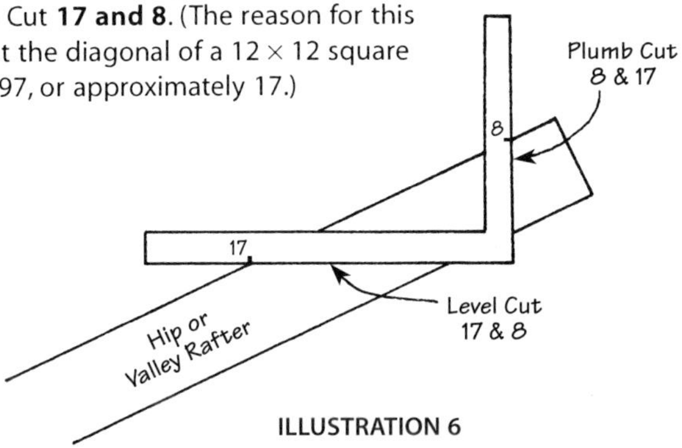


ILLUSTRATION 6

## 3. *Rafter Cuts*

### a. Types of Rafters

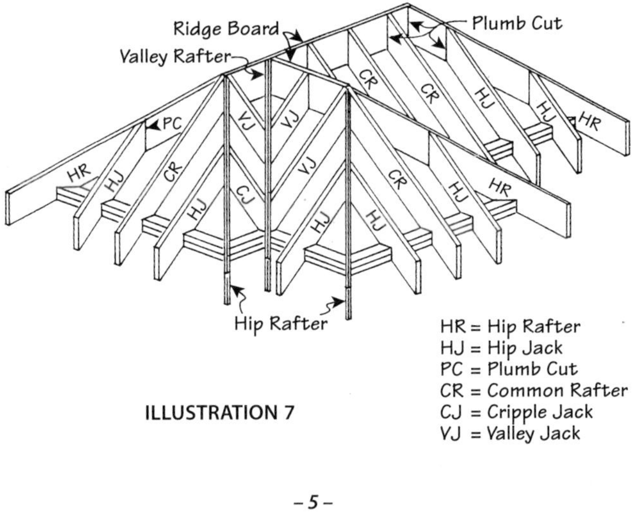


ILLUSTRATION 7

## b. Common Rafters

Common Rafters are those rafters that extend without interruption from the ridge (or ridge board) to the outside wall, or Building Line.

### ■ *Common Rafters, Top Cut*

The **Top Cut** of a Common Rafter is a **Plumb Cut** for that particular roof pitch, laid out with framing square, with saw set at 0°.

### ■ *Common Rafters, Seat Cut*

The Seat Cut consists of two cuts.

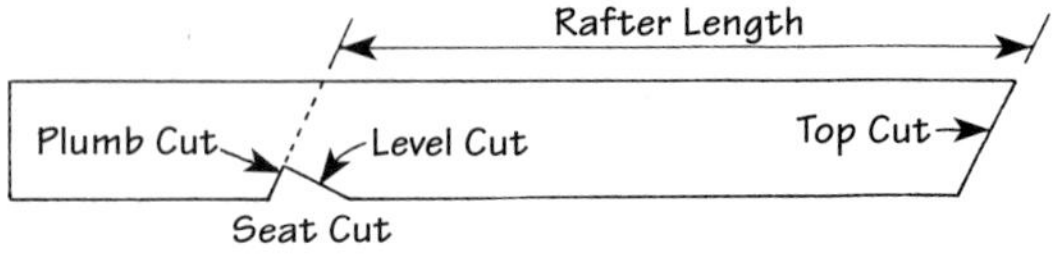


ILLUSTRATION 8

The Plumb Cut portion of the Seat Cut is laid out at the Rafter Length, and the Level Cut portion of the Seat Cut is laid out so that it is long enough for the Rafter to make full contact with the Bearing Wall. Set saw at 0° (see Illustration 9).

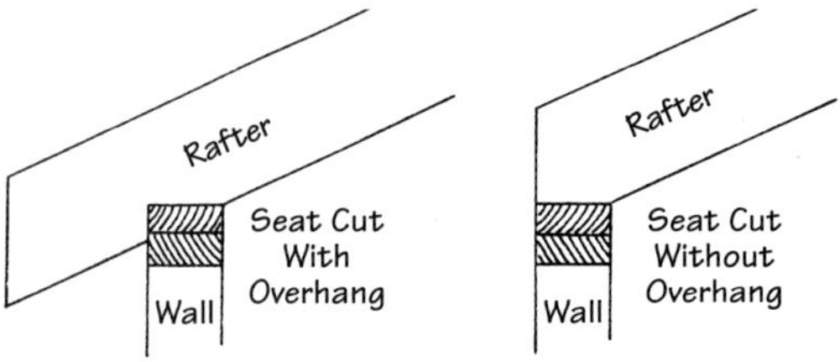


ILLUSTRATION 9

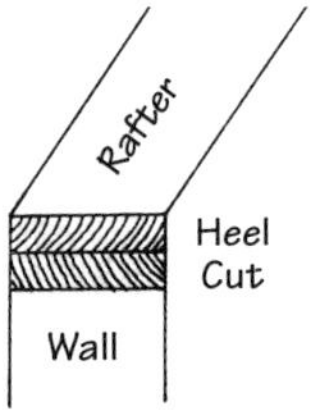


ILLUSTRATION 10

On a steep roof with a wide bearing wall, the plumb portion of the Seat Cut may be eliminated. This type of cut is called a Heel Cut (see Illustration 10).

## ■ *Common Rafters, Tail Cut*

Where the end of the rafter at the overhang is open to view, a wide variety of Tail Cuts may be used. In this book, however, only the two most common Tail Cuts will be discussed: the Plumb Cut and the Square Cut.

The Plumb Tail Cut is laid out with the framing square exactly as the Top Cut, so that the overall configuration of the rafter is that of a parallelogram. Set saw at 0°.

The Square Tail Cut is laid out as the name implies, the lower end of the rafter forming a right angle to the rafter. Set saw at 0°.

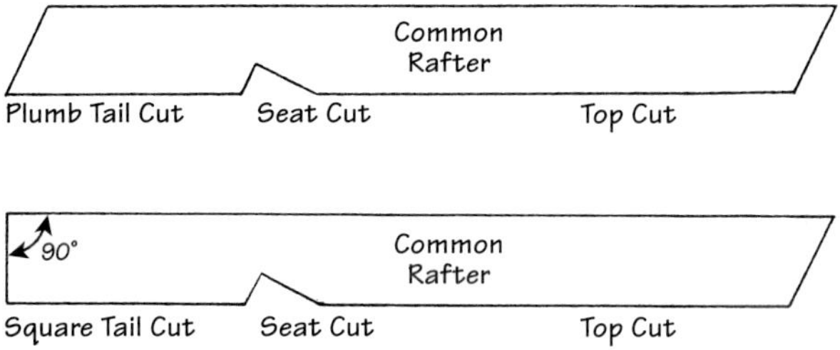


ILLUSTRATION 11

## c. Jack Rafters

Jack Rafters are those rafters that are interrupted between the ridge and the building line by a Hip or Valley Rafter. There are three types of Jack Rafters, as can be seen in Illustration 7 (page 5). The simplified drawing on the following page shows a similar roof as seen from above. The Top Cuts of Hip Jacks must be cut to a bevel to fit against the Hip Rafter. The Bottom Cuts of Valley Jacks must likewise be cut to a bevel to fit against the Valley Rafter. A Cripple Jack (there is one in this drawing) extends from a Hip Rafter to a Valley Rafter and must be cut to a bevel both top and bottom. The bevel cuts are called **Side Cuts** or **Cheek Cuts**. Jack Rafters are usually cut in pairs, for either side of the Hip or Valley Rafter to which they are to be fitted. **Side Cuts for Jack Rafters are always made with the saw set at 45°**.

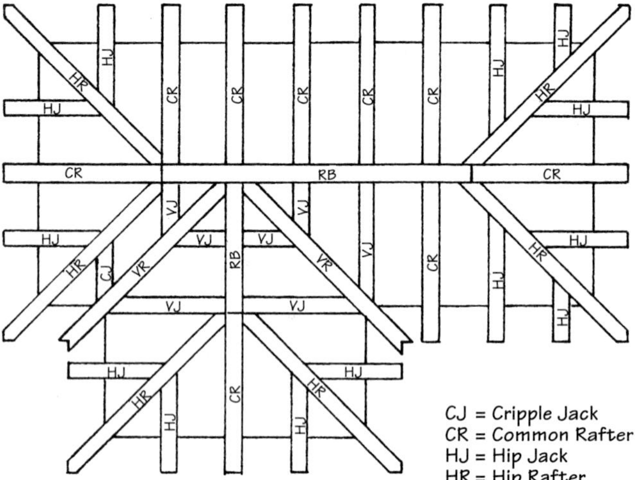


ILLUSTRATION 12

CJ = Cripple Jack
CR = Common Rafter
HJ = Hip Jack
HR = Hip Rafter
RB = Ridge Board
VR = Valley Rafter
VJ = Valley Jack

## ■ *Hip Jack Rafters*

Tail Cut and Seat Cut for Hip Jack Rafters are the same as for Common Rafters. The Top Cut for a Hip Jack Rafter is laid out as a plumb cut, then cut to a bevel, with the saw set at 45°.

## ■ *Valley Jack Rafters*

The Top Cut of a Valley Jack Rafter is the same as for a Common Rafter. The Bottom Cut is laid out as a plumb cut, then cut to a bevel, with the saw set at 45°.

## ■ *Cripple Jack Rafters*

Both Top and Bottom Cuts are laid out as plumb cuts, then cut to a bevel with the saw set at 45°.

## d. Hip and Valley Rafters

### ■ *Plumb and Level Cuts*

A Hip or Valley Rafter must travel the diagonal of the run of a Common Rafter to reach the same amount of rise, and so is not as steep. The diagonal of a 12 × 12 square is 16.97, which means that the **run** of a Hip or Valley Rafter for a 24' - 0" **span** is almost 17 feet. If the pitch of the roof is 6 in 12 (which means that the Common Rafter will rise 6 feet in 12 feet of run), then the pitch of the Hip or Valley Rafter will be **6 in 17**. This general rule applies to roofs of any pitch. **If the pitch of the roof is the number N in 12, then the pitch of the Hip or Valley Rafter is N in 17**. A plumb cut would be N and 17, and a level cut would be 17 and N.

### ■ *Top Cuts for Hip and Valley Rafters*

Hip and Valley Rafters meet the ridge board or other rafters at an angle and so must be cut to a bevel, like Jack Rafters. The Top Cuts are laid out with a framing square as Plumb Cuts; that is, **N and 17**, and **the saw is set at 45°** (Illustration 13 below). When the Hip or Valley Rafter meets the ridge board at its intersection with another ridge board or a Common Rafter, as in Illustration 14 below, the Top Cut must be laid out and cut on both sides of the Rafter to form the necessary point. Set saw at 45° for this cut also.

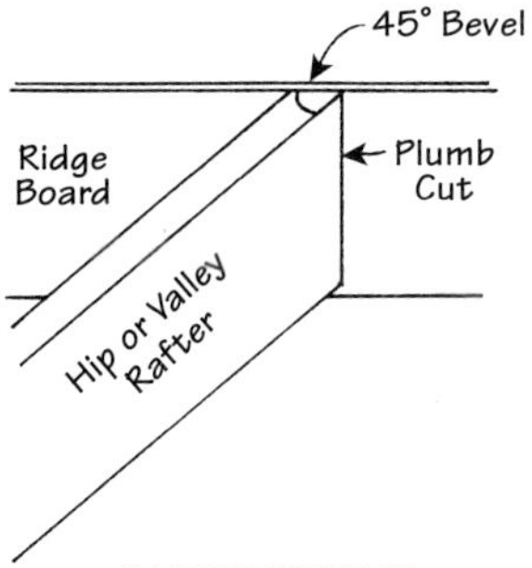


ILLUSTRATION 13
(SIDE VIEW)

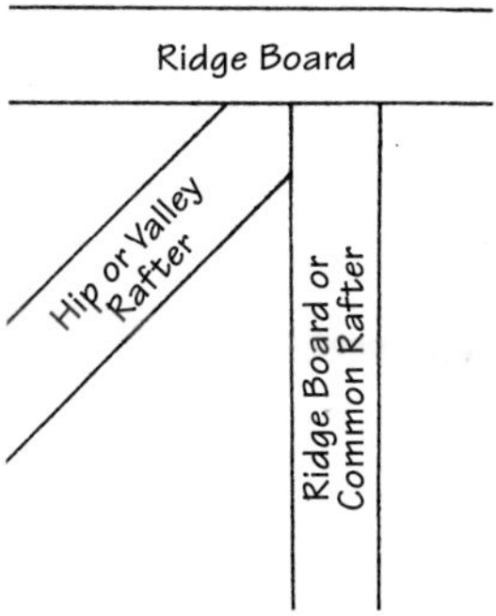


ILLUSTRATION 14
(TOP VIEW)

## Chamfer Cuts for Hip and Valley Rafters

It is generally necessary to cut a chamfer into the top of a Valley Rafter. It is often desirable as well to cut a corresponding chamfer along the top of a Hip Rafter, so that the roof sheathing will get full bearing on the Rafter. The chamfers will appear something like this in cross section:

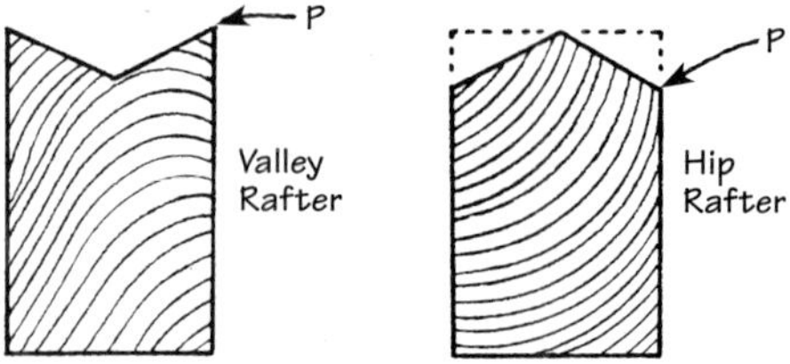


ILLUSTRATION 15

The saw settings to cut these chamfers will vary with the pitch of the roof. The setting is the same for Hip and Valley Rafters for any given pitch. These settings are given in the Tables under the heading, *Chamfer Cut, Hip or Valley Rafters.*

## Seat Cuts for Hip and Valley Rafters

Seat Cuts for Hip and Valley Rafters, like the Seat Cuts for other rafters, consist of plumb and level cuts. Remember, plumb cuts are laid out **N and 17**, and level cuts are **17 and N**.

Hip and Valley Rafters generally protrude from a building at a corner. In order for the Seat Cut to fit snugly against the corner of the bearing walls, the plumb cut must be cut to a point for a Valley Rafter or to a notch for a Hip Rafter. Illustrations 16 and 17 will clarify this. Rafters are shown upside down, as on sawhorses to be cut.

To make these cuts, lay out the Seat Cut on both sides of the rafter. Cut the level cut from either side with the saw set square (0° ), and cut the plumb cuts with the saw set at 45° . Finish the cuts with a chisel. To make these cuts safely, you will need two saws — one right-handed and one left-handed. Worm drive saws generally have the blade on the left, direct drive saws the blade on the right, so if you have one of each, you will be equipped to make these cuts.

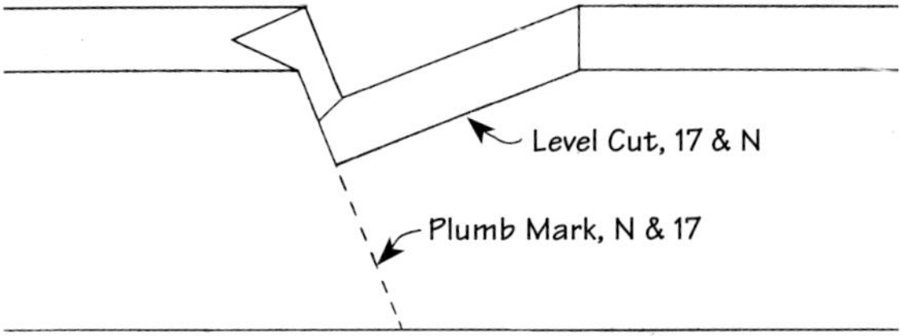


**ILLUSTRATION 16 - HIP RAFTER SEAT CUT**

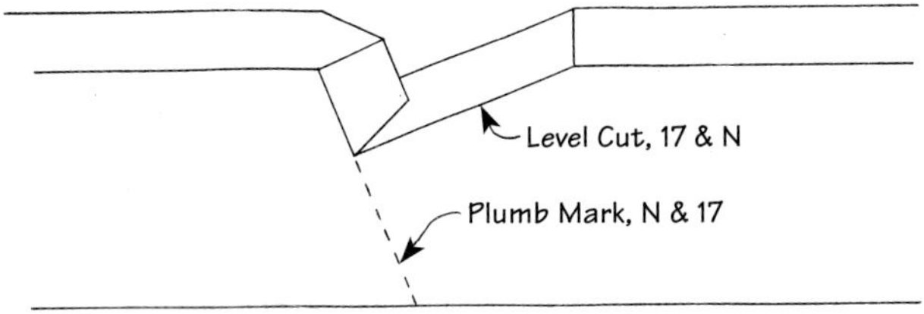


**ILLUSTRATION 17 - VALLEY RAFTER SEAT CUT**

The depth of the seat cut of a Hip or Valley Rafter is adjusted so that height of the rafter as it crosses the building line (the plumb mark) is equal to a Common or Jack Rafter at the same point. In Illustration 18 below, Height "f" would be adjusted to equal Height "d." (If the Hip Rafter is beveled, measure to Point "P," Illustration 15 at left.)

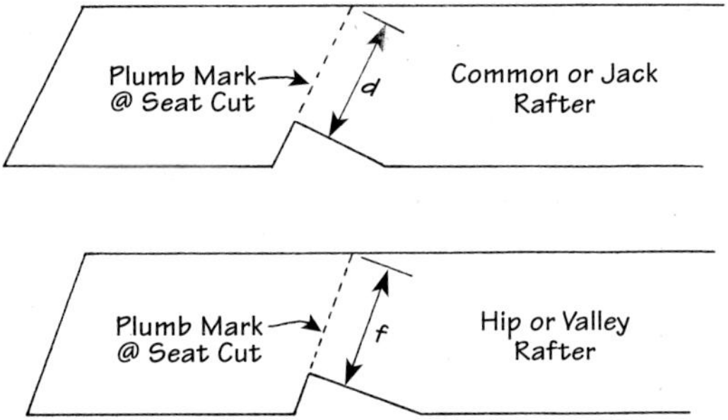


**ILLUSTRATION 18**

If the work is not going to show, Hip and Valley Seat Cuts may be cut as illustrated below. Figure depth of Seat Cut at lines indicated.

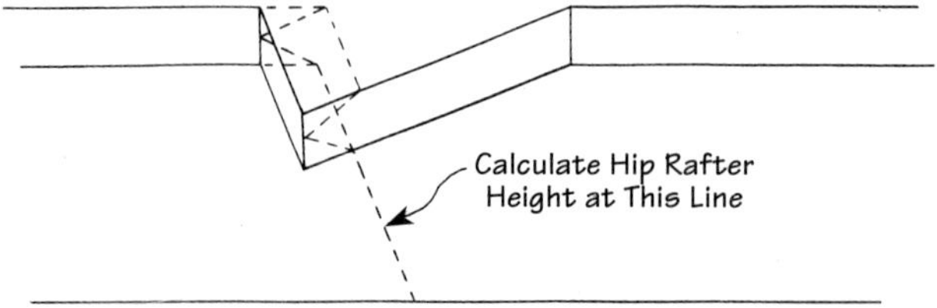


ILLUSTRATION 19 - HIP RAFTER SEAT CUT

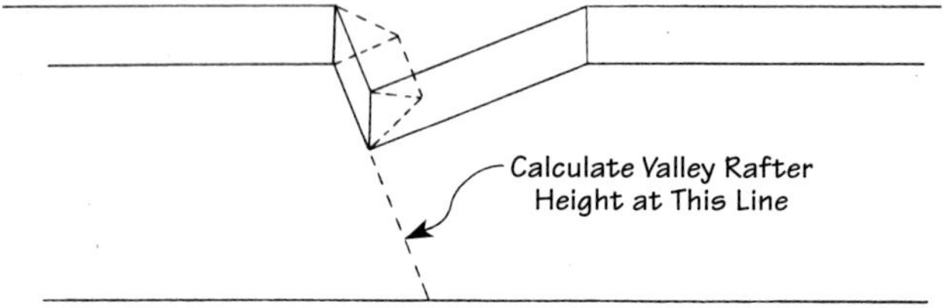


ILLUSTRATION 20 - VALLEY RAFTER SEAT CUT

## ■ *Tail Cuts for Hip and Valley Rafters*

When Hip and Valley Rafters are being cut to receive a Fascia Board, the Tail Cuts of the Hip Rafters will be cut in the form of points and the Tail Cuts of the Valley Rafters in the form of notches. This can be seen in Illustration 12 (page 8).

**With Plumb Tail Cut Common and Jack Rafters** (plumb cut being N and 12), the Hip and Valley Rafters will also be cut plumb at the tail (N and 17). The saw setting for the pointed or notched Tail Cuts is **45**°.

**With Square Tail Cut Common and Jack Rafters**, the angle of the Tail Cuts for Hip and Valley Rafters **and** the saw setting for the pointed or notched Tail Cuts vary with the pitch of the roof. Each roof pitch has its own Hip and Valley Tail Cut angle and its own saw setting for making the Tail Cut. These angles and saw settings are given in the Tables.

***Example:*** Look up 8 in 12 roof pitch in the Tables. Under the heading, *Tail Cut, Hip or Valley Rafters, with Square Cut Common Rafters,* find values: Framing Square: 7 13/16 & 24; set saw at 36°.

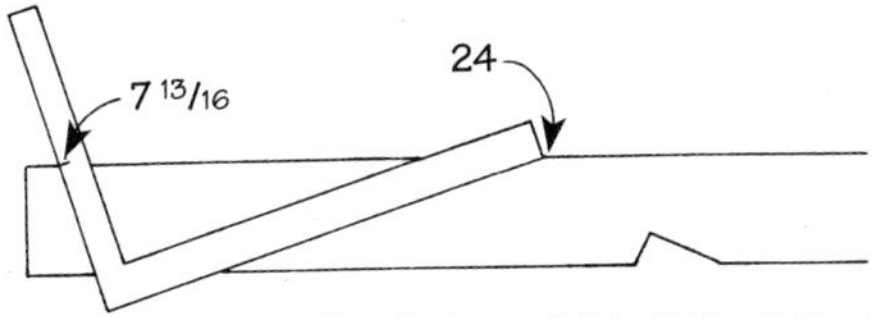


Tail Cut for Hip or Valley Rafter of 8 in 12 Roof Pitch with Square Cut Common Rafters

ILLUSTRATION 21

# 4. Calculating Rafter Lengths From the Tables

## a. Common Rafters

For each roof pitch, the Tables give the length of a rafter for a given span. If the roof is to have no overhang and no ridge board, the length of the rafter will equal the value given in the Tables.

***Example #1:*** The building width is 28' - 0". The roof pitch is 6 in 12. What is the length of a Common Rafter?

Turn to the pages for 6 in 12 pitch. Under the heading *Common Rafters,* in the column marked *Span: Feet,* find 28. Adjacent to it, in the column marked *Length: Feet, Inches,* see 15 - 7 13/16. The Common Rafter, then, is 15 feet, 7 13/16 inches (see Illustration 22).

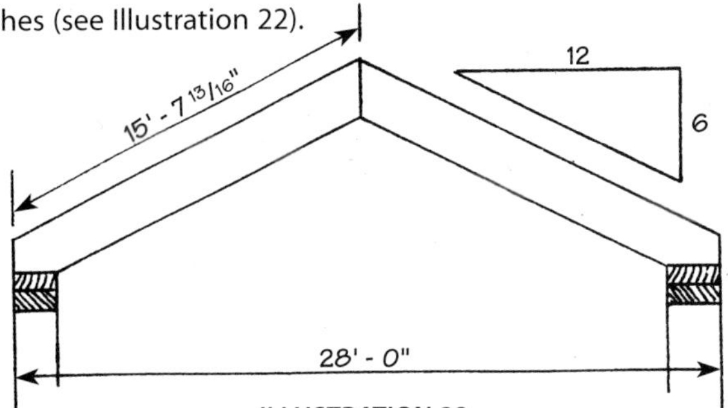


ILLUSTRATION 22

**When the span is not an even feet measurement** calculate the rafter length by adding the values for spans given in feet and spans given in inches.

***Example #2:*** The building width is 28' - 5 1/2". The roof pitch is 6 in 12. What is the length of a Common Rafter?

Turn to the pages for 6 in 12 pitch. Under the heading *Common Rafters,* in the column marked *Span: Feet,* find 28. In the column marked *Length: Feet, Inches,* see 15' - 7 13/16". Now, under the same heading, *Common Rafters,* find the column marked *Span: Inches.* In this column find 5 1/2. Adjacent to it, in the column marked *Length: Inches,* see 3 1/16. Add the two as follows:

| | | |
|---|---|---|
| Span 28' | *gives Common Rafter length* | 15' - 7 13/16" |
| Span 5 1/2" | *gives Common Rafter length* | 3 1/16" |
| Span 28' - 5 1/2" | *gives Common Rafter length* | 15' - 10 7/8" |

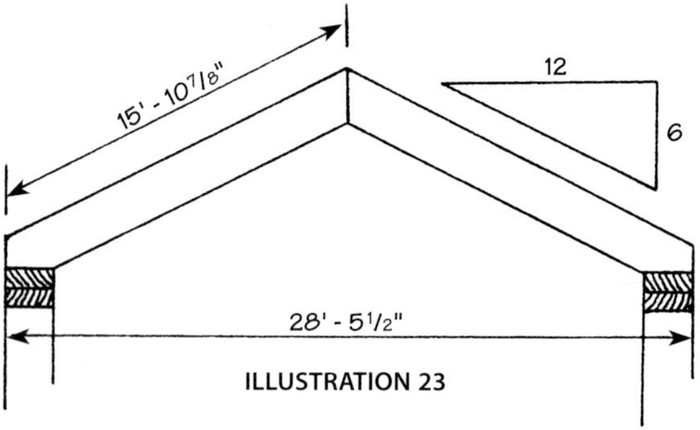

ILLUSTRATION 23

**Calculating the Rafter Length with a Ridge Board**: Generally, carpenters calculate the length of the rafter as if there were to be no ridge board, and then subtract, or cut off at the top cut, sufficient to allow for one-half the thickness of the ridge board (see Illustration 24).

When using these Tables, however, it is simpler to **subtract the thickness of the ridge board from the span. This will give true rafter length.**

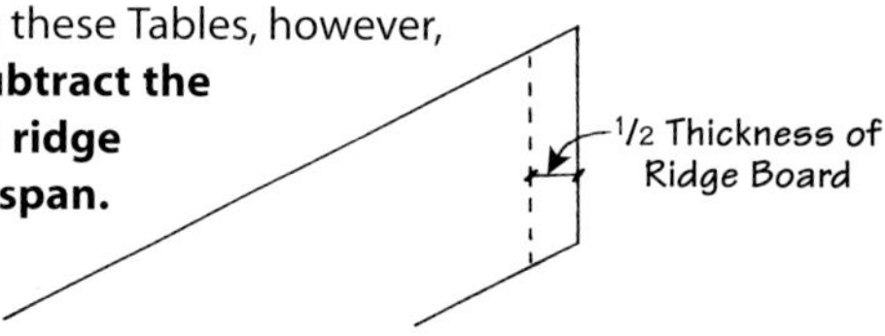

ILLUSTRATION 24

***Example #3:*** The span is 28' - 5 1/2." The roof pitch is 6 in 12. A ridge board 1 1/2" thick is to be used. What is the length of a Common Rafter?

| | | |
|---|---|---|
| | Building Width | 28' - 5 1/2" |
| *less* | Ridge Board | 1 1/2" |
| *equals* | Effective Span | 28' - 4 " |

| | | |
|---|---|---|
| Span 28' | *gives Common Rafter length* | 15' - 7 13/16" |
| Span 4" | *gives Common Rafter length* | 2 1/4" |
| Span 28' - 4" | *gives Common Rafter length* | 15' - 10 1/16" |

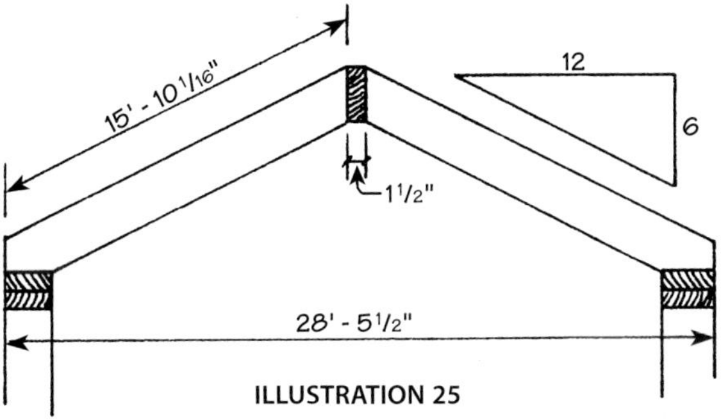


**ILLUSTRATION 25**

**If there is to be an overhang**, the projection of the overhang must be added twice (both sides of the building) to the span to find the total length of the rafter.

***Example #4:*** The building width is 28' - 5 1/2". The roof pitch is 6 in 12. The ridge board is 1 1/2" thick. The overhang is 2' - 6". What is the length of a Common Rafter?

| | | |
|---|---|---|
| | Building Width | 28' - 5 1/2" |
| *less* | Ridge Board | 1 1/2" |
| *plus* | 2 × Overhang | 5' - 0 " |
| *equals* | Effective Overall Span | 33' - 4 " |

| | | |
|---|---|---|
| Span 33' | *gives Common Rafter length* | 18' - 5 3/8" |
| Span 4" | *gives Common Rafter length* | 2 1/4" |
| Span 33' - 4" | *gives Common Rafter length* | 18' - 7 5/8" |

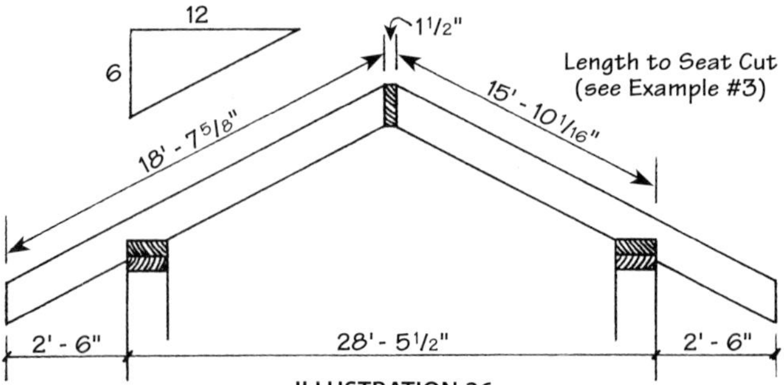


**ILLUSTRATION 26**

Note that the rafter length found in Example #3 (page 15) for the span of 28' - 5 1/2" gives the rafter length from the ridge to the plumb line of the seat cut.

| **Summary and General Rule:** | |
|---|---|
| | Span of building |
| *less* | Thickness of ridge board (if any) |
| *gives* | Length of rafter to plumb line of seat cut |
| | *– and –* |
| | Span of building |
| *less* | Thickness of ridge board (if any) |
| *plus* | 2 × projection of overhang |
| *gives* | Total length of rafter to tail cut |

**Note: To find the length of a rafter for a span greater than 50',** divide the span into portions 50' or less, find the lengths for the corresponding rafters, then add those lengths together.

***Example:*** The span of a 5 in 12 roof is 120' - 10". What will be the length of a Common Rafter?

| | | | |
|---|---|---|---|
| Span | 50' | *gives Common Rafter length* | 27' - 1" |
| Span | 50' | *gives Common Rafter length* | 27' - 1" |
| Span | 20' | *gives Common Rafter length* | 10' - 10" |
| Span | 10" | *gives Common Rafter length* | 5 7/16" |
| Span | 120' - 10" | *gives Common Rafter length* | 65' - 5 7/16" |

## b. Calculating Hip and Valley Rafter Lengths

The true length of a Hip or Valley Rafter is measured along the top center of the rafter, as shown by the dotted line in Illustration 27, below.

Note that the **long side** of the **Hip** and **either side** of the **Valley** are the **same length** as the center line. **This means that the cuts for Hip and Valley Rafters can be laid out in the same manner as the cuts for Common Rafters.** The value of this approach will be quickly apparent to anyone who has tried to calculate the distance from the top center of a Hip or Valley Rafter to the Seat Cut of the same.

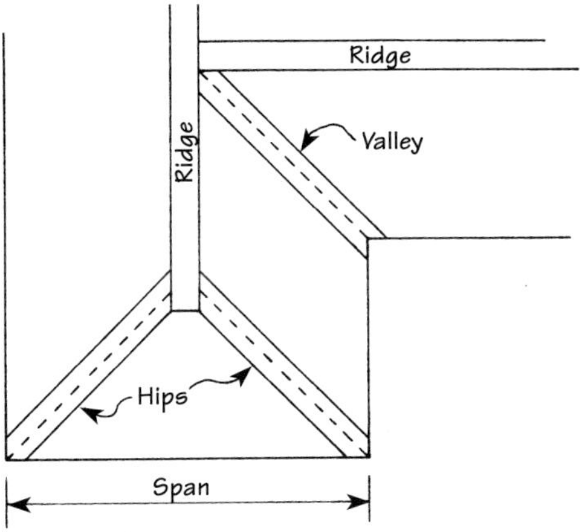


**ILLUSTRATION 27**

To find the length of any Hip or Valley Rafter, use the same rules as those for Common Rafters:

| | |
|---|---|
| | Span of building |
| *less* | Thickness of ridge board (if any) |
| *gives* | Length of rafter to plumb line of seat cut |

*– and –*

| | |
|---|---|
| | Span of building |
| *less* | Thickness of ridge board (if any) |
| *plus* | 2 × projection of overhang |
| *gives* | Total length of rafter to tail cut |

***Example:*** The span of an 8 in 12 roof is 24' - 0". The thickness of the ridge board is 1 1/2", and the projection of the overhang is 2' - 6". What will be the length of a Hip or Valley Rafter? (Note that they will be the same.)

| | | |
|---|---|---|
| | Span of building | 24' - 0" |
| *less* | Thickness of ridge board | 1 1/2" |
| *equals* | | 23' - 10 1/2" |

On the pages for 8 in 12 pitch, under the heading *Hip or Valley Rafters,* find the length of the rafter **to the plumb line of the seat cut**:

| | | |
|---|---|---|
| Span 23' | *gives Rafter length* | 17' - 11 3/4" |
| Span 10 1/2" | *gives Rafter length* | 8 3/16" |
| Span 23' - 10 1/2" | *gives Rafter length* | 18' - 7 15/16" |

**Overall length of Rafter to Tail Cut:**

| | | |
|---|---|---|
| | Span of building | 24' - 0" |
| *less* | Thickness of ridge board | 1 1/2" |
| *plus* | 2 × projection of overhang | 5' - 0" |
| *equals* | | 28' - 10 1/2" |

From the Tables:

| | | |
|---|---|---|
| Span 28' | *gives Rafter length* | 21' - 10 11/16" |
| Span 10 1/2" | *gives Rafter length* | 8 3/16" |
| Span 28' - 10 1/2" | *gives Rafter length* | 22' - 6 7/8" |

The Hip or Valley Rafters would measure as shown in Illustration 28.

***Note:*** With Common and Hip Jack Rafters having square Tail Cuts, the angle of the Hip or Valley Tail Cut — Angle "A" in Illustration 28 — will not be a plumb or a square cut. See page 12, *Tail Cuts for Hip and Valley Rafters.* The Projection of the Common and Hip Jack Rafters should be measured from the Building Line to the top corner of the Common or Hip Jack Tail Cut. Calculate Hip or Valley Tail Cut **length** along the top edge of the Hip or Valley Rafter to Point "P" in Illustration 28.

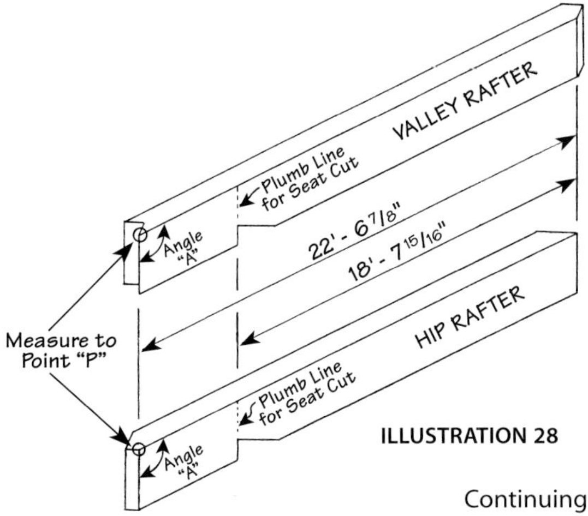


**ILLUSTRATION 28**

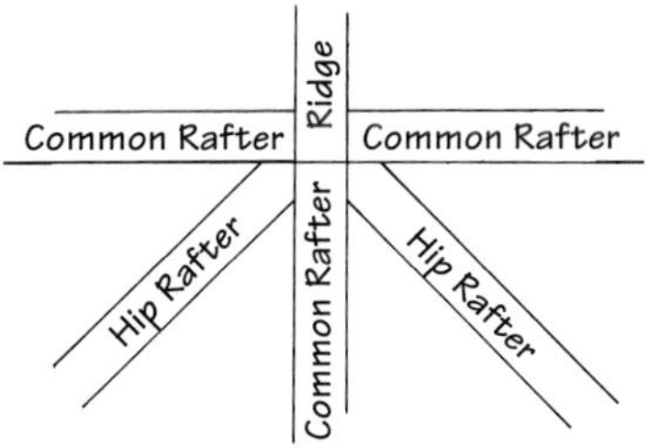


**ILLUSTRATION 29**

Continuing with the example given above, assume the Hip Rafters meet the ridge board in the manner shown in Illustration 29.

Dimensions are the same as in the preceding example. Lay out the cuts on the Hip Rafter as shown in Illustration 30. Double solid line is the first cut, to the length calculated from the Tables. Next step is to square across from the **short** points of each cut, top and bottom (single dotted line). Make a corresponding line on the other side of the rafter and complete the cut (double dotted line).

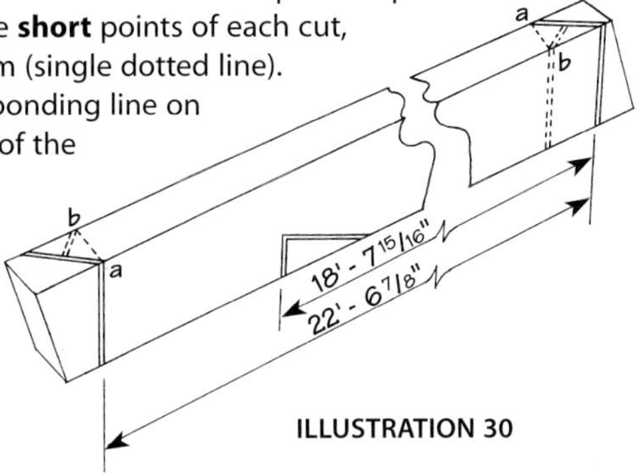


**ILLUSTRATION 30**

### c. Calculating Jack Rafter Lengths

For each roof pitch, the Tables give the **difference** in length for successive Hip or Valley Jack Rafters, for different layout spacings.

Measure Jack Rafters as illustrated below:

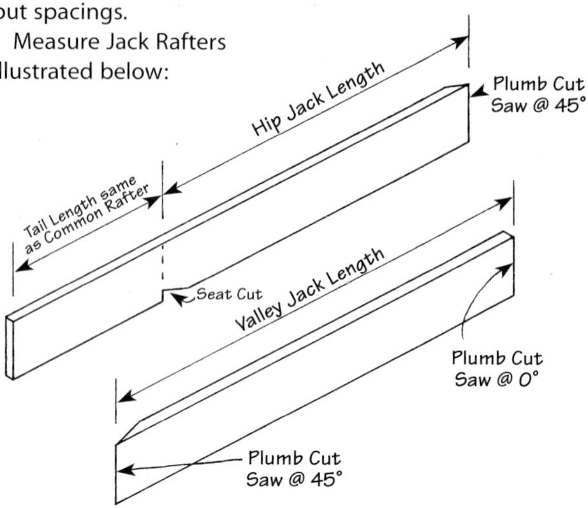


ILLUSTRATION 31

***Example:*** The shortest Hip Jack Rafter on an 8 in 12 pitch roof measures 2' - 4". The rafters are laid out at 24" intervals (24 inches "on center"). What will be the lengths of the next Jack Rafters against that Hip Rafter?

Turn to the pages for 8 in 12 pitch. Under the heading *Jack Rafters,* in the column marked *Spaced: Inches,* find 24. Adjacent to it, in the column marked *Difference: Inches,* find 28 7/8.

| | | | |
|---|---|---|---|
| | Shortest Jack Rafter 2' - 4" | = | 28" |
| *plus* | Difference in length | = | 28 7/8" |
| *equals* | Next Jack Rafter on 24" layout | = | 56 7/8" |
| *plus* | Difference in length | = | 28 7/8" |
| *equals* | Next Jack Rafter on 24" layout | = | 85 3/4" |
| *plus* | Difference in length | = | 28 7/8" |
| *equals* | Next Jack Rafter on 24" layout | = | 114 5/8" |
| | etc. | | |

**Optional:** If you need to **calculate** the length of any Jack Rafter, it may be done as follows:

In Illustration 32 below, **distance "a"** is the distance from the point at which the Hip Rafter crosses the building line to the far side of **any** Jack Rafter on that hip. **Distance "a"** will always equal the **run** of the Hip Jack Rafter "A." To find the length of Hip Jack Rafter "A," double distance **"a"** to find the corresponding span, then look in the Tables for that pitch and span for *Common Rafters*. The value given will equal the length of Hip Jack Rafter "A," measuring the length as in Illustration 31 at left.

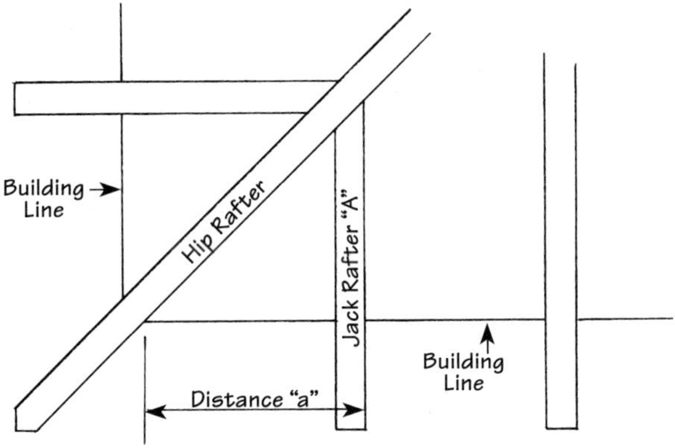


**ILLUSTRATION 32**

***Example:*** Roof pitch is 6 in 12. Distance "a" equals 6' - 2 1/4". What is the length of Hip Jack Rafter "A"?

Distance "a" 6' - 2 1/4" × 2 = 12' - 4 1/2"

Under *Common Rafters* for 6 in 12 roof pitch,

| | | |
|---|---|---|
| Span 12' | *gives Common Rafter length* | 6' - 8 1/2" |
| Span 4 1/2" | *gives Common Rafter length* | 2 1/2" |
| Span 12' - 4 1/2" | *gives Common Rafter length* | 6' - 11" |

**The length of Hip Jack Rafter "A" equals 6' - 11".**

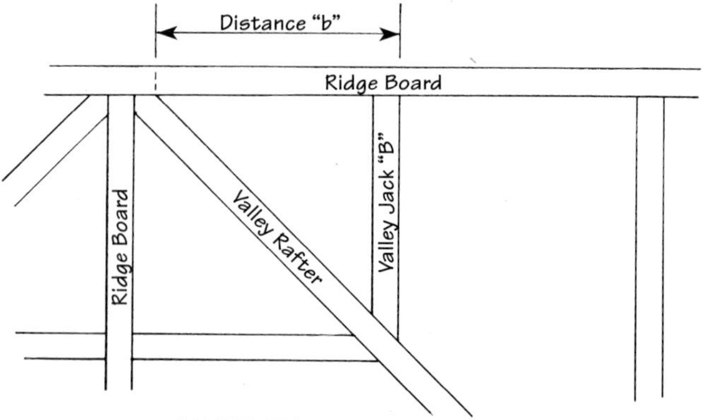


**ILLUSTRATION 33**

Calculations for Valley Jack Rafters are similar. In Illustration 33 above, **distance "b"** is the distance from the point at which the Valley Rafter meets the Ridge Board to the far side of any Valley Jack Rafter on that valley. Proceeding with **distance "b"** as with the example above will give the length of Valley Jack Rafter "B," measuring the length as in Illustration 31 (page 20).

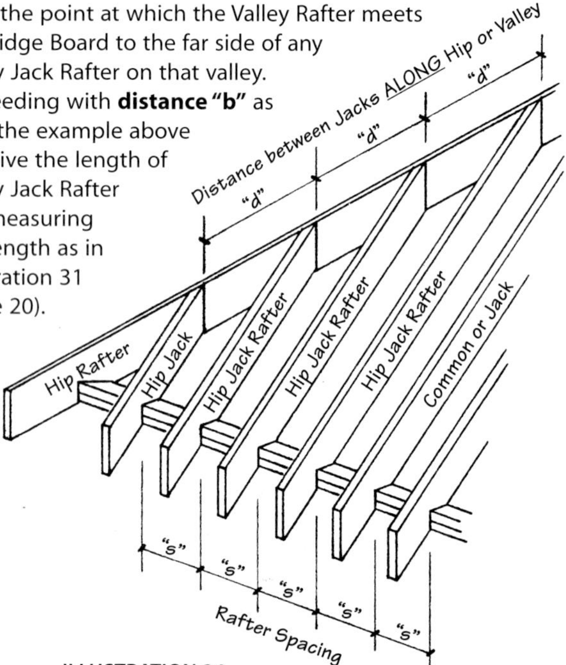


**ILLUSTRATION 34**

In Illustration 34 at left, **distance "s"** equals the Rafter Layout spacing. (For example, if "s" = 24 inches, the rafters are laid out 24 inches "**on center**.") **Distance "d"** represents the Jack Rafter layout spacing **along** the Hip or Valley Rafter. This is a distance that is often handy to know when actually putting together a hip roof. **This distance may be calculated very simply as follows:**

**Multiply rafter layout spacing times 2.** Look up Hip or Valley Rafter length using this value as the **Span.** The result will be distance "d."

***Example:*** An 8 in 12 pitch roof has rafters laid out 16" O.C. What is the Jack Rafter layout spacing (distance "d") **along** a Hip or Valley Rafter?

Layout Spacing, 16 inches × 2 = 32 inches = 2' - 8"

In the Tables, look for pages for 8 in 12 pitch. Under the heading, *Hip or Valley Rafters:*

| | | |
|---|---|---|
| Span 2' | *gives Rafter length* | 1' - 6 3/4" |
| Span 8" | *gives Rafter length* | 6 1/4" |
| Span 2' - 8" | *gives Rafter length* | 2' - 1 " |

The layout spacing **along** the Hip or Valley Rafter equals 2' - 1".

## 5. Fascia Cuts

This book will deal with Fascia Board cuts in two situations only:

1. Rafter Tails Cut Plumb, and
2. (Common) Rafter Tails Cut Square.

### a. Fascia Board Miters When Rafter Tails Are Cut Plumb

**When Rafter Tails are cut plumb**, the Fascia Board will appear as in Illustration 35 on page 24. **For Fascia Board miter cuts at Hip and Valley intersections, lay out cut square on face of Fascia Board, make cut with saw set at 45°.**

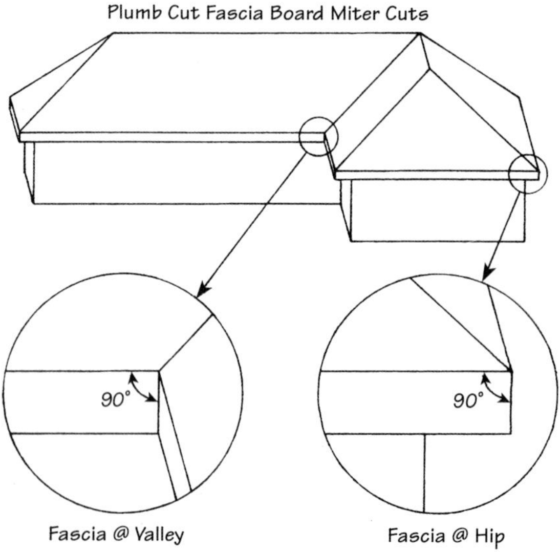


ILLUSTRATION 35

## b. Fascia Board Miters When Rafter Tails Are Cut Square

When Common and Hip Jack Rafter Tails are cut square, Fascia Boards meeting at Hips or Valleys form compound miters. The layout angle on the face of the Fascia Board and the saw setting will vary with the roof pitch. **These angles and saw settings are given in the Tables.** The angles are given for layout with framing square, with saw setting to cut bevel given in degrees. In addition, the setting is given in degrees for cuts to be made with compound miter or radial arm saws.

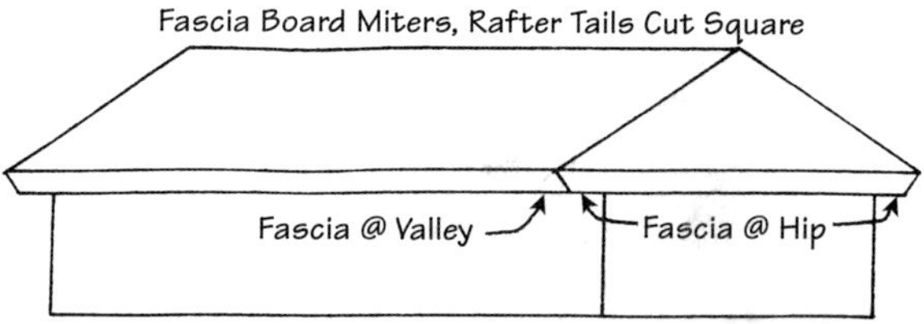


ILLUSTRATION 36

Illustration 36 is an elevation of an L-shaped house with a full hip roof with square cut rafter tails. The long point of the fascia board miter cut is on the bottom of the fascia at a valley, on the top at a hip.

***Example:*** Roof pitch is 8 in 12. Make miter cuts for Fascia Board joints at hips and valleys.

Turn in the Tables to 8 in 12 pitch. Under the heading *Fascia Board Miter Cut at Hip or Valley,* find values: Framing Square: 13 5/16 & 24; set saw at 36°. Lay out cut with framing square as shown:

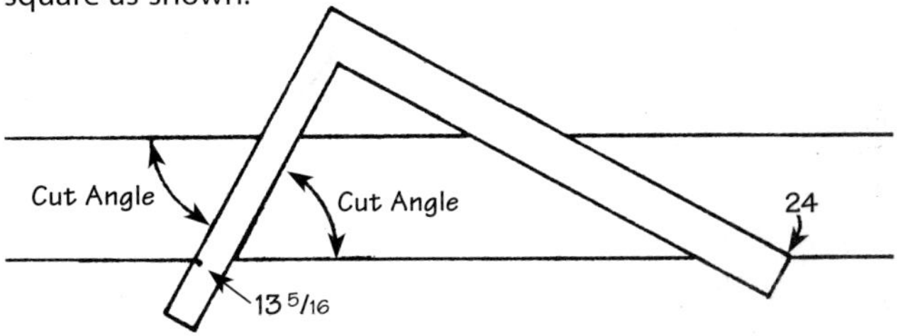


Set saw at 36° and make cuts as follows:

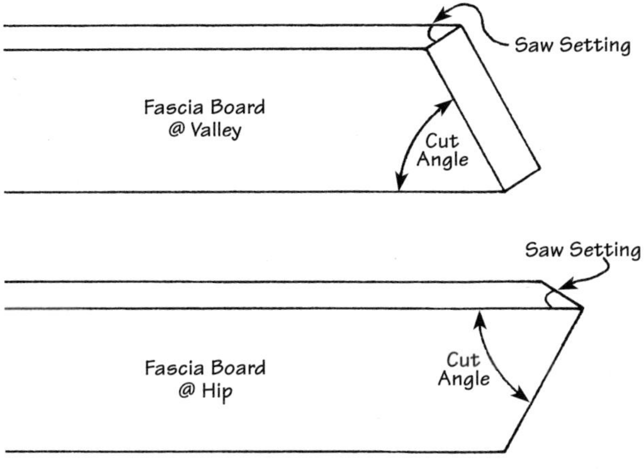


ILLUSTRATION 37

## 6. *Frieze Block Cuts Against Hip & Valley Rafters*

When Frieze Blocks between rafters are to be squared to the rafters as illustrated below, fitting them against a Hip or Valley Rafter requires a cut similar to that for a Fascia Board miter cut as explained in Section 5(b), pages 24–25. **These Frieze Block cuts and saw settings** will vary with the pitch of the roof, and **are given in the Tables**.

**The long point on a Frieze Block meeting a Hip Rafter will be on the face, top edge.**

**The long point on a Frieze Block meeting a Valley Rafter will be on the back, bottom edge.**

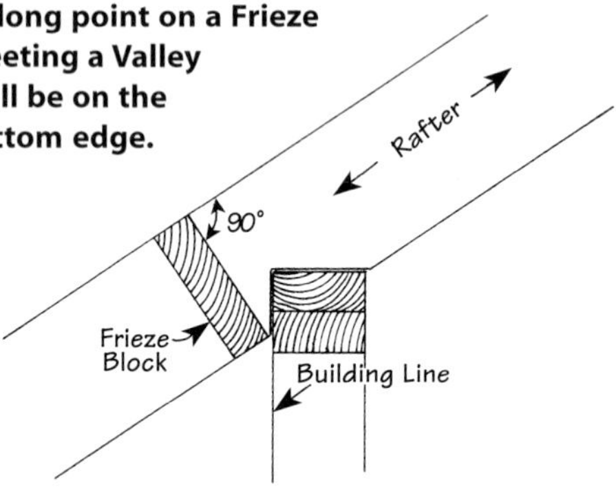


ILLUSTRATION 38

## 7. *Sheathing Cuts*

For roof sheathing to fit properly into a Valley or over a Hip, the end must be cut to an angle and a bevel. **These angles and saw settings** vary with the roof pitch, and **are given in the Tables** under the heading *Roof Sheathing at Hip or Valley.* Values are given for framing square layout, with saw settings in degrees. In addition, settings are given in degrees for cuts to be made with compound miter or radial arm saws. For any given roof pitch, angles and saw settings are the same for both Hips and Valleys.

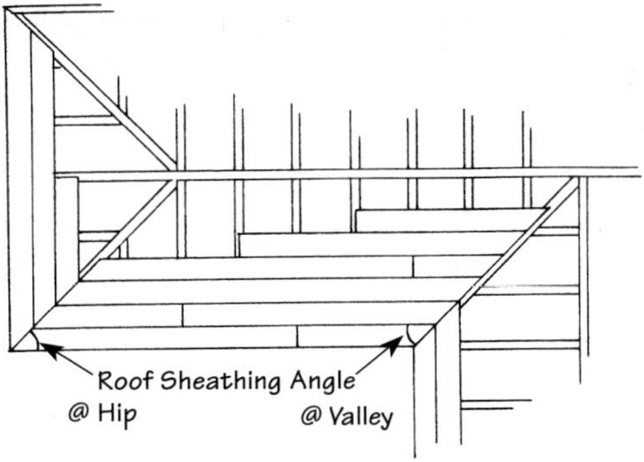


ILLUSTRATION 39 - ROOF SHEATHING ANGLE

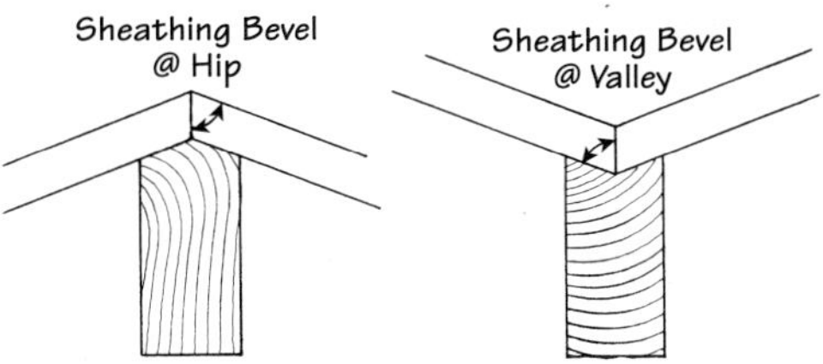


ILLUSTRATION 40 - ROOF SHEATHING BEVEL

**Plywood Roof Sheathing Cuts**: When 48" wide plywood roof sheathing is being used, the proper angle may be found by measuring along one side a given distance.

**Distance "d"** is given for each roof pitch under *Roof Sheathing at Hip or Valley.* Measure **distance "d"** along edge of plywood (top edge for Hip, bottom edge for Valley), find **point "a,"** snap line to nearer opposite corner, cut to same bevel as given for roof sheathing.

(Note: Distance "d" is always less than 48".)

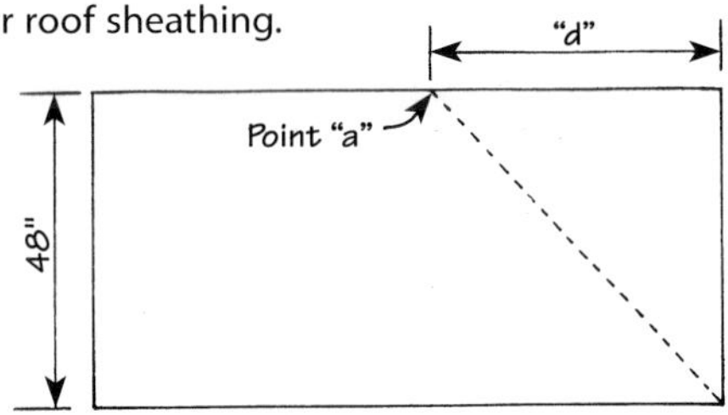


ILLUSTRATION 41

# 1/4 in 12 – 1 1/4°

## *Cuts and Saw Settings*

**Tail Cut, Hip or Valley Rafters,** with Square Cut Common Rafters:
Framing Square: **3/8 & 24**; set saw at **45**°
(Compound Miter or Radial Arm Saws: set bevel at 45°, miter at 3/4°)
Explanation: Section 3(d), pages 12–13.

**Chamfer Cut, Hip or Valley Rafters:**
Set saw at **3/4°**. Explanation: Section 3(d), page 10.

**Fascia Board Miter Cut at Hip or Valley:**
Framing Square: **1/2 & 24**; set saw at **45°**
(Compound Miter or Radial Arm Saws: set bevel at 45°, miter at 1 1/4°)
Explanation: Section 5, pages 24–25.

**Frieze Block against Hip or Valley Rafter:**
Framing Square: **1/2 & 24**; set saw at **45°**
(Compound Miter or Radial Arm Saws: set bevel at 45°, miter at 1 1/4°)
Explanation: Section 6, page 26.

**Roof Sheathing at Hip or Valley:**
Framing Square: **12 & 12**; set saw at **3/4°**
(Compound Miter or Radial Arm Saws: set bevel at 3/4°, miter at 45°)
Distance "d" = **48"**. Explanation: Section 7, pages 26–27.

## *Calculating the Length of a Rafter*

**General Rule:**

| | | |
|---|---|---|
| | | Span of building |
| | less | Thickness of ridge board (if any) |
| | gives | Length of rafter to plumb line of seat cut |
| *and* | | Span of building |
| | less | Thickness of ridge board (if any) |
| | plus | 2 × projection of overhang |
| | gives | Total length of rafter to tail cut |

Explanation: Section 4, pages 13–19.

*Jack Rafters*

| SPACED *Inches* | DIFFERENCE *Inches* | SPACED *Inches* | DIFFERENCE *Inches* | SPACED *Inches* | DIFFERENCE *Inches* |
|---|---|---|---|---|---|
| 1/4 | 1/4 | 15 | 15 | **32** | **32** |
| 1/2 | 1/2 | **16** | **16** | 33 | 33 |
| 3/4 | 3/4 | 17 | 17 | 34 | 34 |
| 1 | 1 | 18 | 18 | 35 | 35 |
| 2 | 2 | 19 | 19 | **36** | **36** |
| 3 | 3 | 20 | 20 | 37 | 37 |
| 4 | 4 | 21 | 21 | 38 | 38 |
| 5 | 5 | 22 | 22 | 39 | 39 |
| 6 | 6 | 23 | 23 | 40 | 40 |
| 7 | 7 | **24** | **24** | 41 | 41 |
| 8 | 8 | 25 | 25 | 42 | 42 |
| 9 | 9 | 26 | 26 | 43 | 43 |
| 10 | 10 | 27 | 27 | 44 | 44 |
| 11 | 11 | 28 | 28 | 45 | 45 |
| **12** | **12** | 29 | 29 | 46 | 46 |
| 13 | 13 | 30 | 30 | 47 | 47 |
| 14 | 14 | 31 | 31 | **48** | **48** |

| Common Rafters<br>1/4 in 12 – 1 1/4° | | | |
|---|---|---|---|
| SPAN<br>*Feet* | LENGTH<br>*Feet Inches* | SPAN<br>*Inches* | LENGTH<br>*Inches* |
| 1 | 0 - 6 | 1/4 | 1/8 |
| 2 | 1 - 0 | 1/2 | 1/4 |
| 3 | 1 - 6 | 3/4 | 3/8 |
| 4 | 2 - 0 | 1 | 1/2 |
| 5 | 2 - 6 | 1 1/4 | 5/8 |
| 6 | 3 - 0 | 1 1/2 | 3/4 |
| 7 | 3 - 6 | 1 3/4 | 7/8 |
| 8 | 4 - 0 | 2 | 1 |
| 9 | 4 - 6 | 2 1/4 | 1 1/8 |
| 10 | 5 - 0 | 2 1/2 | 1 1/4 |
| 11 | 5 - 6 | 2 3/4 | 1 3/8 |
| 12 | 6 - 0 | 3 | 1 1/2 |
| 13 | 6 - 6 | 3 1/4 | 1 5/8 |
| 14 | 7 - 0 | 3 1/2 | 1 3/4 |
| 15 | 7 - 6 | 3 3/4 | 1 7/8 |
| 16 | 8 - 0 | 4 | 2 |
| 17 | 8 - 6 | 4 1/4 | 2 1/8 |
| 18 | 9 - 0 | 4 1/2 | 2 1/4 |
| 19 | 9 - 6 | 4 3/4 | 2 3/8 |
| 20 | 10 - 0 | 5 | 2 1/2 |
| 21 | 10 - 6 | 5 1/4 | 2 5/8 |
| 22 | 11 - 0 | 5 1/2 | 2 3/4 |
| 23 | 11 - 6 | 5 3/4 | 2 7/8 |
| 24 | 12 - 0 | 6 | 3 |
| 25 | 12 - 6 1/16 | 6 1/4 | 3 1/8 |
| 26 | 13 - 0 1/16 | 6 1/2 | 3 1/4 |
| 27 | 13 - 6 1/16 | 6 3/4 | 3 3/8 |
| 28 | 14 - 0 1/16 | 7 | 3 1/2 |
| 29 | 14 - 6 1/16 | 7 1/4 | 3 5/8 |
| 30 | 15 - 0 1/16 | 7 1/2 | 3 3/4 |
| 31 | 15 - 6 1/16 | 7 3/4 | 3 7/8 |
| 32 | 16 - 0 1/16 | 8 | 4 |
| 33 | 16 - 6 1/16 | 8 1/4 | 4 1/8 |
| 34 | 17 - 0 1/16 | 8 1/2 | 4 1/4 |
| 35 | 17 - 6 1/16 | 8 3/4 | 4 3/8 |
| 36 | 18 - 0 1/16 | 9 | 4 1/2 |
| 37 | 18 - 6 1/16 | 9 1/4 | 4 5/8 |
| 38 | 19 - 0 1/16 | 9 1/2 | 4 3/4 |
| 39 | 19 - 6 1/16 | 9 3/4 | 4 7/8 |
| 40 | 20 - 0 1/16 | 10 | 5 |
| 41 | 20 - 6 1/16 | 10 1/4 | 5 1/8 |
| 42 | 21 - 0 1/16 | 10 1/2 | 5 1/4 |
| 43 | 21 - 6 1/16 | 10 3/4 | 5 3/8 |
| 44 | 22 - 0 1/16 | 11 | 5 1/2 |
| 45 | 22 - 6 1/16 | 11 1/4 | 5 5/8 |
| 46 | 23 - 0 1/16 | 11 1/2 | 5 3/4 |
| 47 | 23 - 6 1/16 | 11 3/4 | 5 7/8 |
| 48 | 24 - 0 1/16 | 12 | 6 |
| 49 | 24 - 6 1/16 | | |
| 50 | 25 - 0 1/16 | | |

| Hip or Valley Rafters<br>1/4 in 17 – 3/4° | | | |
|---|---|---|---|
| SPAN<br>*Feet* | LENGTH<br>*Feet Inches* | SPAN<br>*Inches* | LENGTH<br>*Inches* |
| 1 | 0 - 8 1/2 | 1/4 | 3/16 |
| 2 | 1 - 5 | 1/2 | 3/8 |
| 3 | 2 - 1 7/16 | 3/4 | 1/2 |
| 4 | 2 - 9 15/16 | 1 | 11/16 |
| 5 | 3 - 6 7/16 | 1 1/4 | 7/8 |
| 6 | 4 - 2 15/16 | 1 1/2 | 1 1/16 |
| 7 | 4 - 11 3/8 | 1 3/4 | 1 1/4 |
| 8 | 5 - 7 7/8 | 2 | 1 7/16 |
| 9 | 6 - 4 3/8 | 2 1/4 | 1 9/16 |
| 10 | 7 - 0 7/8 | 2 1/2 | 1 3/4 |
| 11 | 7 - 9 3/8 | 2 3/4 | 1 15/16 |
| 12 | 8 - 5 13/16 | 3 | 2 1/8 |
| 13 | 9 - 2 5/16 | 3 1/4 | 2 5/16 |
| 14 | 9 - 10 13/16 | 3 1/2 | 2 1/2 |
| 15 | 10 - 7 5/16 | 3 3/4 | 2 5/8 |
| 16 | 11 - 3 3/4 | 4 | 2 13/16 |
| 17 | 12 - 0 1/4 | 4 1/4 | 3 |
| 18 | 12 - 8 3/4 | 4 1/2 | 3 3/16 |
| 19 | 13 - 5 1/4 | 4 3/4 | 3 3/8 |
| 20 | 14 - 1 3/4 | 5 | 3 9/16 |
| 21 | 14 - 10 3/16 | 5 1/4 | 3 11/16 |
| 22 | 15 - 6 11/16 | 5 1/2 | 3 7/8 |
| 23 | 16 - 3 3/16 | 5 3/4 | 4 1/16 |
| 24 | 16 - 11 11/16 | 6 | 4 1/4 |
| 25 | 17 - 8 1/8 | 6 1/4 | 4 7/16 |
| 26 | 18 - 4 5/8 | 6 1/2 | 4 5/8 |
| 27 | 19 - 1 1/8 | 6 3/4 | 4 3/4 |
| 28 | 19 - 9 5/8 | 7 | 4 15/16 |
| 29 | 20 - 6 1/8 | 7 1/4 | 5 1/8 |
| 30 | 21 - 2 9/16 | 7 1/2 | 5 5/16 |
| 31 | 21 - 11 1/16 | 7 3/4 | 5 1/2 |
| 32 | 22 - 7 9/16 | 8 | 5 11/16 |
| 33 | 23 - 4 1/16 | 8 1/4 | 5 13/16 |
| 34 | 24 - 0 1/2 | 8 1/2 | 6 |
| 35 | 24 - 9 | 8 3/4 | 6 3/16 |
| 36 | 25 - 5 1/2 | 9 | 6 3/8 |
| 37 | 26 - 2 | 9 1/4 | 6 9/16 |
| 38 | 26 - 10 1/2 | 9 1/2 | 6 11/16 |
| 39 | 27 - 6 15/16 | 9 3/4 | 6 7/8 |
| 40 | 28 - 3 7/16 | 10 | 7 1/16 |
| 41 | 28 - 11 15/16 | 10 1/4 | 7 1/4 |
| 42 | 29 - 8 7/16 | 10 1/2 | 7 7/16 |
| 43 | 30 - 4 15/16 | 10 3/4 | 7 5/8 |
| 44 | 31 - 1 3/8 | 11 | 7 3/4 |
| 45 | 31 - 9 7/8 | 11 1/4 | 7 15/16 |
| 46 | 32 - 6 3/8 | 11 1/2 | 8 1/8 |
| 47 | 33 - 2 7/8 | 11 3/4 | 8 5/16 |
| 48 | 33 - 11 5/16 | 12 | 8 1/2 |
| 49 | 34 - 7 13/16 | | |
| 50 | 35 - 4 5/16 | | |

# $1/2$ in 12 – $2\,1/2°$

## *Cuts and Saw Settings*

**Tail Cut, Hip or Valley Rafters,** with Square Cut Common Rafters:
Framing Square: **11/16 & 24**; set saw at **45°**
(Compound Miter or Radial Arm Saws: set bevel at 45°, miter at 1 3/4°)
Explanation: Section 3(d), pages 12–13.

**Chamfer Cut, Hip or Valley Rafters:**
Set saw at **1 3/4°**. Explanation: Section 3(d), page 10.

**Fascia Board Miter Cut at Hip or Valley:**
Framing Square: **1 & 24**; set saw at **45°**
(Compound Miter or Radial Arm Saws: set bevel at 45°, miter at 2 1/2°)
Explanation: Section 5, pages 24–25.

**Frieze Block against Hip or Valley Rafter:**
Framing Square: **1 & 24**; set saw at **45°**
(Compound Miter or Radial Arm Saws: set bevel at 45°, miter at 2 1/2°)
Explanation: Section 6, page 26.

**Roof Sheathing at Hip or Valley:**
Framing Square: **12 & 12**; set saw at **1 3/4°**
(Compound Miter or Radial Arm Saws: set bevel at 1 3/4°, miter at 45°)
Distance "d" = **47 15/16"**. Explanation: Section 7, pages 26–27.

## *Calculating the Length of a Rafter*

**General Rule:** Span of building
less Thickness of ridge board (if any)
gives Length of rafter to plumb line of seat cut

*and* Span of building
less Thickness of ridge board (if any)
plus 2 × projection of overhang
gives Total length of rafter to tail cut

Explanation: Section 4, pages 13–19.

| *Jack Rafters* | | | | | |
|---|---|---|---|---|---|
| SPACED *Inches* | DIFFERENCE *Inches* | SPACED *Inches* | DIFFERENCE *Inches* | SPACED *Inches* | DIFFERENCE *Inches* |
| 1/4 | 1/4 | 15 | 15 | **32** | **32** |
| 1/2 | 1/2 | **16** | **16** | 33 | 33 |
| 3/4 | 3/4 | 17 | 17 | 34 | 34 |
| 1 | 1 | 18 | 18 | 35 | 35 |
| 2 | 2 | 19 | 19 | **36** | **36** |
| 3 | 3 | 20 | 20 | 37 | 37 1/16 |
| 4 | 4 | 21 | 21 | 38 | 38 1/16 |
| 5 | 5 | 22 | 22 | 39 | 39 1/16 |
| 6 | 6 | 23 | 23 | 40 | 40 1/16 |
| 7 | 7 | **24** | **24** | 41 | 41 1/16 |
| 8 | 8 | 25 | 25 | 42 | 42 1/16 |
| 9 | 9 | 26 | 26 | 43 | 43 1/16 |
| 10 | 10 | 27 | 27 | 44 | 44 1/16 |
| 11 | 11 | 28 | 28 | 45 | 45 1/16 |
| **12** | **12** | 29 | 29 | 46 | 46 1/16 |
| 13 | 13 | 30 | 30 | 47 | 47 1/16 |
| 14 | 14 | 31 | 31 | **48** | **48 1/16** |

| ***Common Rafters*** <br> 1/2 in 12 – 2 1/2° | | | |
|---|---|---|---|
| **SPAN** *Feet* | **LENGTH** *Feet Inches* | **SPAN** *Inches* | **LENGTH** *Inches* |
| 1 | 0 - 6 | 1/4 | 1/8 |
| 2 | 1 - 0 | 1/2 | 1/4 |
| 3 | 1 - 6 | 3/4 | 3/8 |
| 4 | 2 - 0 | 1 | 1/2 |
| 5 | 2 - 6 | 1 1/4 | 5/8 |
| 6 | 3 - 0 | 1 1/2 | 3/4 |
| 7 | 3 - 6 1/16 | 1 3/4 | 7/8 |
| 8 | 4 - 0 1/16 | 2 | 1 |
| 9 | 4 - 6 1/16 | 2 1/4 | 1 1/8 |
| 10 | 5 - 0 1/16 | 2 1/2 | 1 1/4 |
| 11 | 5 - 6 1/16 | 2 3/4 | 1 3/8 |
| 12 | 6 - 0 1/16 | 3 | 1 1/2 |
| 13 | 6 - 6 1/16 | 3 1/4 | 1 5/8 |
| 14 | 7 - 0 1/16 | 3 1/2 | 1 3/4 |
| 15 | 7 - 6 1/16 | 3 3/4 | 1 7/8 |
| 16 | 8 - 0 1/16 | 4 | 2 |
| 17 | 8 - 6 1/16 | 4 1/4 | 2 1/8 |
| 18 | 9 - 0 1/16 | 4 1/2 | 2 1/4 |
| 19 | 9 - 6 1/8 | 4 3/4 | 2 3/8 |
| 20 | 10 - 0 1/8 | 5 | 2 1/2 |
| 21 | 10 - 6 1/8 | 5 1/4 | 2 5/8 |
| 22 | 11 - 0 1/8 | 5 1/2 | 2 3/4 |
| 23 | 11 - 6 1/8 | 5 3/4 | 2 7/8 |
| 24 | 12 - 0 1/8 | 6 | 3 |
| 25 | 12 - 6 1/8 | 6 1/4 | 3 1/8 |
| 26 | 13 - 0 1/8 | 6 1/2 | 3 1/4 |
| 27 | 13 - 6 1/8 | 6 3/4 | 3 3/8 |
| 28 | 14 - 0 1/8 | 7 | 3 1/2 |
| 29 | 14 - 6 1/8 | 7 1/4 | 3 5/8 |
| 30 | 15 - 0 1/8 | 7 1/2 | 3 3/4 |
| 31 | 15 - 6 3/16 | 7 3/4 | 3 7/8 |
| 32 | 16 - 0 3/16 | 8 | 4 |
| 33 | 16 - 6 3/16 | 8 1/4 | 4 1/8 |
| 34 | 17 - 0 3/16 | 8 1/2 | 4 1/4 |
| 35 | 17 - 6 3/16 | 8 3/4 | 4 3/8 |
| 36 | 18 - 0 3/16 | 9 | 4 1/2 |
| 37 | 18 - 6 3/16 | 9 1/4 | 4 5/8 |
| 38 | 19 - 0 3/16 | 9 1/2 | 4 3/4 |
| 39 | 19 - 6 3/16 | 9 3/4 | 4 7/8 |
| 40 | 20 - 0 3/16 | 10 | 5 |
| 41 | 20 - 6 3/16 | 10 1/4 | 5 1/8 |
| 42 | 21 - 0 3/16 | 10 1/2 | 5 1/4 |
| 43 | 21 - 6 1/4 | 10 3/4 | 5 3/8 |
| 44 | 22 - 0 1/4 | 11 | 5 1/2 |
| 45 | 22 - 6 1/4 | 11 1/4 | 5 5/8 |
| 46 | 23 - 0 1/4 | 11 1/2 | 5 3/4 |
| 47 | 23 - 6 1/4 | 11 3/4 | 5 7/8 |
| 48 | 24 - 0 1/4 | 12 | 6 |
| 49 | 24 - 6 1/4 | | |
| 50 | 25 - 0 1/4 | | |

| ***Hip or Valley Rafters*** <br> 1/2 in 17 – 1 3/4° | | | |
|---|---|---|---|
| **SPAN** *Feet* | **LENGTH** *Feet Inches* | **SPAN** *Inches* | **LENGTH** *Inches* |
| 1 | 0- 8 1/2 | 1/4 | 3/16 |
| 2 | 1- 5 | 1/2 | 3/8 |
| 3 | 2- 1 7/16 | 3/4 | 1/2 |
| 4 | 2- 9 15/16 | 1 | 11/16 |
| 5 | 3- 6 7/16 | 1 1/4 | 7/8 |
| 6 | 4- 2 15/16 | 1 1/2 | 1 1/16 |
| 7 | 4-11 7/16 | 1 3/4 | 1 1/4 |
| 8 | 5- 7 15/16 | 2 | 1 7/16 |
| 9 | 6- 4 3/8 | 2 1/4 | 1 9/16 |
| 10 | 7- 0 7/8 | 2 1/2 | 1 3/4 |
| 11 | 7- 9 3/8 | 2 3/4 | 1 15/16 |
| 12 | 8- 5 7/8 | 3 | 2 1/8 |
| 13 | 9- 2 3/8 | 3 1/4 | 2 5/16 |
| 14 | 9-10 7/8 | 3 1/2 | 2 1/2 |
| 15 | 10- 7 5/16 | 3 3/4 | 2 5/8 |
| 16 | 11- 3 13/16 | 4 | 2 13/16 |
| 17 | 12- 0 5/16 | 4 1/4 | 3 |
| 18 | 12- 8 13/16 | 4 1/2 | 3 3/16 |
| 19 | 13- 5 5/16 | 4 3/4 | 3 3/8 |
| 20 | 14- 1 3/4 | 5 | 3 9/16 |
| 21 | 14-10 1/4 | 5 1/4 | 3 11/16 |
| 22 | 15- 6 3/4 | 5 1/2 | 3 7/8 |
| 23 | 16- 3 1/4 | 5 3/4 | 4 1/16 |
| 24 | 16-11 3/4 | 6 | 4 1/4 |
| 25 | 17- 8 1/4 | 6 1/4 | 4 7/16 |
| 26 | 18- 4 11/16 | 6 1/2 | 4 5/8 |
| 27 | 19- 1 3/16 | 6 3/4 | 4 3/4 |
| 28 | 19- 9 11/16 | 7 | 4 15/16 |
| 29 | 20- 6 3/16 | 7 1/4 | 5 1/8 |
| 30 | 21- 2 11/16 | 7 1/2 | 5 5/16 |
| 31 | 21-11 3/16 | 7 3/4 | 5 1/2 |
| 32 | 22- 7 5/8 | 8 | 5 11/16 |
| 33 | 23- 4 1/8 | 8 1/4 | 5 13/16 |
| 34 | 24- 0 5/8 | 8 1/2 | 6 |
| 35 | 24- 9 1/8 | 8 3/4 | 6 3/16 |
| 36 | 25- 5 5/8 | 9 | 6 3/8 |
| 37 | 26- 2 1/16 | 9 1/4 | 6 9/16 |
| 38 | 26-10 9/16 | 9 1/2 | 6 3/4 |
| 39 | 27- 7 1/16 | 9 3/4 | 6 7/8 |
| 40 | 28- 3 9/16 | 10 | 7 1/16 |
| 41 | 29- 0 1/16 | 10 1/4 | 7 1/4 |
| 42 | 29- 8 9/16 | 10 1/2 | 7 7/16 |
| 43 | 30- 5 | 10 3/4 | 7 5/8 |
| 44 | 31- 1 1/2 | 11 | 7 13/16 |
| 45 | 31-10 | 11 1/4 | 7 15/16 |
| 46 | 32- 6 1/2 | 11 1/2 | 8 1/8 |
| 47 | 33- 3 | 11 3/4 | 8 5/16 |
| 48 | 33-11 1/2 | 12 | 8 1/2 |
| 49 | 34- 7 15/16 | | |
| 50 | 35- 4 7/16 | | |

# 1 in 12 – 4 3/4°

## *Cuts and Saw Settings*

**Tail Cut, Hip or Valley Rafters,** with Square Cut Common Rafters:
Framing Square: **1 3/8 & 24**; set saw at **44 3/4°**
(Compound Miter or Radial Arm Saws: set bevel at 44 3/4°, miter at 3 1/4°)
Explanation: Section 3(d), pages 12–13.

**Chamfer Cut, Hip or Valley Rafters:**
Set saw at **3 1/4°**. Explanation: Section 3(d), page 10.

**Fascia Board Miter Cut at Hip or Valley:**
Framing Square: **2 & 24**; set saw at **44 3/4°**
(Compound Miter or Radial Arm Saws: set bevel at 44 3/4°, miter at 4 3/4°)
Explanation: Section 5, pages 24–25.

**Frieze Block against Hip or Valley Rafter:**
Framing Square: **2 & 24**; set saw at **44 3/4°**
(Compound Miter or Radial Arm Saws: set bevel at 44 3/4°, miter at 4 3/4°)
Explanation: Section 6, page 26.

**Roof Sheathing at Hip or Valley:**
Framing Square: **12 & 12 1/16**; set saw at **3 1/4°**
(Compound Miter or Radial Arm Saws: set bevel at 3 1/4°, miter at 45°)
Distance "d" = **47 13/16"**. Explanation: Section 7, pages 26–27.

## *Calculating the Length of a Rafter*

**General Rule:**

| | | |
|---|---|---|
| | | Span of building |
| | less | Thickness of ridge board (if any) |
| | gives | Length of rafter to plumb line of seat cut |
| *and* | | Span of building |
| | less | Thickness of ridge board (if any) |
| | plus | 2 × projection of overhang |
| | gives | Total length of rafter to tail cut |

Explanation: Section 4, pages 13–19.

| *Jack Rafters* | | | | | |
|---|---|---|---|---|---|
| SPACED *Inches* | DIFFERENCE *Inches* | SPACED *Inches* | DIFFERENCE *Inches* | SPACED *Inches* | DIFFERENCE *Inches* |
| 1/4 | 1/4 | 15 | 15 1/16 | **32** | **32 1/8** |
| 1/2 | 1/2 | **16** | **16 1/16** | 33 | 33 1/8 |
| 3/4 | 3/4 | 17 | 17 1/16 | 34 | 34 1/8 |
| 1 | 1 | 18 | 18 1/16 | 35 | 35 1/8 |
| 2 | 2 | 19 | 19 1/16 | **36** | **36 1/8** |
| 3 | 3 | 20 | 20 1/16 | 37 | 37 1/8 |
| 4 | 4 | 21 | 21 1/16 | 38 | 38 1/8 |
| 5 | 5 | 22 | 22 1/16 | 39 | 39 1/8 |
| 6 | 6 | 23 | 23 1/16 | 40 | 40 1/8 |
| 7 | 7 | **24** | **24 1/16** | 41 | 41 1/8 |
| 8 | 8 | 25 | 25 1/16 | 42 | 42 1/8 |
| 9 | 9 | 26 | 26 1/16 | 43 | 43 1/8 |
| 10 | 10 1/16 | 27 | 27 1/16 | 44 | 44 1/8 |
| 11 | 11 1/16 | 28 | 28 1/8 | 45 | 45 1/8 |
| **12** | **12 1/16** | 29 | 29 1/8 | 46 | 46 3/16 |
| 13 | 13 1/16 | 30 | 30 1/8 | 47 | 47 3/16 |
| 14 | 14 1/16 | 31 | 31 1/8 | **48** | **48 3/16** |

| **Common Rafters 1 in 12 – 4 3/4°** | | | |
|---|---|---|---|
| SPAN *Feet* | LENGTH *Feet Inches* | SPAN *Inches* | LENGTH *Inches* |
| 1 | 0 - 6 | 1/4 | 1/8 |
| 2 | 1 - 0 1/16 | 1/2 | 1/4 |
| 3 | 1 - 6 1/16 | 3/4 | 3/8 |
| 4 | 2 - 0 1/16 | 1 | 1/2 |
| 5 | 2 - 6 1/8 | 1 1/4 | 5/8 |
| 6 | 3 - 0 1/8 | 1 1/2 | 3/4 |
| 7 | 3 - 6 1/8 | 1 3/4 | 7/8 |
| 8 | 4 - 0 3/16 | 2 | 1 |
| 9 | 4 - 6 3/16 | 2 1/4 | 1 1/8 |
| 10 | 5 - 0 3/16 | 2 1/2 | 1 1/4 |
| 11 | 5 - 6 1/4 | 2 3/4 | 1 3/8 |
| 12 | 6 - 0 1/4 | 3 | 1 1/2 |
| 13 | 6 - 6 1/4 | 3 1/4 | 1 5/8 |
| 14 | 7 - 0 5/16 | 3 1/2 | 1 3/4 |
| 15 | 7 - 6 5/16 | 3 3/4 | 1 7/8 |
| 16 | 8 - 0 5/16 | 4 | 2 |
| 17 | 8 - 6 3/8 | 4 1/4 | 2 1/8 |
| 18 | 9 - 0 3/8 | 4 1/2 | 2 1/4 |
| 19 | 9 - 6 3/8 | 4 3/4 | 2 3/8 |
| 20 | 10 - 0 7/16 | 5 | 2 1/2 |
| 21 | 10 - 6 7/16 | 5 1/4 | 2 5/8 |
| 22 | 11 - 0 7/16 | 5 1/2 | 2 3/4 |
| 23 | 11 - 6 1/2 | 5 3/4 | 2 7/8 |
| 24 | 12 - 0 1/2 | 6 | 3 |
| 25 | 12 - 6 1/2 | 6 1/4 | 3 1/8 |
| 26 | 13 - 0 9/16 | 6 1/2 | 3 1/4 |
| 27 | 13 - 6 9/16 | 6 3/4 | 3 3/8 |
| 28 | 14 - 0 9/16 | 7 | 3 1/2 |
| 29 | 14 - 6 5/8 | 7 1/4 | 3 5/8 |
| 30 | 15 - 0 5/8 | 7 1/2 | 3 3/4 |
| 31 | 15 - 6 5/8 | 7 3/4 | 3 7/8 |
| 32 | 16 - 0 11/16 | 8 | 4 |
| 33 | 16 - 6 11/16 | 8 1/4 | 4 1/8 |
| 34 | 17 - 0 11/16 | 8 1/2 | 4 1/4 |
| 35 | 17 - 6 3/4 | 8 3/4 | 4 3/8 |
| 36 | 18 - 0 3/4 | 9 | 4 1/2 |
| 37 | 18 - 6 3/4 | 9 1/4 | 4 5/8 |
| 38 | 19 - 0 13/16 | 9 1/2 | 4 3/4 |
| 39 | 19 - 6 13/16 | 9 3/4 | 4 7/8 |
| 40 | 20 - 0 13/16 | 10 | 5 |
| 41 | 20 - 6 7/8 | 10 1/4 | 5 1/8 |
| 42 | 21 - 0 7/8 | 10 1/2 | 5 1/4 |
| 43 | 21 - 6 7/8 | 10 3/4 | 5 3/8 |
| 44 | 22 - 0 15/16 | 11 | 5 1/2 |
| 45 | 22 - 6 15/16 | 11 1/4 | 5 5/8 |
| 46 | 23 - 0 15/16 | 11 1/2 | 5 3/4 |
| 47 | 23 - 7 | 11 3/4 | 5 7/8 |
| 48 | 24 - 1 | 12 | 6 |
| 49 | 24 - 7 | | |
| 50 | 25 - 1 1/16 | | |

| **Hip or Valley Rafters 1 in 17 – 3 1/4°** | | | |
|---|---|---|---|
| SPAN *Feet* | LENGTH *Feet Inches* | SPAN *Inches* | LENGTH *Inches* |
| 1 | 0- 8 1/2 | 1/4 | 3/16 |
| 2 | 1- 5 | 1/2 | 3/8 |
| 3 | 2- 1 1/2 | 3/4 | 9/16 |
| 4 | 2-10 | 1 | 11/16 |
| 5 | 3- 6 1/2 | 1 1/4 | 7/8 |
| 6 | 4- 3 | 1 1/2 | 1 1/16 |
| 7 | 4-11 1/2 | 1 3/4 | 1 1/4 |
| 8 | 5- 8 | 2 | 1 7/16 |
| 9 | 6- 4 1/2 | 2 1/4 | 1 5/8 |
| 10 | 7- 1 | 2 1/2 | 1 3/4 |
| 11 | 7- 9 1/2 | 2 3/4 | 1 15/16 |
| 12 | 8- 6 | 3 | 2 1/8 |
| 13 | 9- 2 1/2 | 3 1/4 | 2 5/16 |
| 14 | 9-11 | 3 1/2 | 2 1/2 |
| 15 | 10- 7 1/2 | 3 3/4 | 2 11/16 |
| 16 | 11- 4 | 4 | 2 13/16 |
| 17 | 12- 0 1/2 | 4 1/4 | 3 |
| 18 | 12- 9 | 4 1/2 | 3 3/16 |
| 19 | 13- 5 1/2 | 4 3/4 | 3 3/8 |
| 20 | 14- 2 | 5 | 3 9/16 |
| 21 | 14-10 1/2 | 5 1/4 | 3 3/4 |
| 22 | 15- 7 | 5 1/2 | 3 7/8 |
| 23 | 16- 3 1/2 | 5 3/4 | 4 1/16 |
| 24 | 17- 0 | 6 | 4 1/4 |
| 25 | 17- 8 1/2 | 6 1/4 | 4 7/16 |
| 26 | 18- 5 | 6 1/2 | 4 5/8 |
| 27 | 19- 1 1/2 | 6 3/4 | 4 13/16 |
| 28 | 19-10 | 7 | 4 15/16 |
| 29 | 20- 6 1/2 | 7 1/4 | 5 1/8 |
| 30 | 21- 3 | 7 1/2 | 5 5/16 |
| 31 | 21-11 1/2 | 7 3/4 | 5 1/2 |
| 32 | 22- 8 | 8 | 5 11/16 |
| 33 | 23- 4 1/2 | 8 1/4 | 5 7/8 |
| 34 | 24- 1 | 8 1/2 | 6 |
| 35 | 24- 9 1/2 | 8 3/4 | 6 3/16 |
| 36 | 25- 6 | 9 | 6 3/8 |
| 37 | 26- 2 1/2 | 9 1/4 | 6 9/16 |
| 38 | 26-11 | 9 1/2 | 6 3/4 |
| 39 | 27- 7 1/2 | 9 3/4 | 6 15/16 |
| 40 | 28- 4 | 10 | 7 1/16 |
| 41 | 29- 0 1/2 | 10 1/4 | 7 1/4 |
| 42 | 29- 9 | 10 1/2 | 7 7/16 |
| 43 | 30- 5 1/2 | 10 3/4 | 7 5/8 |
| 44 | 31- 2 | 11 | 7 13/16 |
| 45 | 31-10 1/2 | 11 1/4 | 8 |
| 46 | 32- 7 | 11 1/2 | 8 1/8 |
| 47 | 33- 3 1/2 | 11 3/4 | 8 5/16 |
| 48 | 34- 0 | 12 | 8 1/2 |
| 49 | 34- 8 1/2 | | |
| 50 | 35- 5 | | |

# $1\frac{1}{2}$ in 12 – $7\frac{1}{4}$°

## *Cuts and Saw Settings*

**Tail Cut, Hip or Valley Rafters,** with Square Cut Common Rafters:
Framing Square: **2 1/16 & 24**; set saw at **44 1/2°**
(Compound Miter or Radial Arm Saws: set bevel at 44 1/2°, miter at 5°)
Explanation: Section 3(d), pages 12–13.

**Chamfer Cut, Hip or Valley Rafters:**
Set saw at **5°**. Explanation: Section 3(d), page 10.

**Fascia Board Miter Cut at Hip or Valley:**
Framing Square: **3 & 24**; set saw at **44 1/2°**
(Compound Miter or Radial Arm Saws: set bevel at 44 1/2°, miter at 7°)
Explanation: Section 5, pages 24–25.

**Frieze Block against Hip or Valley Rafter:**
Framing Square: **3 & 24**; set saw at **44 1/2°**
(Compound Miter or Radial Arm Saws: set bevel at 44 1/2°, miter at 7°)
Explanation: Section 6, page 26.

**Roof Sheathing at Hip or Valley:**
Framing Square: **12 & 12 1/16**; set saw at **5°**
(Compound Miter or Radial Arm Saws: set bevel at 5°, miter at 44 3/4°)
Distance "d" = **47 5/8"**. Explanation: Section 7, pages 26–27.

## *Calculating the Length of a Rafter*

| **General Rule:** | | Span of building |
|---|---|---|
| | less | Thickness of ridge board (if any) |
| | gives | Length of rafter to plumb line of seat cut |
| *and* | | Span of building |
| | less | Thickness of ridge board (if any) |
| | plus | 2 × projection of overhang |
| | gives | Total length of rafter to tail cut |

Explanation: Section 4, pages 13–19.

| *Jack Rafters* | | | | | |
|---|---|---|---|---|---|
| SPACED *Inches* | DIFFERENCE *Inches* | SPACED *Inches* | DIFFERENCE *Inches* | SPACED *Inches* | DIFFERENCE *Inches* |
| 1/4 | 1/4 | 15 | 15 1/8 | **32** | **32 1/4** |
| 1/2 | 1/2 | **16** | **16 1/8** | 33 | 33 1/4 |
| 3/4 | 3/4 | 17 | 17 1/8 | 34 | 34 1/4 |
| 1 | 1 | 18 | 18 1/8 | 35 | 35 1/4 |
| 2 | 2 | 19 | 19 1/8 | **36** | **36 1/4** |
| 3 | 3 | 20 | 20 1/8 | 37 | 37 5/16 |
| 4 | 4 | 21 | 21 3/16 | 38 | 38 5/16 |
| 5 | 5 1/16 | 22 | 22 3/16 | 39 | 39 5/16 |
| 6 | 6 1/16 | 23 | 23 3/16 | 40 | 40 5/16 |
| 7 | 7 1/16 | **24** | **24 3/16** | 41 | 41 5/16 |
| 8 | 8 1/16 | 25 | 25 3/16 | 42 | 42 5/16 |
| 9 | 9 1/16 | 26 | 26 3/16 | 43 | 43 5/16 |
| 10 | 10 1/16 | 27 | 27 3/16 | 44 | 44 5/16 |
| 11 | 11 1/16 | 28 | 28 3/16 | 45 | 45 3/8 |
| **12** | **12 1/16** | 29 | 29 1/4 | 46 | 46 3/8 |
| 13 | 13 1/8 | 30 | 30 1/4 | 47 | 47 3/8 |
| 14 | 14 1/8 | 31 | 31 1/4 | **48** | **48 3/8** |

**Common Rafters**
**1½ in 12 – 7¼°**

| SPAN Feet | LENGTH Feet Inches | SPAN Inches | LENGTH Inches |
|---|---|---|---|
| 1 | 0 - 6 1/16 | 1/4 | 1/8 |
| 2 | 1 - 0 1/16 | 1/2 | 1/4 |
| 3 | 1 - 6 1/8 | 3/4 | 3/8 |
| 4 | 2 - 0 3/16 | 1 | 1/2 |
| 5 | 2 - 6 1/4 | 1 1/4 | 5/8 |
| 6 | 3 - 0 1/4 | 1 1/2 | 3/4 |
| 7 | 3 - 6 5/16 | 1 3/4 | 7/8 |
| 8 | 4 - 0 3/8 | 2 | 1 |
| 9 | 4 - 6 7/16 | 2 1/4 | 1 1/8 |
| 10 | 5 - 0 7/16 | 2 1/2 | 1 1/4 |
| 11 | 5 - 6 1/2 | 2 3/4 | 1 3/8 |
| 12 | 6 - 0 9/16 | 3 | 1 1/2 |
| 13 | 6 - 6 5/8 | 3 1/4 | 1 5/8 |
| 14 | 7 - 0 5/8 | 3 1/2 | 1 3/4 |
| 15 | 7 - 6 11/16 | 3 3/4 | 1 7/8 |
| 16 | 8 - 0 3/4 | 4 | 2 |
| 17 | 8 - 6 13/16 | 4 1/4 | 2 1/8 |
| 18 | 9 - 0 13/16 | 4 1/2 | 2 1/4 |
| 19 | 9 - 6 7/8 | 4 3/4 | 2 3/8 |
| 20 | 10 - 0 15/16 | 5 | 2 1/2 |
| 21 | 10 - 7 | 5 1/4 | 2 5/8 |
| 22 | 11 - 1 | 5 1/2 | 2 3/4 |
| 23 | 11 - 7 1/16 | 5 3/4 | 2 7/8 |
| 24 | 12 - 1 1/8 | 6 | 3 |
| 25 | 12 - 7 3/16 | 6 1/4 | 3 1/8 |
| 26 | 13 - 1 3/16 | 6 1/2 | 3 1/4 |
| 27 | 13 - 7 1/4 | 6 3/4 | 3 3/8 |
| 28 | 14 - 1 5/16 | 7 | 3 1/2 |
| 29 | 14 - 7 3/8 | 7 1/4 | 3 5/8 |
| 30 | 15 - 1 3/8 | 7 1/2 | 3 3/4 |
| 31 | 15 - 7 7/16 | 7 3/4 | 3 7/8 |
| 32 | 16 - 1 1/2 | 8 | 4 |
| 33 | 16 - 7 9/16 | 8 1/4 | 4 3/16 |
| 34 | 17 - 1 9/16 | 8 1/2 | 4 5/16 |
| 35 | 17 - 7 5/8 | 8 3/4 | 4 7/16 |
| 36 | 18 - 1 11/16 | 9 | 4 9/16 |
| 37 | 18 - 7 3/4 | 9 1/4 | 4 11/16 |
| 38 | 19 - 1 3/4 | 9 1/2 | 4 13/16 |
| 39 | 19 - 7 13/16 | 9 3/4 | 4 15/16 |
| 40 | 20 - 1 7/8 | 10 | 5 1/16 |
| 41 | 20 - 7 15/16 | 10 1/4 | 5 3/16 |
| 42 | 21 - 1 15/16 | 10 1/2 | 5 5/16 |
| 43 | 21 - 8 | 10 3/4 | 5 7/16 |
| 44 | 22 - 2 1/16 | 11 | 5 9/16 |
| 45 | 22 - 8 1/8 | 11 1/4 | 5 11/16 |
| 46 | 23 - 2 1/8 | 11 1/2 | 5 13/16 |
| 47 | 23 - 8 3/16 | 11 3/4 | 5 15/16 |
| 48 | 24 - 2 1/4 | 12 | 6 1/16 |
| 49 | 24 - 8 5/16 | | |
| 50 | 25 - 2 5/16 | | |

**Hip or Valley Rafters**
**1½ in 17 – 5°**

| SPAN Feet | LENGTH Feet Inches | SPAN Inches | LENGTH Inches |
|---|---|---|---|
| 1 | 0 - 8 1/2 | 1/4 | 3/16 |
| 2 | 1 - 5 1/16 | 1/2 | 3/8 |
| 3 | 2 - 1 9/16 | 3/4 | 9/16 |
| 4 | 2 - 10 1/16 | 1 | 11/16 |
| 5 | 3 - 6 9/16 | 1 1/4 | 7/8 |
| 6 | 4 - 3 1/8 | 1 1/2 | 1 1/16 |
| 7 | 4 - 11 5/8 | 1 3/4 | 1 1/4 |
| 8 | 5 - 8 1/8 | 2 | 1 7/16 |
| 9 | 6 - 4 11/16 | 2 1/4 | 1 5/8 |
| 10 | 7 - 1 3/16 | 2 1/2 | 1 3/4 |
| 11 | 7 - 9 11/16 | 2 3/4 | 1 15/16 |
| 12 | 8 - 6 1/4 | 3 | 2 1/8 |
| 13 | 9 - 2 3/4 | 3 1/4 | 2 5/16 |
| 14 | 9 - 11 1/4 | 3 1/2 | 2 1/2 |
| 15 | 10 - 7 3/4 | 3 3/4 | 2 11/16 |
| 16 | 11 - 4 5/16 | 4 | 2 13/16 |
| 17 | 12 - 0 13/16 | 4 1/4 | 3 |
| 18 | 12 - 9 5/16 | 4 1/2 | 3 3/16 |
| 19 | 13 - 5 7/8 | 4 3/4 | 3 3/8 |
| 20 | 14 - 2 3/8 | 5 | 3 9/16 |
| 21 | 14 - 10 7/8 | 5 1/4 | 3 3/4 |
| 22 | 15 - 7 3/8 | 5 1/2 | 3 7/8 |
| 23 | 16 - 3 15/16 | 5 3/4 | 4 1/16 |
| 24 | 17 - 0 7/16 | 6 | 4 1/4 |
| 25 | 17 - 8 15/16 | 6 1/4 | 4 7/16 |
| 26 | 18 - 5 1/2 | 6 1/2 | 4 5/8 |
| 27 | 19 - 2 | 6 3/4 | 4 13/16 |
| 28 | 19 - 10 1/2 | 7 | 5 |
| 29 | 20 - 7 1/16 | 7 1/4 | 5 1/8 |
| 30 | 21 - 3 9/16 | 7 1/2 | 5 5/16 |
| 31 | 22 - 0 1/16 | 7 3/4 | 5 1/2 |
| 32 | 22 - 8 9/16 | 8 | 5 11/16 |
| 33 | 23 - 5 1/8 | 8 1/4 | 5 7/8 |
| 34 | 24 - 1 5/8 | 8 1/2 | 6 1/16 |
| 35 | 24 - 10 1/8 | 8 3/4 | 6 3/16 |
| 36 | 25 - 6 11/16 | 9 | 6 3/8 |
| 37 | 26 - 3 3/16 | 9 1/4 | 6 9/16 |
| 38 | 26 - 11 11/16 | 9 1/2 | 6 3/4 |
| 39 | 27 - 8 3/16 | 9 3/4 | 6 15/16 |
| 40 | 28 - 4 3/4 | 10 | 7 1/8 |
| 41 | 29 - 1 1/4 | 10 1/4 | 7 1/4 |
| 42 | 29 - 9 3/4 | 10 1/2 | 7 7/16 |
| 43 | 30 - 6 5/16 | 10 3/4 | 7 5/8 |
| 44 | 31 - 2 13/16 | 11 | 7 13/16 |
| 45 | 31 - 11 5/16 | 11 1/4 | 8 |
| 46 | 32 - 7 7/8 | 11 1/2 | 8 3/16 |
| 47 | 33 - 4 3/8 | 11 3/4 | 8 5/16 |
| 48 | 34 - 0 7/8 | 12 | 8 1/2 |
| 49 | 34 - 9 3/8 | | |
| 50 | 35 - 5 15/16 | | |

# 2 in 12 – 9 1/2°

## *Cuts and Saw Settings*

**Tail Cut, Hip or Valley Rafters,** with Square Cut Common Rafters:
Framing Square: **2 3/4 & 24**; set saw at **44 1/4°**
(Compound Miter or Radial Arm Saws: set bevel at 44 1/4°, miter at 6 1/2°)
Explanation: Section 3(d), pages 12–13.

**Chamfer Cut, Hip or Valley Rafters:**
Set saw at **6 3/4°**. Explanation: Section 3(d), page 10.

**Fascia Board Miter Cut at Hip or Valley:**
Framing Square: **3 15/16 & 24**; set saw at **44 1/4°**
(Compound Miter or Radial Arm Saws: set bevel at 44 1/4°, miter at 9 1/4°)
Explanation: Section 5, pages 24–25.

**Frieze Block against Hip or Valley Rafter:**
Framing Square: **3 15/16 & 24**; set saw at **44 1/4°**
(Compound Miter or Radial Arm Saws: set bevel at 44 1/4°, miter at 9 1/4°)
Explanation: Section 6, page 26.

**Roof Sheathing at Hip or Valley:**
Framing Square: **12 & 12 3/16**; set saw at **6 3/4°**
(Compound Miter or Radial Arm Saws: set bevel at 6 3/4°, miter at 44 1/2°)
Distance "d" = **47 3/8"**. Explanation: Section 7, pages 26–27.

## *Calculating the Length of a Rafter*

**General Rule:**

| | | |
|---|---|---|
| | | Span of building |
| | less | Thickness of ridge board (if any) |
| | gives | Length of rafter to plumb line of seat cut |
| *and* | | Span of building |
| | less | Thickness of ridge board (if any) |
| | plus | 2 × projection of overhang |
| | gives | Total length of rafter to tail cut |

Explanation: Section 4, pages 13–19.

| *Jack Rafters* | | | | | |
|---|---|---|---|---|---|
| SPACED *Inches* | DIFFERENCE *Inches* | SPACED *Inches* | DIFFERENCE *Inches* | SPACED *Inches* | DIFFERENCE *Inches* |
| 1/4 | 1/4 | 15 | 15 3/16 | **32** | **32 7/16** |
| 1/2 | 1/2 | **16** | **16 1/4** | 33 | 33 7/16 |
| 3/4 | 3/4 | 17 | 17 1/4 | 34 | 34 1/2 |
| 1 | 1 | 18 | 18 1/4 | 35 | 35 1/2 |
| 2 | 2 | 19 | 19 1/4 | **36** | **36 1/2** |
| 3 | 3 1/16 | 20 | 20 1/4 | 37 | 37 1/2 |
| 4 | 4 1/16 | 21 | 21 5/16 | 38 | 38 1/2 |
| 5 | 5 1/16 | 22 | 22 5/16 | 39 | 39 9/16 |
| 6 | 6 1/16 | 23 | 23 5/16 | 40 | 40 9/16 |
| 7 | 7 1/8 | **24** | **24 5/16** | 41 | 41 9/16 |
| 8 | 8 1/8 | 25 | 25 3/8 | 42 | 42 9/16 |
| 9 | 9 1/8 | 26 | 26 3/8 | 43 | 43 9/16 |
| 10 | 10 1/8 | 27 | 27 3/8 | 44 | 44 5/8 |
| 11 | 11 1/8 | 28 | 28 3/8 | 45 | 45 5/8 |
| **12** | **12 3/16** | 29 | 29 3/8 | 46 | 46 5/8 |
| 13 | 13 3/16 | 30 | 30 7/16 | 47 | 47 5/8 |
| 14 | 14 3/16 | 31 | 31 7/16 | **48** | **48 11/16** |

**Common Rafters**
**2 in 12 – 9 1/2°**

| SPAN Feet | LENGTH Feet Inches | SPAN Inches | LENGTH Inches |
|---|---|---|---|
| 1 | 0 - 6 1/16 | 1/4 | 1/8 |
| 2 | 1 - 0 3/16 | 1/2 | 1/4 |
| 3 | 1 - 6 1/4 | 3/4 | 3/8 |
| 4 | 2 - 0 5/16 | 1 | 1/2 |
| 5 | 2 - 6 7/16 | 1 1/4 | 5/8 |
| 6 | 3 - 0 1/2 | 1 1/2 | 3/4 |
| 7 | 3 - 6 9/16 | 1 3/4 | 7/8 |
| 8 | 4 - 0 11/16 | 2 | 1 |
| 9 | 4 - 6 3/4 | 2 1/4 | 1 1/8 |
| 10 | 5 - 0 13/16 | 2 1/2 | 1 1/4 |
| 11 | 5 - 6 15/16 | 2 3/4 | 1 3/8 |
| 12 | 6 - 1 | 3 | 1 1/2 |
| 13 | 6 - 7 1/16 | 3 1/4 | 1 5/8 |
| 14 | 7 - 1 3/16 | 3 1/2 | 1 3/4 |
| 15 | 7 - 7 1/4 | 3 3/4 | 1 7/8 |
| 16 | 8 - 1 5/16 | 4 | 2 |
| 17 | 8 - 7 7/16 | 4 1/4 | 2 1/8 |
| 18 | 9 - 1 1/2 | 4 1/2 | 2 1/4 |
| 19 | 9 - 7 9/16 | 4 3/4 | 2 7/16 |
| 20 | 10 - 1 5/8 | 5 | 2 9/16 |
| 21 | 10 - 7 3/4 | 5 1/4 | 2 11/16 |
| 22 | 11 - 1 13/16 | 5 1/2 | 2 13/16 |
| 23 | 11 - 7 7/8 | 5 3/4 | 2 15/16 |
| 24 | 12 - 2 | 6 | 3 1/16 |
| 25 | 12 - 8 1/16 | 6 1/4 | 3 3/16 |
| 26 | 13 - 2 1/8 | 6 1/2 | 3 5/16 |
| 27 | 13 - 8 1/4 | 6 3/4 | 3 7/16 |
| 28 | 14 - 2 5/16 | 7 | 3 9/16 |
| 29 | 14 - 8 3/8 | 7 1/4 | 3 11/16 |
| 30 | 15 - 2 1/2 | 7 1/2 | 3 13/16 |
| 31 | 15 - 8 9/16 | 7 3/4 | 3 15/16 |
| 32 | 16 - 2 5/8 | 8 | 4 1/16 |
| 33 | 16 - 8 3/4 | 8 1/4 | 4 3/16 |
| 34 | 17 - 2 13/16 | 8 1/2 | 4 5/16 |
| 35 | 17 - 8 7/8 | 8 3/4 | 4 7/16 |
| 36 | 18 - 3 | 9 | 4 9/16 |
| 37 | 18 - 9 1/16 | 9 1/4 | 4 11/16 |
| 38 | 19 - 3 1/8 | 9 1/2 | 4 13/16 |
| 39 | 19 - 9 1/4 | 9 3/4 | 4 15/16 |
| 40 | 20 - 3 5/16 | 10 | 5 1/16 |
| 41 | 20 - 9 3/8 | 10 1/4 | 5 3/16 |
| 42 | 21 - 3 1/2 | 10 1/2 | 5 5/16 |
| 43 | 21 - 9 9/16 | 10 3/4 | 5 7/16 |
| 44 | 22 - 3 5/8 | 11 | 5 9/16 |
| 45 | 22 - 9 3/4 | 11 1/4 | 5 11/16 |
| 46 | 23 - 3 13/16 | 11 1/2 | 5 13/16 |
| 47 | 23 - 9 7/8 | 11 3/4 | 5 15/16 |
| 48 | 24 - 4 | 12 | 6 1/16 |
| 49 | 24 -10 1/16 | | |
| 50 | 25 - 4 1/8 | | |

**Hip or Valley Rafters**
**2 in 17 – 6 3/4°**

| SPAN Feet | LENGTH Feet Inches | SPAN Inches | LENGTH Inches |
|---|---|---|---|
| 1 | 0 - 8 9/16 | 1/4 | 3/16 |
| 2 | 1 - 5 1/16 | 1/2 | 3/8 |
| 3 | 2 - 1 5/8 | 3/4 | 9/16 |
| 4 | 2 -10 3/16 | 1 | 11/16 |
| 5 | 3 - 6 3/4 | 1 1/4 | 7/8 |
| 6 | 4 - 3 1/4 | 1 1/2 | 1 1/16 |
| 7 | 4 -11 13/16 | 1 3/4 | 1 1/4 |
| 8 | 5 - 8 3/8 | 2 | 1 7/16 |
| 9 | 6 - 4 7/8 | 2 1/4 | 1 5/8 |
| 10 | 7 - 1 7/16 | 2 1/2 | 1 3/4 |
| 11 | 7 -10 | 2 3/4 | 1 15/16 |
| 12 | 8 - 6 1/2 | 3 | 2 1/8 |
| 13 | 9 - 3 1/16 | 3 1/4 | 2 5/16 |
| 14 | 9 -11 5/8 | 3 1/2 | 2 1/2 |
| 15 | 10 - 8 3/16 | 3 3/4 | 2 11/16 |
| 16 | 11 - 4 11/16 | 4 | 2 7/8 |
| 17 | 12 - 1 1/4 | 4 1/4 | 3 |
| 18 | 12 - 9 13/16 | 4 1/2 | 3 3/16 |
| 19 | 13 - 6 5/16 | 4 3/4 | 3 3/8 |
| 20 | 14 - 2 7/8 | 5 | 3 9/16 |
| 21 | 14 -11 7/16 | 5 1/4 | 3 3/4 |
| 22 | 15 - 7 15/16 | 5 1/2 | 3 15/16 |
| 23 | 16 - 4 1/2 | 5 3/4 | 4 1/8 |
| 24 | 17 - 1 1/16 | 6 | 4 1/4 |
| 25 | 17 - 9 5/8 | 6 1/4 | 4 7/16 |
| 26 | 18 - 6 1/8 | 6 1/2 | 4 5/8 |
| 27 | 19 - 2 11/16 | 6 3/4 | 4 13/16 |
| 28 | 19 -11 1/4 | 7 | 5 |
| 29 | 20 - 7 3/4 | 7 1/4 | 5 3/16 |
| 30 | 21 - 4 5/16 | 7 1/2 | 5 5/16 |
| 31 | 22 - 0 7/8 | 7 3/4 | 5 1/2 |
| 32 | 22 - 9 7/16 | 8 | 5 11/16 |
| 33 | 23 - 5 15/16 | 8 1/4 | 5 7/8 |
| 34 | 24 - 2 1/2 | 8 1/2 | 6 1/16 |
| 35 | 24 -11 1/16 | 8 3/4 | 6 1/4 |
| 36 | 25 - 7 9/16 | 9 | 6 7/16 |
| 37 | 26 - 4 1/8 | 9 1/4 | 6 9/16 |
| 38 | 27 - 0 11/16 | 9 1/2 | 6 3/4 |
| 39 | 27 - 9 3/16 | 9 3/4 | 6 15/16 |
| 40 | 28 - 5 3/4 | 10 | 7 1/8 |
| 41 | 29 - 2 5/16 | 10 1/4 | 7 5/16 |
| 42 | 29 -10 7/8 | 10 1/2 | 7 1/2 |
| 43 | 30 - 7 3/8 | 10 3/4 | 7 5/8 |
| 44 | 31 - 3 15/16 | 11 | 7 13/16 |
| 45 | 32 - 0 1/2 | 11 1/4 | 8 |
| 46 | 32 - 9 | 11 1/2 | 8 3/16 |
| 47 | 33 - 5 9/16 | 11 3/4 | 8 3/8 |
| 48 | 34 - 2 1/8 | 12 | 8 9/16 |
| 49 | 34 -10 5/8 | | |
| 50 | 35 - 7 3/16 | | |

## $2^1/_2$ in 12 – $11^3/_4$°

### *Cuts and Saw Settings*

**Tail Cut, Hip or Valley Rafters,** with Square Cut Common Rafters:
Framing Square: **3 3/8 & 24**; set saw at **43 3/4°**
(Compound Miter or Radial Arm Saws: set bevel at 43 3/4°, miter at 8°)
Explanation: Section 3(d), pages 12–13.

**Chamfer Cut, Hip or Valley Rafters:**
Set saw at **8 1/4°**. Explanation: Section 3(d), page 10.

**Fascia Board Miter Cut at Hip or Valley:**
Framing Square: **4 7/8 & 24**; set saw at **43 3/4°**
(Compound Miter or Radial Arm Saws: set bevel at 43 3/4°, miter at 11 1/2°)
Explanation: Section 5, pages 24–25.

**Frieze Block against Hip or Valley Rafter:**
Framing Square: **4 7/8 & 24**; set saw at **43 3/4°**
(Compound Miter or Radial Arm Saws: set bevel at 43 3/4°, miter at 11 1/2°)
Explanation: Section 6, page 26.

**Roof Sheathing at Hip or Valley:**
Framing Square: **12 & 12 1/4**; set saw at **8 1/4°**
(Compound Miter or Radial Arm Saws: set bevel at 8 1/4°, miter at 44 1/2°)
Distance "d" = **47"**. Explanation: Section 7, pages 26–27.

### *Calculating the Length of a Rafter*

**General Rule:** Span of building
less Thickness of ridge board (if any)
gives Length of rafter to plumb line of seat cut

*and* Span of building
less Thickness of ridge board (if any)
plus 2 × projection of overhang
gives Total length of rafter to tail cut

Explanation: Section 4, pages 13–19.

| *Jack Rafters* | | | | | |
|---|---|---|---|---|---|
| SPACED *Inches* | DIFFERENCE *Inches* | SPACED *Inches* | DIFFERENCE *Inches* | SPACED *Inches* | DIFFERENCE *Inches* |
| 1/4 | 1/4 | 15 | 15 5/16 | **32** | **32 11/16** |
| 1/2 | 1/2 | **16** | **16 5/16** | 33 | 33 11/16 |
| 3/4 | 3/4 | 17 | 17 3/8 | 34 | 34 3/4 |
| 1 | 1 | 18 | 18 3/8 | 35 | 35 3/4 |
| 2 | 2 1/16 | 19 | 19 7/16 | **36** | **36 3/4** |
| 3 | 3 1/16 | 20 | 20 7/16 | 37 | 37 13/16 |
| 4 | 4 1/16 | 21 | 21 7/16 | 38 | 38 13/16 |
| 5 | 5 1/8 | 22 | 22 1/2 | 39 | 39 13/16 |
| 6 | 6 1/8 | 23 | 23 1/2 | 40 | 40 7/8 |
| 7 | 7 1/8 | **24** | **24 1/2** | 41 | 41 7/8 |
| 8 | 8 3/16 | 25 | 25 9/16 | 42 | 42 7/8 |
| 9 | 9 3/16 | 26 | 26 9/16 | 43 | 43 15/16 |
| 10 | 10 3/16 | 27 | 27 9/16 | 44 | 44 15/16 |
| 11 | 11 1/4 | 28 | 28 5/8 | 45 | 45 15/16 |
| **12** | **12 1/4** | 29 | 29 5/8 | 46 | 47 |
| 13 | 13 1/4 | 30 | 30 5/8 | 47 | 48 |
| 14 | 14 5/16 | 31 | 31 11/16 | **48** | **49** |

| ***Common Rafters*** **2 1/2 in 12 – 11 3/4°** | | | |
|---|---|---|---|
| **SPAN *Feet*** | **LENGTH *Feet Inches*** | **SPAN *Inches*** | **LENGTH *Inches*** |
| 1 | 0- 6 1/8 | 1/4 | 1/8 |
| 2 | 1- 0 1/4 | 1/2 | 1/4 |
| 3 | 1- 6 3/8 | 3/4 | 3/8 |
| 4 | 2- 0 1/2 | 1 | 1/2 |
| 5 | 2- 6 5/8 | 1 1/4 | 5/8 |
| 6 | 3- 0 3/4 | 1 1/2 | 3/4 |
| 7 | 3- 6 7/8 | 1 3/4 | 7/8 |
| 8 | 4- 1 | 2 | 1 |
| 9 | 4- 7 3/16 | 2 1/4 | 1 1/8 |
| 10 | 5- 1 5/16 | 2 1/2 | 1 1/4 |
| 11 | 5- 7 7/16 | 2 3/4 | 1 3/8 |
| 12 | 6- 1 9/16 | 3 | 1 9/16 |
| 13 | 6- 7 11/16 | 3 1/4 | 1 11/16 |
| 14 | 7- 1 13/16 | 3 1/2 | 1 13/16 |
| 15 | 7- 7 15/16 | 3 3/4 | 1 15/16 |
| 16 | 8- 2 1/16 | 4 | 2 1/16 |
| 17 | 8- 8 3/16 | 4 1/4 | 2 3/16 |
| 18 | 9- 2 5/16 | 4 1/2 | 2 5/16 |
| 19 | 9- 8 7/16 | 4 3/4 | 2 7/16 |
| 20 | 10- 2 9/16 | 5 | 2 9/16 |
| 21 | 10- 8 11/16 | 5 1/4 | 2 11/16 |
| 22 | 11- 2 13/16 | 5 1/2 | 2 13/16 |
| 23 | 11- 8 15/16 | 5 3/4 | 2 15/16 |
| 24 | 12- 3 1/16 | 6 | 3 1/16 |
| 25 | 12- 9 1/4 | 6 1/4 | 3 3/16 |
| 26 | 13- 3 3/8 | 6 1/2 | 3 5/16 |
| 27 | 13- 9 1/2 | 6 3/4 | 3 7/16 |
| 28 | 14- 3 5/8 | 7 | 3 9/16 |
| 29 | 14- 9 3/4 | 7 1/4 | 3 11/16 |
| 30 | 15- 3 7/8 | 7 1/2 | 3 13/16 |
| 31 | 15-10 | 7 3/4 | 3 15/16 |
| 32 | 16- 4 1/8 | 8 | 4 1/16 |
| 33 | 16-10 1/4 | 8 1/4 | 4 3/16 |
| 34 | 17- 4 3/8 | 8 1/2 | 4 5/16 |
| 35 | 17-10 1/2 | 8 3/4 | 4 1/2 |
| 36 | 18- 4 5/8 | 9 | 4 5/8 |
| 37 | 18-10 3/4 | 9 1/4 | 4 3/4 |
| 38 | 19- 4 7/8 | 9 1/2 | 4 7/8 |
| 39 | 19-11 | 9 3/4 | 5 |
| 40 | 20- 5 1/8 | 10 | 5 1/8 |
| 41 | 20-11 5/16 | 10 1/4 | 5 1/4 |
| 42 | 21- 5 7/16 | 10 1/2 | 5 3/8 |
| 43 | 21-11 9/16 | 10 3/4 | 5 1/2 |
| 44 | 22- 5 11/16 | 11 | 5 5/8 |
| 45 | 22-11 13/16 | 11 1/4 | 5 3/4 |
| 46 | 23- 5 15/16 | 11 1/2 | 5 7/8 |
| 47 | 24- 0 1/16 | 11 3/4 | 6 |
| 48 | 24- 6 3/16 | 12 | 6 1/8 |
| 49 | 25- 0 5/16 | | |
| 50 | 25- 6 7/16 | | |

| ***Hip or Valley Rafters*** **2 1/2 in 17 – 8 1/2°** | | | |
|---|---|---|---|
| **SPAN *Feet*** | **LENGTH *Feet Inches*** | **SPAN *Inches*** | **LENGTH *Inches*** |
| 1 | 0- 8 9/16 | 1/4 | 3/16 |
| 2 | 1- 5 1/8 | 1/2 | 3/8 |
| 3 | 2- 1 3/4 | 3/4 | 9/16 |
| 4 | 2-10 5/16 | 1 | 11/16 |
| 5 | 3- 6 7/8 | 1 1/4 | 7/8 |
| 6 | 4- 3 7/16 | 1 1/2 | 1 1/16 |
| 7 | 5- 0 1/16 | 1 3/4 | 1 1/4 |
| 8 | 5- 8 5/8 | 2 | 1 7/16 |
| 9 | 6- 5 3/16 | 2 1/4 | 1 5/8 |
| 10 | 7- 1 3/4 | 2 1/2 | 1 13/16 |
| 11 | 7-10 3/8 | 2 3/4 | 1 15/16 |
| 12 | 8- 6 15/16 | 3 | 2 1/8 |
| 13 | 9- 3 1/2 | 3 1/4 | 2 5/16 |
| 14 | 10- 0 1/16 | 3 1/2 | 2 1/2 |
| 15 | 10- 8 5/8 | 3 3/4 | 2 11/16 |
| 16 | 11- 5 1/4 | 4 | 2 7/8 |
| 17 | 12- 1 13/16 | 4 1/4 | 3 1/16 |
| 18 | 12-10 3/8 | 4 1/2 | 3 3/16 |
| 19 | 13- 6 15/16 | 4 3/4 | 3 3/8 |
| 20 | 14- 3 9/16 | 5 | 3 9/16 |
| 21 | 15- 0 1/8 | 5 1/4 | 3 3/4 |
| 22 | 15- 8 11/16 | 5 1/2 | 3 15/16 |
| 23 | 16- 5 1/4 | 5 3/4 | 4 1/8 |
| 24 | 17- 1 7/8 | 6 | 4 5/16 |
| 25 | 17-10 7/16 | 6 1/4 | 4 7/16 |
| 26 | 18- 7 | 6 1/2 | 4 5/8 |
| 27 | 19- 3 9/16 | 6 3/4 | 4 13/16 |
| 28 | 20- 0 1/8 | 7 | 5 |
| 29 | 20- 8 3/4 | 7 1/4 | 5 3/16 |
| 30 | 21- 5 5/16 | 7 1/2 | 5 3/8 |
| 31 | 22- 1 7/8 | 7 3/4 | 5 9/16 |
| 32 | 22-10 7/16 | 8 | 5 11/16 |
| 33 | 23- 7 1/16 | 8 1/4 | 5 7/8 |
| 34 | 24- 3 5/8 | 8 1/2 | 6 1/16 |
| 35 | 25- 0 3/16 | 8 3/4 | 6 1/4 |
| 36 | 25- 8 3/4 | 9 | 6 7/16 |
| 37 | 26- 5 3/8 | 9 1/4 | 6 5/8 |
| 38 | 27- 1 15/16 | 9 1/2 | 6 13/16 |
| 39 | 27-10 1/2 | 9 3/4 | 6 15/16 |
| 40 | 28- 7 1/16 | 10 | 7 1/8 |
| 41 | 29- 3 5/8 | 10 1/4 | 7 5/16 |
| 42 | 30- 0 1/4 | 10 1/2 | 7 1/2 |
| 43 | 30- 8 13/16 | 10 3/4 | 7 11/16 |
| 44 | 31- 5 3/8 | 11 | 7 7/8 |
| 45 | 32- 1 15/16 | 11 1/4 | 8 1/16 |
| 46 | 32-10 9/16 | 11 1/2 | 8 1/4 |
| 47 | 33- 7 1/8 | 11 3/4 | 8 3/8 |
| 48 | 34- 3 11/16 | 12 | 8 9/16 |
| 49 | 35- 0 1/4 | | |
| 50 | 35- 8 13/16 | | |

# 3 in 12 – 14°

## *Cuts and Saw Settings*

**Tail Cut, Hip or Valley Rafters,** with Square Cut Common Rafters:
Framing Square: **4 & 24**; set saw at **43 1/4°**
(Compound Miter or Radial Arm Saws: set bevel at 43 1/4°, miter at 9 1/2°)
Explanation: Section 3(d), pages 12–13.

**Chamfer Cut, Hip or Valley Rafters:**
Set saw at **9 3/4°**. Explanation: Section 3(d), page 10.

**Fascia Board Miter Cut at Hip or Valley:**
Framing Square: **5 13/16 & 24**; set saw at **43 1/4°**
(Compound Miter or Radial Arm Saws: set bevel at 43 1/4°, miter at 13 3/4°)
Explanation: Section 5, pages 24–25.

**Frieze Block against Hip or Valley Rafter:**
Framing Square: **5 13/16 & 24**; set saw at **43 1/4°**
(Compound Miter or Radial Arm Saws: set bevel at 43 1/4°, miter at 13 3/4°)
Explanation: Section 6, page 26.

**Roof Sheathing at Hip or Valley:**
Framing Square: **12 & 12 3/8**; set saw at **9 3/4°**
(Compound Miter or Radial Arm Saws: set bevel at 9 3/4°, miter at 44 1/4°)
Distance "d" = **46 9/16"**. Explanation: Section 7, pages 26–27.

## *Calculating the Length of a Rafter*

| **General Rule:** | | Span of building |
|---|---|---|
| | less | Thickness of ridge board (if any) |
| | gives | Length of rafter to plumb line of seat cut |
| *and* | | Span of building |
| | less | Thickness of ridge board (if any) |
| | plus | 2 × projection of overhang |
| | gives | Total length of rafter to tail cut |

Explanation: Section 4, pages 13–19.

| *Jack Rafters* | | | | | |
|---|---|---|---|---|---|
| SPACED *Inches* | DIFFERENCE *Inches* | SPACED *Inches* | DIFFERENCE *Inches* | SPACED *Inches* | DIFFERENCE *Inches* |
| 1/4 | 1/4 | 15 | 15 7/16 | **32** | **33** |
| 1/2 | 1/2 | **16** | **16 1/2** | 33 | 34 |
| 3/4 | 3/4 | 17 | 17 1/2 | 34 | 35 1/16 |
| 1 | 1 | 18 | 18 9/16 | 35 | 36 1/16 |
| 2 | 2 1/16 | 19 | 19 9/16 | **36** | **37 1/8** |
| 3 | 3 1/16 | 20 | 20 5/8 | 37 | 38 1/8 |
| 4 | 4 1/8 | 21 | 21 5/8 | 38 | 39 3/16 |
| 5 | 5 1/8 | 22 | 22 11/16 | 39 | 40 3/16 |
| 6 | 6 3/16 | 23 | 23 11/16 | 40 | 41 1/4 |
| 7 | 7 3/16 | **24** | **24 3/4** | 41 | 42 1/4 |
| 8 | 8 1/4 | 25 | 25 3/4 | 42 | 43 5/16 |
| 9 | 9 1/4 | 26 | 26 13/16 | 43 | 44 5/16 |
| 10 | 10 5/16 | 27 | 27 13/16 | 44 | 45 3/8 |
| 11 | 11 5/16 | 28 | 28 7/8 | 45 | 46 3/8 |
| **12** | **12 3/8** | 29 | 29 7/8 | 46 | 47 7/16 |
| 13 | 13 3/8 | 30 | 30 15/16 | 47 | 48 7/16 |
| 14 | 14 7/16 | 31 | 31 15/16 | **48** | **49 1/2** |

## *Common Rafters*
## 3 in 12 – 14°

| SPAN *Feet* | LENGTH *Feet Inches* | SPAN *Inches* | LENGTH *Inches* |
|---|---|---|---|
| 1 | 0 - 6 3/16 | 1/4 | 1/8 |
| 2 | 1 - 0 3/8 | 1/2 | 1/4 |
| 3 | 1 - 6 9/16 | 3/4 | 3/8 |
| 4 | 2 - 0 3/4 | 1 | 1/2 |
| 5 | 2 - 6 15/16 | 1 1/4 | 5/8 |
| 6 | 3 - 1 1/8 | 1 1/2 | 3/4 |
| 7 | 3 - 7 5/16 | 1 3/4 | 7/8 |
| 8 | 4 - 1 1/2 | 2 | 1 |
| 9 | 4 - 7 11/16 | 2 1/4 | 1 3/16 |
| 10 | 5 - 1 7/8 | 2 1/2 | 1 5/16 |
| 11 | 5 - 8 | 2 3/4 | 1 7/16 |
| 12 | 6 - 2 3/16 | 3 | 1 9/16 |
| 13 | 6 - 8 3/8 | 3 1/4 | 1 11/16 |
| 14 | 7 - 2 9/16 | 3 1/2 | 1 13/16 |
| 15 | 7 - 8 3/4 | 3 3/4 | 1 15/16 |
| 16 | 8 - 2 15/16 | 4 | 2 1/16 |
| 17 | 8 - 9 1/8 | 4 1/4 | 2 3/16 |
| 18 | 9 - 3 5/16 | 4 1/2 | 2 5/16 |
| 19 | 9 - 9 1/2 | 4 3/4 | 2 7/16 |
| 20 | 10 - 3 11/16 | 5 | 2 9/16 |
| 21 | 10 - 9 7/8 | 5 1/4 | 2 11/16 |
| 22 | 11 - 4 1/16 | 5 1/2 | 2 13/16 |
| 23 | 11 -10 1/4 | 5 3/4 | 2 15/16 |
| 24 | 12 - 4 7/16 | 6 | 3 1/16 |
| 25 | 12 -10 5/8 | 6 1/4 | 3 1/4 |
| 26 | 13 - 4 13/16 | 6 1/2 | 3 3/8 |
| 27 | 13 -11 | 6 3/4 | 3 1/2 |
| 28 | 14 - 5 3/16 | 7 | 3 5/8 |
| 29 | 14 -11 3/8 | 7 1/4 | 3 3/4 |
| 30 | 15 - 5 9/16 | 7 1/2 | 3 7/8 |
| 31 | 15 -11 3/4 | 7 3/4 | 4 |
| 32 | 16 - 5 15/16 | 8 | 4 1/8 |
| 33 | 17 - 0 1/16 | 8 1/4 | 4 1/4 |
| 34 | 17 - 6 1/4 | 8 1/2 | 4 3/8 |
| 35 | 18 - 0 7/16 | 8 3/4 | 4 1/2 |
| 36 | 18 - 6 5/8 | 9 | 4 5/8 |
| 37 | 19 - 0 13/16 | 9 1/4 | 4 3/4 |
| 38 | 19 - 7 | 9 1/2 | 4 7/8 |
| 39 | 20 - 1 3/16 | 9 3/4 | 5 |
| 40 | 20 - 7 3/8 | 10 | 5 1/8 |
| 41 | 21 - 1 9/16 | 10 1/4 | 5 5/16 |
| 42 | 21 - 7 3/4 | 10 1/2 | 5 7/16 |
| 43 | 22 - 1 15/16 | 10 3/4 | 5 9/16 |
| 44 | 22 - 8 1/8 | 11 | 5 11/16 |
| 45 | 23 - 2 5/16 | 11 1/4 | 5 13/16 |
| 46 | 23 - 8 1/2 | 11 1/2 | 5 15/16 |
| 47 | 24 - 2 11/16 | 11 3/4 | 6 1/16 |
| 48 | 24 - 8 7/8 | 12 | 6 3/16 |
| 49 | 25 - 3 1/16 | | |
| 50 | 25 - 9 1/4 | | |

## *Hip or Valley Rafters*
## 3 in 17 – 10°

| SPAN *Feet* | LENGTH *Feet Inches* | SPAN *Inches* | LENGTH *Inches* |
|---|---|---|---|
| 1 | 0 - 8 5/8 | 1/4 | 3/16 |
| 2 | 1 - 5 1/4 | 1/2 | 3/8 |
| 3 | 2 - 1 7/8 | 3/4 | 9/16 |
| 4 | 2 -10 7/16 | 1 | 11/16 |
| 5 | 3 - 7 1/16 | 1 1/4 | 7/8 |
| 6 | 4 - 3 11/16 | 1 1/2 | 1 1/16 |
| 7 | 5 - 0 5/16 | 1 3/4 | 1 1/4 |
| 8 | 5 - 8 15/16 | 2 | 1 7/16 |
| 9 | 6 - 5 9/16 | 2 1/4 | 1 5/8 |
| 10 | 7 - 2 3/16 | 2 1/2 | 1 13/16 |
| 11 | 7 -10 13/16 | 2 3/4 | 2 |
| 12 | 8 - 7 3/8 | 3 | 2 1/8 |
| 13 | 9 - 4 | 3 1/4 | 2 5/16 |
| 14 | 10 - 0 5/8 | 3 1/2 | 2 1/2 |
| 15 | 10 - 9 1/4 | 3 3/4 | 2 11/16 |
| 16 | 11 - 5 7/8 | 4 | 2 7/8 |
| 17 | 12 - 2 1/2 | 4 1/4 | 3 1/16 |
| 18 | 12 -11 1/8 | 4 1/2 | 3 1/4 |
| 19 | 13 - 7 3/4 | 4 3/4 | 3 7/16 |
| 20 | 14 - 4 5/16 | 5 | 3 9/16 |
| 21 | 15 - 0 15/16 | 5 1/4 | 3 3/4 |
| 22 | 15 - 9 9/16 | 5 1/2 | 3 15/16 |
| 23 | 16 - 6 3/16 | 5 3/4 | 4 1/8 |
| 24 | 17 - 2 13/16 | 6 | 4 5/16 |
| 25 | 17 -11 7/16 | 6 1/4 | 4 1/2 |
| 26 | 18 - 8 1/16 | 6 1/2 | 4 11/16 |
| 27 | 19 - 4 5/8 | 6 3/4 | 4 7/8 |
| 28 | 20 - 1 1/4 | 7 | 5 |
| 29 | 20 - 9 7/8 | 7 1/4 | 5 3/16 |
| 30 | 21 - 6 1/2 | 7 1/2 | 5 3/8 |
| 31 | 22 - 3 1/8 | 7 3/4 | 5 9/16 |
| 32 | 22 -11 3/4 | 8 | 5 3/4 |
| 33 | 23 - 8 3/8 | 8 1/4 | 5 15/16 |
| 34 | 24 - 5 | 8 1/2 | 6 1/8 |
| 35 | 25 - 1 9/16 | 8 3/4 | 6 5/16 |
| 36 | 25 -10 3/16 | 9 | 6 7/16 |
| 37 | 26 - 6 13/16 | 9 1/4 | 6 5/8 |
| 38 | 27 - 3 7/16 | 9 1/2 | 6 13/16 |
| 39 | 28 - 0 1/16 | 9 3/4 | 7 |
| 40 | 28 - 8 11/16 | 10 | 7 3/16 |
| 41 | 29 - 5 5/16 | 10 1/4 | 7 3/8 |
| 42 | 30 - 1 15/16 | 10 1/2 | 7 9/16 |
| 43 | 30 -10 1/2 | 10 3/4 | 7 3/4 |
| 44 | 31 - 7 1/8 | 11 | 7 7/8 |
| 45 | 32 - 3 3/4 | 11 1/4 | 8 1/16 |
| 46 | 33 - 0 3/8 | 11 1/2 | 8 1/4 |
| 47 | 33 - 9 | 11 3/4 | 8 7/16 |
| 48 | 34 - 5 5/8 | 12 | 8 5/8 |
| 49 | 35 - 2 1/4 | | |
| 50 | 35 -10 13/16 | | |

# 3 1/2 in 12 – 16 1/4°

### *Cuts and Saw Settings*

**Tail Cut, Hip or Valley Rafters,** with Square Cut Common Rafters:
Framing Square: **4 9/16 & 24**; set saw at **42 3/4°**
(Compound Miter or Radial Arm Saws: set bevel at 42 3/4°, miter at 10 3/4°)
Explanation: Section 3(d), pages 12–13.

**Chamfer Cut, Hip or Valley Rafters:**
Set saw at **11 1/2°**. Explanation: Section 3(d), page 10.

**Fascia Board Miter Cut at Hip or Valley:**
Framing Square: **6 3/4 & 24**; set saw at **42 3/4°**
(Compound Miter or Radial Arm Saws: set bevel at 42 3/4°, miter at 15 3/4°)
Explanation: Section 5, pages 24–25.

**Frieze Block against Hip or Valley Rafter:**
Framing Square: **6 3/4 & 24**; set saw at **42 3/4°**
(Compound Miter or Radial Arm Saws: set bevel at 42 3/4°, miter at 15 3/4°)
Explanation: Section 6, page 26.

**Roof Sheathing at Hip or Valley:**
Framing Square: **12 & 12 1/2**; set saw at **11 1/2°**
(Compound Miter or Radial Arm Saws: set bevel at 11 1/2°, miter at 43 3/4°)
Distance "d" = **46 1/16"**. Explanation: Section 7, pages 26–27.

### *Calculating the Length of a Rafter*

**General Rule:**

| | | |
|---|---|---|
| | | Span of building |
| | less | Thickness of ridge board (if any) |
| | gives | Length of rafter to plumb line of seat cut |
| *and* | | Span of building |
| | less | Thickness of ridge board (if any) |
| | plus | 2 × projection of overhang |
| | gives | Total length of rafter to tail cut |

Explanation: Section 4, pages 13–19.

| *Jack Rafters* | | | | | |
|---|---|---|---|---|---|
| SPACED *Inches* | DIFFERENCE *Inches* | SPACED *Inches* | DIFFERENCE *Inches* | SPACED *Inches* | DIFFERENCE *Inches* |
| 1/4 | 1/4 | 15 | 15 5/8 | **32** | **33 5/16** |
| 1/2 | 1/2 | **16** | **16 11/16** | 33 | 34 3/8 |
| 3/4 | 13/16 | 17 | 17 11/16 | 34 | 35 7/16 |
| 1 | 1 1/16 | 18 | 18 3/4 | 35 | 36 7/16 |
| 2 | 2 1/16 | 19 | 19 13/16 | **36** | **37 1/2** |
| 3 | 3 1/8 | 20 | 20 13/16 | 37 | 38 9/16 |
| 4 | 4 3/16 | 21 | 21 7/8 | 38 | 39 9/16 |
| 5 | 5 3/16 | 22 | 22 15/16 | 39 | 40 5/8 |
| 6 | 6 1/4 | 23 | 23 15/16 | 40 | 41 11/16 |
| 7 | 7 5/16 | **24** | **25** | 41 | 42 11/16 |
| 8 | 8 5/16 | 25 | 26 1/16 | 42 | 43 3/4 |
| 9 | 9 3/8 | 26 | 27 1/16 | 43 | 44 13/16 |
| 10 | 10 7/16 | 27 | 28 1/8 | 44 | 45 13/16 |
| 11 | 11 7/16 | 28 | 29 3/16 | 45 | 46 7/8 |
| **12** | **12 1/2** | 29 | 30 3/16 | 46 | 47 15/16 |
| 13 | 13 9/16 | 30 | 31 1/4 | 47 | 48 15/16 |
| 14 | 14 9/16 | 31 | 32 5/16 | **48** | **50** |

| **Common Rafters**<br>**3 1/2 in 12 – 16 1/4°** | | | |
|---|---|---|---|
| **SPAN** *Feet* | **LENGTH** *Feet Inches* | **SPAN** *Inches* | **LENGTH** *Inches* |
| 1 | 0- 6 1/4 | 1/4 | 1/8 |
| 2 | 1- 0 1/2 | 1/2 | 1/4 |
| 3 | 1- 6 3/4 | 3/4 | 3/8 |
| 4 | 2- 1 | 1 | 1/2 |
| 5 | 2- 7 1/4 | 1 1/4 | 5/8 |
| 6 | 3- 1 1/2 | 1 1/2 | 13/16 |
| 7 | 3- 7 3/4 | 1 3/4 | 15/16 |
| 8 | 4- 2 | 2 | 1 1/16 |
| 9 | 4- 8 1/4 | 2 1/4 | 1 3/16 |
| 10 | 5- 2 1/2 | 2 1/2 | 1 5/16 |
| 11 | 5- 8 3/4 | 2 3/4 | 1 7/16 |
| 12 | 6- 3 | 3 | 1 9/16 |
| 13 | 6- 9 1/4 | 3 1/4 | 1 11/16 |
| 14 | 7- 3 1/2 | 3 1/2 | 1 13/16 |
| 15 | 7- 9 3/4 | 3 3/4 | 1 15/16 |
| 16 | 8- 4 | 4 | 2 1/16 |
| 17 | 8-10 1/4 | 4 1/4 | 2 3/16 |
| 18 | 9- 4 1/2 | 4 1/2 | 2 3/8 |
| 19 | 9-10 3/4 | 4 3/4 | 2 1/2 |
| 20 | 10- 5 | 5 | 2 5/8 |
| 21 | 10-11 1/4 | 5 1/4 | 2 3/4 |
| 22 | 11- 5 1/2 | 5 1/2 | 2 7/8 |
| 23 | 11-11 3/4 | 5 3/4 | 3 |
| 24 | 12- 6 | 6 | 3 1/8 |
| 25 | 13- 0 1/4 | 6 1/4 | 3 1/4 |
| 26 | 13- 6 1/2 | 6 1/2 | 3 3/8 |
| 27 | 14- 0 3/4 | 6 3/4 | 3 1/2 |
| 28 | 14- 7 | 7 | 3 5/8 |
| 29 | 15- 1 1/4 | 7 1/4 | 3 3/4 |
| 30 | 15- 7 1/2 | 7 1/2 | 3 15/16 |
| 31 | 16- 1 3/4 | 7 3/4 | 4 1/16 |
| 32 | 16- 8 | 8 | 4 3/16 |
| 33 | 17- 2 1/4 | 8 1/4 | 4 5/16 |
| 34 | 17- 8 1/2 | 8 1/2 | 4 7/16 |
| 35 | 18- 2 3/4 | 8 3/4 | 4 9/16 |
| 36 | 18- 9 | 9 | 4 11/16 |
| 37 | 19- 3 1/4 | 9 1/4 | 4 13/16 |
| 38 | 19- 9 1/2 | 9 1/2 | 4 15/16 |
| 39 | 20- 3 3/4 | 9 3/4 | 5 1/16 |
| 40 | 20-10 | 10 | 5 3/16 |
| 41 | 21- 4 1/4 | 10 1/4 | 5 5/16 |
| 42 | 21-10 1/2 | 10 1/2 | 5 1/2 |
| 43 | 22- 4 3/4 | 10 3/4 | 5 5/8 |
| 44 | 22-11 | 11 | 5 3/4 |
| 45 | 23- 5 1/4 | 11 1/4 | 5 7/8 |
| 46 | 23-11 1/2 | 11 1/2 | 6 |
| 47 | 24- 5 3/4 | 11 3/4 | 6 1/8 |
| 48 | 25- 0 | 12 | 6 1/4 |
| 49 | 25- 6 1/4 | | |
| 50 | 26- 0 1/2 | | |

| **Hip or Valley Rafters**<br>**3 1/2 in 17 – 11 3/4°** | | | |
|---|---|---|---|
| **SPAN** *Feet* | **LENGTH** *Feet Inches* | **SPAN** *Inches* | **LENGTH** *Inches* |
| 1 | 0- 8 11/16 | 1/4 | 3/16 |
| 2 | 1- 5 5/16 | 1/2 | 3/8 |
| 3 | 2- 2 | 3/4 | 9/16 |
| 4 | 2-10 5/8 | 1 | 3/4 |
| 5 | 3- 7 5/16 | 1 1/4 | 7/8 |
| 6 | 4- 4 | 1 1/2 | 1 1/16 |
| 7 | 5- 0 5/8 | 1 3/4 | 1 1/4 |
| 8 | 5- 9 5/16 | 2 | 1 7/16 |
| 9 | 6- 6 | 2 1/4 | 1 5/8 |
| 10 | 7- 2 5/8 | 2 1/2 | 1 13/16 |
| 11 | 7-11 5/16 | 2 3/4 | 2 |
| 12 | 8- 7 15/16 | 3 | 2 3/16 |
| 13 | 9- 4 5/8 | 3 1/4 | 2 3/8 |
| 14 | 10- 1 5/16 | 3 1/2 | 2 1/2 |
| 15 | 10- 9 15/16 | 3 3/4 | 2 11/16 |
| 16 | 11- 6 5/8 | 4 | 2 7/8 |
| 17 | 12- 3 5/16 | 4 1/4 | 3 1/16 |
| 18 | 12-11 15/16 | 4 1/2 | 3 1/4 |
| 19 | 13- 8 5/8 | 4 3/4 | 3 7/16 |
| 20 | 14- 5 1/4 | 5 | 3 5/8 |
| 21 | 15- 1 15/16 | 5 1/4 | 3 13/16 |
| 22 | 15-10 5/8 | 5 1/2 | 4 |
| 23 | 16- 7 1/4 | 5 3/4 | 4 1/8 |
| 24 | 17- 3 15/16 | 6 | 4 5/16 |
| 25 | 18- 0 5/8 | 6 1/4 | 4 1/2 |
| 26 | 18- 9 1/4 | 6 1/2 | 4 11/16 |
| 27 | 19- 5 15/16 | 6 3/4 | 4 7/8 |
| 28 | 20- 2 9/16 | 7 | 5 1/16 |
| 29 | 20-11 1/4 | 7 1/4 | 5 1/4 |
| 30 | 21- 7 15/16 | 7 1/2 | 5 7/16 |
| 31 | 22- 4 9/16 | 7 3/4 | 5 5/8 |
| 32 | 23- 1 1/4 | 8 | 5 3/4 |
| 33 | 23- 9 15/16 | 8 1/4 | 5 15/16 |
| 34 | 24- 6 9/16 | 8 1/2 | 6 1/8 |
| 35 | 25- 3 1/4 | 8 3/4 | 6 5/16 |
| 36 | 25-11 7/8 | 9 | 6 1/2 |
| 37 | 26- 8 9/16 | 9 1/4 | 6 11/16 |
| 38 | 27- 5 1/4 | 9 1/2 | 6 7/8 |
| 39 | 28- 1 7/8 | 9 3/4 | 7 1/16 |
| 40 | 28-10 9/16 | 10 | 7 1/4 |
| 41 | 29- 7 3/16 | 10 1/4 | 7 3/8 |
| 42 | 30- 3 7/8 | 10 1/2 | 7 9/16 |
| 43 | 31- 0 9/16 | 10 3/4 | 7 3/4 |
| 44 | 31- 9 3/16 | 11 | 7 15/16 |
| 45 | 32- 5 7/8 | 11 1/4 | 8 1/8 |
| 46 | 33- 2 9/16 | 11 1/2 | 8 5/16 |
| 47 | 33-11 3/16 | 11 3/4 | 8 1/2 |
| 48 | 34- 7 7/8 | 12 | 8 11/16 |
| 49 | 35- 4 1/2 | | |
| 50 | 36- 1 3/16 | | |

# 4 in 12 – 18½°

## *Cuts and Saw Settings*

**Tail Cut, Hip or Valley Rafters,** with Square Cut Common Rafters:
Framing Square: **5 1/16 & 24**; set saw at **42 1/4°**
(Compound Miter or Radial Arm Saws: set bevel at 42 1/4°, miter at 12°)
Explanation: Section 3(d), pages 12–13.

**Chamfer Cut, Hip or Valley Rafters:**
Set saw at **13°**. Explanation: Section 3(d), page 10.

**Fascia Board Miter Cut at Hip or Valley:**
Framing Square: **7 9/16 & 24**; set saw at **42 1/4°**
(Compound Miter or Radial Arm Saws: set bevel at 42 1/4°, miter at 17 1/2°)
Explanation: Section 5, pages 24–25.

**Frieze Block against Hip or Valley Rafter:**
Framing Square: **7 9/16 & 24**; set saw at **42 1/4°**
(Compound Miter or Radial Arm Saws: set bevel at 42 1/4°, miter at 17 1/2°)
Explanation: Section 6, page 26.

**Roof Sheathing at Hip or Valley:**
Framing Square: **12 & 12 5/8**; set saw at **13°**
(Compound Miter or Radial Arm Saws: set bevel at 13°, miter at 43 1/2°)
Distance "d" = **45 9/16"**. Explanation: Section 7, pages 26–27.

## *Calculating the Length of a Rafter*

**General Rule:**

| | | |
|---|---|---|
| **General Rule:** | | Span of building |
| | less | Thickness of ridge board (if any) |
| | gives | Length of rafter to plumb line of seat cut |
| *and* | | Span of building |
| | less | Thickness of ridge board (if any) |
| | plus | 2 × projection of overhang |
| | gives | Total length of rafter to tail cut |

Explanation: Section 4, pages 13–19.

| *Jack Rafters* | | | | | |
|---|---|---|---|---|---|
| SPACED *Inches* | DIFFERENCE *Inches* | SPACED *Inches* | DIFFERENCE *Inches* | SPACED *Inches* | DIFFERENCE *Inches* |
| 1/4 | 1/4 | 15 | 15 13/16 | **32** | **33 3/4** |
| 1/2 | 1/2 | **16** | **16 7/8** | 33 | 34 13/16 |
| 3/4 | 13/16 | 17 | 17 15/16 | 34 | 35 13/16 |
| 1 | 1 1/16 | 18 | 19 | 35 | 36 7/8 |
| 2 | 2 1/8 | 19 | 20 | **36** | **37 15/16** |
| 3 | 3 3/16 | 20 | 21 1/16 | 37 | 39 |
| 4 | 4 3/16 | 21 | 22 1/8 | 38 | 40 1/16 |
| 5 | 5 1/4 | 22 | 23 3/16 | 39 | 41 1/8 |
| 6 | 6 5/16 | 23 | 24 1/4 | 40 | 42 3/16 |
| 7 | 7 3/8 | **24** | **25 5/16** | 41 | 43 3/16 |
| 8 | 8 7/16 | 25 | 26 3/8 | 42 | 44 1/4 |
| 9 | 9 1/2 | 26 | 27 7/16 | 43 | 45 5/16 |
| 10 | 10 9/16 | 27 | 28 7/16 | 44 | 46 3/8 |
| 11 | 11 5/8 | 28 | 29 1/2 | 45 | 47 7/16 |
| **12** | **12 5/8** | 29 | 30 9/16 | 46 | 48 1/2 |
| 13 | 13 11/16 | 30 | 31 5/8 | 47 | 49 9/16 |
| 14 | 14 3/4 | 31 | 32 11/16 | **48** | **50 5/8** |

| **Common Rafters**<br>**4 in 12 – 18 1/2°** | | | |
|---|---|---|---|
| SPAN<br>*Feet* | LENGTH<br>*Feet Inches* | SPAN<br>*Inches* | LENGTH<br>*Inches* |
| 1 | 0- 6 5/16 | 1/4 | 1/8 |
| 2 | 1- 0 5/8 | 1/2 | 1/4 |
| 3 | 1- 7 | 3/4 | 3/8 |
| 4 | 2- 1 5/16 | 1 | 1/2 |
| 5 | 2- 7 5/8 | 1 1/4 | 11/16 |
| 6 | 3- 1 15/16 | 1 1/2 | 13/16 |
| 7 | 3- 8 1/4 | 1 3/4 | 15/16 |
| 8 | 4- 2 5/8 | 2 | 1 1/16 |
| 9 | 4- 8 15/16 | 2 1/4 | 1 3/16 |
| 10 | 5- 3 1/4 | 2 1/2 | 1 5/16 |
| 11 | 5- 9 9/16 | 2 3/4 | 1 7/16 |
| 12 | 6- 3 7/8 | 3 | 1 9/16 |
| 13 | 6-10 1/4 | 3 1/4 | 1 11/16 |
| 14 | 7- 4 9/16 | 3 1/2 | 1 7/8 |
| 15 | 7-10 7/8 | 3 3/4 | 2 |
| 16 | 8- 5 3/16 | 4 | 2 1/8 |
| 17 | 8-11 1/2 | 4 1/4 | 2 1/4 |
| 18 | 9- 5 13/16 | 4 1/2 | 2 3/8 |
| 19 | 10- 0 3/16 | 4 3/4 | 2 1/2 |
| 20 | 10- 6 1/2 | 5 | 2 5/8 |
| 21 | 11- 0 13/16 | 5 1/4 | 2 3/4 |
| 22 | 11- 7 1/8 | 5 1/2 | 2 7/8 |
| 23 | 12- 1 7/16 | 5 3/4 | 3 |
| 24 | 12- 7 13/16 | 6 | 3 3/16 |
| 25 | 13- 2 1/8 | 6 1/4 | 3 5/16 |
| 26 | 13- 8 7/16 | 6 1/2 | 3 7/16 |
| 27 | 14- 2 3/4 | 6 3/4 | 3 9/16 |
| 28 | 14- 9 1/16 | 7 | 3 11/16 |
| 29 | 15- 3 7/16 | 7 1/4 | 3 13/16 |
| 30 | 15- 9 3/4 | 7 1/2 | 3 15/16 |
| 31 | 16- 4 1/16 | 7 3/4 | 4 1/16 |
| 32 | 16-10 3/8 | 8 | 4 3/16 |
| 33 | 17- 4 11/16 | 8 1/4 | 4 3/8 |
| 34 | 17-11 1/16 | 8 1/2 | 4 1/2 |
| 35 | 18- 5 3/8 | 8 3/4 | 4 5/8 |
| 36 | 18-11 11/16 | 9 | 4 3/4 |
| 37 | 19- 6 | 9 1/4 | 4 7/8 |
| 38 | 20- 0 5/16 | 9 1/2 | 5 |
| 39 | 20- 6 11/16 | 9 3/4 | 5 1/8 |
| 40 | 21- 1 | 10 | 5 1/4 |
| 41 | 21- 7 5/16 | 10 1/4 | 5 3/8 |
| 42 | 22- 1 5/8 | 10 1/2 | 5 9/16 |
| 43 | 22- 7 15/16 | 10 3/4 | 5 11/16 |
| 44 | 23- 2 1/4 | 11 | 5 13/16 |
| 45 | 23- 8 5/8 | 11 1/4 | 5 15/16 |
| 46 | 24- 2 15/16 | 11 1/2 | 6 1/16 |
| 47 | 24- 9 1/4 | 11 3/4 | 6 3/16 |
| 48 | 25- 3 9/16 | 12 | 6 5/16 |
| 49 | 25- 9 7/8 | | |
| 50 | 26- 4 1/4 | | |

| **Hip or Valley Rafters**<br>**4 in 17 – 13 1/4°** | | | |
|---|---|---|---|
| SPAN<br>*Feet* | LENGTH<br>*Feet Inches* | SPAN<br>*Inches* | LENGTH<br>*Inches* |
| 1 | 0- 8 11/16 | 1/4 | 3/16 |
| 2 | 1- 5 7/16 | 1/2 | 3/8 |
| 3 | 2- 2 1/8 | 3/4 | 9/16 |
| 4 | 2-10 7/8 | 1 | 3/4 |
| 5 | 3- 7 9/16 | 1 1/4 | 15/16 |
| 6 | 4- 4 5/16 | 1 1/2 | 1 1/16 |
| 7 | 5- 1 | 1 3/4 | 1 1/4 |
| 8 | 5- 9 3/4 | 2 | 1 7/16 |
| 9 | 6- 6 7/16 | 2 1/4 | 1 5/8 |
| 10 | 7- 3 3/16 | 2 1/2 | 1 13/16 |
| 11 | 7-11 7/8 | 2 3/4 | 2 |
| 12 | 8- 8 5/8 | 3 | 2 3/16 |
| 13 | 9- 5 5/16 | 3 1/4 | 2 3/8 |
| 14 | 10- 2 1/16 | 3 1/2 | 2 9/16 |
| 15 | 10-10 3/4 | 3 3/4 | 2 3/4 |
| 16 | 11- 7 1/2 | 4 | 2 7/8 |
| 17 | 12- 4 3/16 | 4 1/4 | 3 1/16 |
| 18 | 13- 0 15/16 | 4 1/2 | 3 1/4 |
| 19 | 13- 9 5/8 | 4 3/4 | 3 7/16 |
| 20 | 14- 6 3/8 | 5 | 3 5/8 |
| 21 | 15- 3 1/16 | 5 1/4 | 3 13/16 |
| 22 | 15-11 13/16 | 5 1/2 | 4 |
| 23 | 16- 8 1/2 | 5 3/4 | 4 3/16 |
| 24 | 17- 5 1/4 | 6 | 4 3/8 |
| 25 | 18- 1 15/16 | 6 1/4 | 4 9/16 |
| 26 | 18-10 11/16 | 6 1/2 | 4 3/4 |
| 27 | 19- 7 3/8 | 6 3/4 | 4 7/8 |
| 28 | 20- 4 1/8 | 7 | 5 1/16 |
| 29 | 21- 0 13/16 | 7 1/4 | 5 1/4 |
| 30 | 21- 9 9/16 | 7 1/2 | 5 7/16 |
| 31 | 22- 6 1/4 | 7 3/4 | 5 5/8 |
| 32 | 23- 3 | 8 | 5 13/16 |
| 33 | 23-11 11/16 | 8 1/4 | 6 |
| 34 | 24- 8 3/8 | 8 1/2 | 6 3/16 |
| 35 | 25- 5 1/8 | 8 3/4 | 6 3/8 |
| 36 | 26- 1 13/16 | 9 | 6 9/16 |
| 37 | 26-10 9/16 | 9 1/4 | 6 3/4 |
| 38 | 27- 7 1/4 | 9 1/2 | 6 7/8 |
| 39 | 28- 4 | 9 3/4 | 7 1/16 |
| 40 | 29- 0 11/16 | 10 | 7 1/4 |
| 41 | 29- 9 7/16 | 10 1/4 | 7 7/16 |
| 42 | 30- 6 1/8 | 10 1/2 | 7 5/8 |
| 43 | 31- 2 7/8 | 10 3/4 | 7 13/16 |
| 44 | 31-11 9/16 | 11 | 8 |
| 45 | 32- 8 5/16 | 11 1/4 | 8 3/16 |
| 46 | 33- 5 | 11 1/2 | 8 3/8 |
| 47 | 34- 1 3/4 | 11 3/4 | 8 9/16 |
| 48 | 34-10 7/16 | 12 | 8 11/16 |
| 49 | 35- 7 3/16 | | |
| 50 | 36- 3 7/8 | | |

# 4 1/2 in 12 – 20 1/2°

## *Cuts and Saw Settings*

**Tail Cut, Hip or Valley Rafters,** with Square Cut Common Rafters:
Framing Square: **5 9/16 & 24**; set saw at **41 1/2°**
(Compound Miter or Radial Arm Saws: set bevel at 41 1/2°, miter at 13°)
Explanation: Section 3(d), pages 12–13.

**Chamfer Cut, Hip or Valley Rafters:**
Set saw at **14 1/2°**. Explanation: Section 3(d), page 10.

**Fascia Board Miter Cut at Hip or Valley:**
Framing Square: **8 7/16 & 24**; set saw at **41 1/2°**
(Compound Miter or Radial Arm Saws: set bevel at 41 1/2°, miter at 19 1/4°)
Explanation: Section 5, pages 24–25.

**Frieze Block against Hip or Valley Rafter:**
Framing Square: **8 7/16 & 24**; set saw at **41 1/2°**
(Compound Miter or Radial Arm Saws: set bevel at 41 1/2°, miter at 19 1/4°)
Explanation: Section 6, page 26.

**Roof Sheathing at Hip or Valley:**
Framing Square: **12 & 12 13/16**; set saw at **14 1/2°**
(Compound Miter or Radial Arm Saws: set bevel at 14 1/2°, miter at 43°)
Distance "d" = **44 15/16"**. Explanation: Section 7, pages 26–27.

## *Calculating the Length of a Rafter*

| **General Rule:** | | Span of building |
|---|---|---|
| | less | Thickness of ridge board (if any) |
| | gives | Length of rafter to plumb line of seat cut |
| *and* | | Span of building |
| | less | Thickness of ridge board (if any) |
| | plus | 2 × projection of overhang |
| | gives | Total length of rafter to tail cut |

Explanation: Section 4, pages 13–19.

| *Jack Rafters* | | | | | |
|---|---|---|---|---|---|
| SPACED *Inches* | DIFFERENCE *Inches* | SPACED *Inches* | DIFFERENCE *Inches* | SPACED *Inches* | DIFFERENCE *Inches* |
| 1/4 | 1/4 | 15 | 16 | **32** | **34 3/16** |
| 1/2 | 9/16 | **16** | **17 1/16** | 33 | 35 1/4 |
| 3/4 | 13/16 | 17 | 18 1/8 | 34 | 36 5/16 |
| 1 | 1 1/16 | 18 | 19 1/4 | 35 | 37 3/8 |
| 2 | 2 1/8 | 19 | 20 5/16 | **36** | **38 7/16** |
| 3 | 3 3/16 | 20 | 21 3/8 | 37 | 39 1/2 |
| 4 | 4 1/4 | 21 | 22 7/16 | 38 | 40 9/16 |
| 5 | 5 5/16 | 22 | 23 1/2 | 39 | 41 5/8 |
| 6 | 6 7/16 | 23 | 24 9/16 | 40 | 42 3/4 |
| 7 | 7 1/2 | **24** | **25 5/8** | 41 | 43 13/16 |
| 8 | 8 9/16 | 25 | 26 11/16 | 42 | 44 7/8 |
| 9 | 9 5/8 | 26 | 27 3/4 | 43 | 45 15/16 |
| 10 | 10 11/16 | 27 | 28 13/16 | 44 | 47 |
| 11 | 11 3/4 | 28 | 29 7/8 | 45 | 48 1/16 |
| **12** | **12 13/16** | 29 | 31 | 46 | 49 1/8 |
| 13 | 13 7/8 | 30 | 32 1/16 | 47 | 50 3/16 |
| 14 | 14 15/16 | 31 | 33 1/8 | **48** | **51 1/4** |

## Common Rafters
### 4 1/2 in 12 – 20 1/2°

| SPAN Feet | LENGTH Feet Inches | SPAN Inches | LENGTH Inches |
|---|---|---|---|
| 1 | 0- 6 7/16 | 1/4 | 1/8 |
| 2 | 1- 0 13/16 | 1/2 | 1/4 |
| 3 | 1- 7 1/4 | 3/4 | 3/8 |
| 4 | 2- 1 5/8 | 1 | 9/16 |
| 5 | 2- 8 1/16 | 1 1/4 | 11/16 |
| 6 | 3- 2 7/16 | 1 1/2 | 13/16 |
| 7 | 3- 8 7/8 | 1 3/4 | 15/16 |
| 8 | 4- 3 1/4 | 2 | 1 1/16 |
| 9 | 4- 9 11/16 | 2 1/4 | 1 3/16 |
| 10 | 5- 4 1/16 | 2 1/2 | 1 5/16 |
| 11 | 5-10 1/2 | 2 3/4 | 1 7/16 |
| 12 | 6- 4 7/8 | 3 | 1 5/8 |
| 13 | 6-11 5/16 | 3 1/4 | 1 3/4 |
| 14 | 7- 5 11/16 | 3 1/2 | 1 7/8 |
| 15 | 8- 0 1/8 | 3 3/4 | 2 |
| 16 | 8- 6 1/2 | 4 | 2 1/8 |
| 17 | 9- 0 15/16 | 4 1/4 | 2 1/4 |
| 18 | 9- 7 3/8 | 4 1/2 | 2 3/8 |
| 19 | 10- 1 3/4 | 4 3/4 | 2 9/16 |
| 20 | 10- 8 3/16 | 5 | 2 11/16 |
| 21 | 11- 2 9/16 | 5 1/4 | 2 13/16 |
| 22 | 11- 9 | 5 1/2 | 2 15/16 |
| 23 | 12- 3 3/8 | 5 3/4 | 3 1/16 |
| 24 | 12- 9 13/16 | 6 | 3 3/16 |
| 25 | 13- 4 3/16 | 6 1/4 | 3 5/16 |
| 26 | 13-10 5/8 | 6 1/2 | 3 1/2 |
| 27 | 14- 5 | 6 3/4 | 3 5/8 |
| 28 | 14-11 7/16 | 7 | 3 3/4 |
| 29 | 15- 5 13/16 | 7 1/4 | 3 7/8 |
| 30 | 16- 0 1/4 | 7 1/2 | 4 |
| 31 | 16- 6 5/8 | 7 3/4 | 4 1/8 |
| 32 | 17- 1 1/16 | 8 | 4 1/4 |
| 33 | 17- 7 7/16 | 8 1/4 | 4 3/8 |
| 34 | 18- 1 7/8 | 8 1/2 | 4 9/16 |
| 35 | 18- 8 1/4 | 8 3/4 | 4 11/16 |
| 36 | 19- 2 11/16 | 9 | 4 13/16 |
| 37 | 19- 9 1/8 | 9 1/4 | 4 15/16 |
| 38 | 20- 3 1/2 | 9 1/2 | 5 1/16 |
| 39 | 20- 9 15/16 | 9 3/4 | 5 3/16 |
| 40 | 21- 4 5/16 | 10 | 5 5/16 |
| 41 | 21-10 3/4 | 10 1/4 | 5 1/2 |
| 42 | 22- 5 1/8 | 10 1/2 | 5 5/8 |
| 43 | 22-11 9/16 | 10 3/4 | 5 3/4 |
| 44 | 23- 5 15/16 | 11 | 5 7/8 |
| 45 | 24- 0 3/8 | 11 1/4 | 6 |
| 46 | 24- 6 3/4 | 11 1/2 | 6 1/8 |
| 47 | 25- 1 3/16 | 11 3/4 | 6 1/4 |
| 48 | 25- 7 9/16 | 12 | 6 7/16 |
| 49 | 26- 2 | | |
| 50 | 26- 8 3/8 | | |

## Hip or Valley Rafters
### 4 1/2 in 17 – 14 3/4°

| SPAN Feet | LENGTH Feet Inches | SPAN Inches | LENGTH Inches |
|---|---|---|---|
| 1 | 0- 8 3/4 | 1/4 | 3/16 |
| 2 | 1- 5 9/16 | 1/2 | 3/8 |
| 3 | 2- 2 5/16 | 3/4 | 9/16 |
| 4 | 2-11 1/8 | 1 | 3/4 |
| 5 | 3- 7 7/8 | 1 1/4 | 15/16 |
| 6 | 4- 4 11/16 | 1 1/2 | 1 1/8 |
| 7 | 5- 1 7/16 | 1 3/4 | 1 1/4 |
| 8 | 5-10 1/4 | 2 | 1 7/16 |
| 9 | 6- 7 | 2 1/4 | 1 5/8 |
| 10 | 7- 3 13/16 | 2 1/2 | 1 13/16 |
| 11 | 8- 0 9/16 | 2 3/4 | 2 |
| 12 | 8- 9 5/16 | 3 | 2 3/16 |
| 13 | 9- 6 1/8 | 3 1/4 | 2 3/8 |
| 14 | 10- 2 7/8 | 3 1/2 | 2 9/16 |
| 15 | 10-11 11/16 | 3 3/4 | 2 3/4 |
| 16 | 11- 8 7/16 | 4 | 2 15/16 |
| 17 | 12- 5 1/4 | 4 1/4 | 3 1/8 |
| 18 | 13- 2 | 4 1/2 | 3 5/16 |
| 19 | 13-10 13/16 | 4 3/4 | 3 1/2 |
| 20 | 14- 7 9/16 | 5 | 3 11/16 |
| 21 | 15- 4 3/8 | 5 1/4 | 3 13/16 |
| 22 | 16- 1 1/8 | 5 1/2 | 4 |
| 23 | 16- 9 7/8 | 5 3/4 | 4 3/16 |
| 24 | 17- 6 11/16 | 6 | 4 3/8 |
| 25 | 18- 3 7/16 | 6 1/4 | 4 9/16 |
| 26 | 19- 0 1/4 | 6 1/2 | 4 3/4 |
| 27 | 19- 9 | 6 3/4 | 4 15/16 |
| 28 | 20- 5 13/16 | 7 | 5 1/8 |
| 29 | 21- 2 9/16 | 7 1/4 | 5 5/16 |
| 30 | 21-11 3/8 | 7 1/2 | 5 1/2 |
| 31 | 22- 8 1/8 | 7 3/4 | 5 11/16 |
| 32 | 23- 4 15/16 | 8 | 5 7/8 |
| 33 | 24- 1 11/16 | 8 1/4 | 6 1/16 |
| 34 | 24-10 1/2 | 8 1/2 | 6 3/16 |
| 35 | 25- 7 1/4 | 8 3/4 | 6 3/8 |
| 36 | 26- 4 | 9 | 6 9/16 |
| 37 | 27- 0 13/16 | 9 1/4 | 6 3/4 |
| 38 | 27- 9 9/16 | 9 1/2 | 6 15/16 |
| 39 | 28- 6 3/8 | 9 3/4 | 7 1/8 |
| 40 | 29- 3 1/8 | 10 | 7 5/16 |
| 41 | 29-11 15/16 | 10 1/4 | 7 1/2 |
| 42 | 30- 8 11/16 | 10 1/2 | 7 11/16 |
| 43 | 31- 5 1/2 | 10 3/4 | 7 7/8 |
| 44 | 32- 2 1/4 | 11 | 8 1/16 |
| 45 | 32-11 1/16 | 11 1/4 | 8 1/4 |
| 46 | 33- 7 13/16 | 11 1/2 | 8 7/16 |
| 47 | 34- 4 9/16 | 11 3/4 | 8 5/8 |
| 48 | 35- 1 3/8 | 12 | 8 3/4 |
| 49 | 35-10 1/8 | | |
| 50 | 36- 6 15/16 | | |

# 5 in 12 – $22^1/_2°$

## *Cuts and Saw Settings*

**Tail Cut, Hip or Valley Rafters,** with Square Cut Common Rafters:
Framing Square: **6 & 24**; set saw at **40 3/4°**
(Compound Miter or Radial Arm Saws: set bevel at 40 3/4°, miter at 14°)
Explanation: Section 3(d), pages 12–13.

**Chamfer Cut, Hip or Valley Rafters:**
Set saw at **15 3/4°**. Explanation: Section 3(d), page 10.

**Fascia Board Miter Cut at Hip or Valley:**
Framing Square: **9 1/4 & 24**; set saw at **40 3/4°**
(Compound Miter or Radial Arm Saws: set bevel at 40 3/4°, miter at 21°)
Explanation: Section 5, pages 24–25.

**Frieze Block against Hip or Valley Rafter:**
Framing Square: **9 1/4 & 24**; set saw at **40 3/4°**
(Compound Miter or Radial Arm Saws: set bevel at 40 3/4°, miter at 21°)
Explanation: Section 6, page 26.

**Roof Sheathing at Hip or Valley:**
Framing Square: **12 & 13**; set saw at **15 3/4°**
(Compound Miter or Radial Arm Saws: set bevel at 15 3/4°, miter at 42 3/4°)
Distance "d" = **44 5/16"**. Explanation: Section 7, pages 26–27.

## *Calculating the Length of a Rafter*

| | | |
|---|---|---|
| **General Rule:** | | Span of building |
| | less | Thickness of ridge board (if any) |
| | gives | Length of rafter to plumb line of seat cut |
| *and* | | Span of building |
| | less | Thickness of ridge board (if any) |
| | plus | 2 × projection of overhang |
| | gives | Total length of rafter to tail cut |

Explanation: Section 4, pages 13–19.

| *Jack Rafters* | | | | | |
|---|---|---|---|---|---|
| SPACED *Inches* | DIFFERENCE *Inches* | SPACED *Inches* | DIFFERENCE *Inches* | SPACED *Inches* | DIFFERENCE *Inches* |
| 1/4 | 1/4 | 15 | 16 1/4 | **32** | **34 11/16** |
| 1/2 | 9/16 | **16** | **17 5/16** | 33 | 35 3/4 |
| 3/4 | 13/16 | 17 | 18 7/16 | 34 | 36 13/16 |
| 1 | 1 1/16 | 18 | 19 1/2 | 35 | 37 15/16 |
| 2 | 2 3/16 | 19 | 20 9/16 | **36** | **39** |
| 3 | 3 1/4 | 20 | 21 11/16 | 37 | 40 1/16 |
| 4 | 4 5/16 | 21 | 22 3/4 | 38 | 41 3/16 |
| 5 | 5 7/16 | 22 | 23 13/16 | 39 | 42 1/4 |
| 6 | 6 1/2 | 23 | 24 15/16 | 40 | 43 5/16 |
| 7 | 7 9/16 | **24** | **26** | 41 | 44 7/16 |
| 8 | 8 11/16 | 25 | 27 1/16 | 42 | 45 1/2 |
| 9 | 9 3/4 | 26 | 28 3/16 | 43 | 46 9/16 |
| 10 | 10 13/16 | 27 | 29 1/4 | 44 | 47 11/16 |
| 11 | 11 15/16 | 28 | 30 5/16 | 45 | 48 3/4 |
| **12** | **13** | 29 | 31 7/16 | 46 | 49 13/16 |
| 13 | 14 1/16 | 30 | 32 1/2 | 47 | 50 15/16 |
| 14 | 15 3/16 | 31 | 33 9/16 | **48** | **52** |

## Common Rafters
### 5 in 12 – 22 1/2°

| SPAN *Feet* | LENGTH *Feet Inches* | SPAN *Inches* | LENGTH *Inches* |
|---|---|---|---|
| 1 | 0- 6 1/2 | 1/4 | 1/8 |
| 2 | 1- 1 | 1/2 | 1/4 |
| 3 | 1- 7 1/2 | 3/4 | 7/16 |
| 4 | 2- 2 | 1 | 9/16 |
| 5 | 2- 8 1/2 | 1 1/4 | 11/16 |
| 6 | 3- 3 | 1 1/2 | 13/16 |
| 7 | 3- 9 1/2 | 1 3/4 | 15/16 |
| 8 | 4- 4 | 2 | 1 1/16 |
| 9 | 4-10 1/2 | 2 1/4 | 1 1/4 |
| 10 | 5- 5 | 2 1/2 | 1 3/8 |
| 11 | 5-11 1/2 | 2 3/4 | 1 1/2 |
| 12 | 6- 6 | 3 | 1 5/8 |
| 13 | 7- 0 1/2 | 3 1/4 | 1 3/4 |
| 14 | 7- 7 | 3 1/2 | 1 7/8 |
| 15 | 8- 1 1/2 | 3 3/4 | 2 1/16 |
| 16 | 8- 8 | 4 | 2 3/16 |
| 17 | 9- 2 1/2 | 4 1/4 | 2 5/16 |
| 18 | 9- 9 | 4 1/2 | 2 7/16 |
| 19 | 10- 3 1/2 | 4 3/4 | 2 9/16 |
| 20 | 10-10 | 5 | 2 11/16 |
| 21 | 11- 4 1/2 | 5 1/4 | 2 7/8 |
| 22 | 11-11 | 5 1/2 | 3 |
| 23 | 12- 5 1/2 | 5 3/4 | 3 1/8 |
| 24 | 13- 0 | 6 | 3 1/4 |
| 25 | 13- 6 1/2 | 6 1/4 | 3 3/8 |
| 26 | 14- 1 | 6 1/2 | 3 1/2 |
| 27 | 14- 7 1/2 | 6 3/4 | 3 11/16 |
| 28 | 15- 2 | 7 | 3 13/16 |
| 29 | 15- 8 1/2 | 7 1/4 | 3 15/16 |
| 30 | 16- 3 | 7 1/2 | 4 1/16 |
| 31 | 16- 9 1/2 | 7 3/4 | 4 3/16 |
| 32 | 17- 4 | 8 | 4 5/16 |
| 33 | 17-10 1/2 | 8 1/4 | 4 1/2 |
| 34 | 18- 5 | 8 1/2 | 4 5/8 |
| 35 | 18-11 1/2 | 8 3/4 | 4 3/4 |
| 36 | 19- 6 | 9 | 4 7/8 |
| 37 | 20- 0 1/2 | 9 1/4 | 5 |
| 38 | 20- 7 | 9 1/2 | 5 1/8 |
| 39 | 21- 1 1/2 | 9 3/4 | 5 5/16 |
| 40 | 21- 8 | 10 | 5 7/16 |
| 41 | 22- 2 1/2 | 10 1/4 | 5 9/16 |
| 42 | 22- 9 | 10 1/2 | 5 11/16 |
| 43 | 23- 3 1/2 | 10 3/4 | 5 13/16 |
| 44 | 23-10 | 11 | 5 15/16 |
| 45 | 24- 4 1/2 | 11 1/4 | 6 1/8 |
| 46 | 24-11 | 11 1/2 | 6 1/4 |
| 47 | 25- 5 1/2 | 11 3/4 | 6 3/8 |
| 48 | 26- 0 | 12 | 6 1/2 |
| 49 | 26- 6 1/2 | | |
| 50 | 27- 1 | | |

## Hip or Valley Rafters
### 5 in 17 – 16 1/2°

| SPAN *Feet* | LENGTH *Feet Inches* | SPAN *Inches* | LENGTH *Inches* |
|---|---|---|---|
| 1 | 0- 8 7/8 | 1/4 | 3/16 |
| 2 | 1- 5 11/16 | 1/2 | 3/8 |
| 3 | 2- 2 9/16 | 3/4 | 9/16 |
| 4 | 2-11 3/8 | 1 | 3/4 |
| 5 | 3- 8 1/4 | 1 1/4 | 15/16 |
| 6 | 4- 5 1/16 | 1 1/2 | 1 1/8 |
| 7 | 5- 1 15/16 | 1 3/4 | 1 5/16 |
| 8 | 5-10 3/4 | 2 | 1 1/2 |
| 9 | 6- 7 5/8 | 2 1/4 | 1 11/16 |
| 10 | 7- 4 7/16 | 2 1/2 | 1 13/16 |
| 11 | 8- 1 5/16 | 2 3/4 | 2 |
| 12 | 8-10 1/8 | 3 | 2 3/16 |
| 13 | 9- 7 | 3 1/4 | 2 3/8 |
| 14 | 10- 3 13/16 | 3 1/2 | 2 9/16 |
| 15 | 11- 0 11/16 | 3 3/4 | 2 3/4 |
| 16 | 11- 9 9/16 | 4 | 2 15/16 |
| 17 | 12- 6 3/8 | 4 1/4 | 3 1/8 |
| 18 | 13- 3 1/4 | 4 1/2 | 3 5/16 |
| 19 | 14- 0 1/16 | 4 3/4 | 3 1/2 |
| 20 | 14- 8 15/16 | 5 | 3 11/16 |
| 21 | 15- 5 3/4 | 5 1/4 | 3 7/8 |
| 22 | 16- 2 5/8 | 5 1/2 | 4 1/16 |
| 23 | 16-11 7/16 | 5 3/4 | 4 1/4 |
| 24 | 17- 8 5/16 | 6 | 4 7/16 |
| 25 | 18- 5 1/8 | 6 1/4 | 4 5/8 |
| 26 | 19- 2 | 6 1/2 | 4 13/16 |
| 27 | 19-10 13/16 | 6 3/4 | 5 |
| 28 | 20- 7 11/16 | 7 | 5 3/16 |
| 29 | 21- 4 1/2 | 7 1/4 | 5 3/8 |
| 30 | 22- 1 3/8 | 7 1/2 | 5 1/2 |
| 31 | 22-10 1/4 | 7 3/4 | 5 11/16 |
| 32 | 23- 7 1/16 | 8 | 5 7/8 |
| 33 | 24- 3 15/16 | 8 1/4 | 6 1/16 |
| 34 | 25- 0 3/4 | 8 1/2 | 6 1/4 |
| 35 | 25- 9 5/8 | 8 3/4 | 6 7/16 |
| 36 | 26- 6 7/16 | 9 | 6 5/8 |
| 37 | 27- 3 5/16 | 9 1/4 | 6 13/16 |
| 38 | 28- 0 1/8 | 9 1/2 | 7 |
| 39 | 28- 9 | 9 3/4 | 7 3/16 |
| 40 | 29- 5 13/16 | 10 | 7 3/8 |
| 41 | 30- 2 11/16 | 10 1/4 | 7 9/16 |
| 42 | 30-11 1/2 | 10 1/2 | 7 3/4 |
| 43 | 31- 8 3/8 | 10 3/4 | 7 15/16 |
| 44 | 32- 5 1/4 | 11 | 8 1/8 |
| 45 | 33- 2 1/16 | 11 1/4 | 8 5/16 |
| 46 | 33-10 15/16 | 11 1/2 | 8 1/2 |
| 47 | 34- 7 3/4 | 11 3/4 | 8 11/16 |
| 48 | 35- 4 5/8 | 12 | 8 7/8 |
| 49 | 36- 1 7/16 | | |
| 50 | 36-10 5/16 | | |

# 5½ in 12 – 24½°

## *Cuts and Saw Settings*

**Tail Cut, Hip or Valley Rafters,** with Square Cut Common Rafters:
Framing Square: **6 7/16 & 24**; set saw at **40°**
(Compound Miter or Radial Arm Saws: set bevel at 40°, miter at 15°)
Explanation: Section 3(d), pages 12–13.

**Chamfer Cut, Hip or Valley Rafters:**
Set saw at **17 1/4°**. Explanation: Section 3(d), page 10.

**Fascia Board Miter Cut at Hip or Valley:**
Framing Square: **10 & 24**; set saw at **40°**
(Compound Miter or Radial Arm Saws: set bevel at 40°, miter at 22 1/2°)
Explanation: Section 5, pages 24–25.

**Frieze Block against Hip or Valley Rafter:**
Framing Square: **10 & 24**; set saw at **40°**
(Compound Miter or Radial Arm Saws: set bevel at 40°, miter at 22 1/2°)
Explanation: Section 6, page 26.

**Roof Sheathing at Hip or Valley:**
Framing Square: **12 & 13 3/16**; set saw at **17 1/4°**
(Compound Miter or Radial Arm Saws: set bevel at 17 1/4°, miter at 42 1/4°)
Distance "d" = **43 5/8"**. Explanation: Section 7, pages 26–27.

## *Calculating the Length of a Rafter*

| **General Rule:** | | Span of building |
|---|---|---|
| | less | Thickness of ridge board (if any) |
| | gives | Length of rafter to plumb line of seat cut |
| *and* | | Span of building |
| | less | Thickness of ridge board (if any) |
| | plus | 2 × projection of overhang |
| | gives | Total length of rafter to tail cut |

Explanation: Section 4, pages 13–19.

### *Jack Rafters*

| SPACED *Inches* | DIFFERENCE *Inches* | SPACED *Inches* | DIFFERENCE *Inches* | SPACED *Inches* | DIFFERENCE *Inches* |
|---|---|---|---|---|---|
| 1/4 | 1/4 | 15 | 16 1/2 | **32** | **35 3/16** |
| 1/2 | 9/16 | **16** | **17 5/8** | 33 | 36 5/16 |
| 3/4 | 13/16 | 17 | 18 11/16 | 34 | 37 3/8 |
| 1 | 1 1/8 | 18 | 19 13/16 | 35 | 38 1/2 |
| 2 | 2 3/16 | 19 | 20 7/8 | **36** | **39 5/8** |
| 3 | 3 5/16 | 20 | 22 | 37 | 40 11/16 |
| 4 | 4 3/8 | 21 | 23 1/8 | 38 | 41 13/16 |
| 5 | 5 1/2 | 22 | 24 3/16 | 39 | 42 7/8 |
| 6 | 6 5/8 | 23 | 25 5/16 | 40 | 44 |
| 7 | 7 11/16 | **24** | **26 3/8** | 41 | 45 1/8 |
| 8 | 8 13/16 | 25 | 27 1/2 | 42 | 46 3/16 |
| 9 | 9 7/8 | 26 | 28 5/8 | 43 | 47 5/16 |
| 10 | 11 | 27 | 29 11/16 | 44 | 48 3/8 |
| 11 | 12 1/8 | 28 | 30 13/16 | 45 | 49 1/2 |
| **12** | **13 3/16** | 29 | 31 7/8 | 46 | 50 5/8 |
| 13 | 14 5/16 | 30 | 33 | 47 | 51 11/16 |
| 14 | 15 3/8 | 31 | 34 1/8 | **48** | **52 13/16** |

**Common Rafters**
**5 1/2 in 12 – 24 1/2°**

| SPAN Feet | LENGTH Feet Inches | SPAN Inches | LENGTH Inches |
|---|---|---|---|
| 1 | 0- 6 5/8 | 1/4 | 1/8 |
| 2 | 1- 1 3/16 | 1/2 | 1/4 |
| 3 | 1- 7 13/16 | 3/4 | 7/16 |
| 4 | 2- 2 3/8 | 1 | 9/16 |
| 5 | 2- 9 | 1 1/4 | 11/16 |
| 6 | 3- 3 5/8 | 1 1/2 | 13/16 |
| 7 | 3-10 3/16 | 1 3/4 | 15/16 |
| 8 | 4- 4 13/16 | 2 | 1 1/8 |
| 9 | 4-11 3/8 | 2 1/4 | 1 1/4 |
| 10 | 5- 6 | 2 1/2 | 1 3/8 |
| 11 | 6- 0 5/8 | 2 3/4 | 1 1/2 |
| 12 | 6- 7 3/16 | 3 | 1 5/8 |
| 13 | 7- 1 13/16 | 3 1/4 | 1 13/16 |
| 14 | 7- 8 3/8 | 3 1/2 | 1 15/16 |
| 15 | 8- 3 | 3 3/4 | 2 1/16 |
| 16 | 8- 9 5/8 | 4 | 2 3/16 |
| 17 | 9- 4 3/16 | 4 1/4 | 2 5/16 |
| 18 | 9-10 13/16 | 4 1/2 | 2 1/2 |
| 19 | 10- 5 3/8 | 4 3/4 | 2 5/8 |
| 20 | 11- 0 | 5 | 2 3/4 |
| 21 | 11- 6 5/8 | 5 1/4 | 2 7/8 |
| 22 | 12- 1 3/16 | 5 1/2 | 3 |
| 23 | 12- 7 13/16 | 5 3/4 | 3 3/16 |
| 24 | 13- 2 3/8 | 6 | 3 5/16 |
| 25 | 13- 9 | 6 1/4 | 3 7/16 |
| 26 | 14- 3 5/8 | 6 1/2 | 3 9/16 |
| 27 | 14-10 3/16 | 6 3/4 | 3 11/16 |
| 28 | 15- 4 13/16 | 7 | 3 7/8 |
| 29 | 15-11 3/8 | 7 1/4 | 4 |
| 30 | 16- 6 | 7 1/2 | 4 1/8 |
| 31 | 17- 0 5/8 | 7 3/4 | 4 1/4 |
| 32 | 17- 7 3/16 | 8 | 4 3/8 |
| 33 | 18- 1 13/16 | 8 1/4 | 4 9/16 |
| 34 | 18- 8 7/16 | 8 1/2 | 4 11/16 |
| 35 | 19- 3 | 8 3/4 | 4 13/16 |
| 36 | 19- 9 5/8 | 9 | 4 15/16 |
| 37 | 20- 4 3/16 | 9 1/4 | 5 1/16 |
| 38 | 20-10 13/16 | 9 1/2 | 5 1/4 |
| 39 | 21- 5 7/16 | 9 3/4 | 5 3/8 |
| 40 | 22- 0 | 10 | 5 1/2 |
| 41 | 22- 6 5/8 | 10 1/4 | 5 5/8 |
| 42 | 23- 1 3/16 | 10 1/2 | 5 3/4 |
| 43 | 23- 7 13/16 | 10 3/4 | 5 15/16 |
| 44 | 24- 2 7/16 | 11 | 6 1/16 |
| 45 | 24- 9 | 11 1/4 | 6 3/16 |
| 46 | 25- 3 5/8 | 11 1/2 | 6 5/16 |
| 47 | 25-10 3/16 | 11 3/4 | 6 7/16 |
| 48 | 26- 4 13/16 | 12 | 6 5/8 |
| 49 | 26-11 7/16 | | |
| 50 | 27- 6 | | |

**Hip or Valley Rafters**
**5 1/2 in 17 – 18°**

| SPAN Feet | LENGTH Feet Inches | SPAN Inches | LENGTH Inches |
|---|---|---|---|
| 1 | 0- 8 15/16 | 1/4 | 3/16 |
| 2 | 1- 5 13/16 | 1/2 | 3/8 |
| 3 | 2- 2 3/4 | 3/4 | 9/16 |
| 4 | 2-11 11/16 | 1 | 3/4 |
| 5 | 3- 8 5/8 | 1 1/4 | 15/16 |
| 6 | 4- 5 1/2 | 1 1/2 | 1 1/8 |
| 7 | 5- 2 7/16 | 1 3/4 | 1 5/16 |
| 8 | 5-11 3/8 | 2 | 1 1/2 |
| 9 | 6- 8 1/4 | 2 1/4 | 1 11/16 |
| 10 | 7- 5 3/16 | 2 1/2 | 1 7/8 |
| 11 | 8- 2 1/8 | 2 3/4 | 2 1/16 |
| 12 | 8-11 1/16 | 3 | 2 1/4 |
| 13 | 9- 7 15/16 | 3 1/4 | 2 7/16 |
| 14 | 10- 4 7/8 | 3 1/2 | 2 5/8 |
| 15 | 11- 1 13/16 | 3 3/4 | 2 13/16 |
| 16 | 11-10 11/16 | 4 | 3 |
| 17 | 12- 7 5/8 | 4 1/4 | 3 3/16 |
| 18 | 13- 4 9/16 | 4 1/2 | 3 3/8 |
| 19 | 14- 1 1/2 | 4 3/4 | 3 1/2 |
| 20 | 14-10 3/8 | 5 | 3 11/16 |
| 21 | 15- 7 5/16 | 5 1/4 | 3 7/8 |
| 22 | 16- 4 1/4 | 5 1/2 | 4 1/16 |
| 23 | 17- 1 1/8 | 5 3/4 | 4 1/4 |
| 24 | 17-10 1/16 | 6 | 4 7/16 |
| 25 | 18- 7 | 6 1/4 | 4 5/8 |
| 26 | 19- 3 15/16 | 6 1/2 | 4 13/16 |
| 27 | 20- 0 13/16 | 6 3/4 | 5 |
| 28 | 20- 9 3/4 | 7 | 5 3/16 |
| 29 | 21- 6 11/16 | 7 1/4 | 5 3/8 |
| 30 | 22- 3 9/16 | 7 1/2 | 5 9/16 |
| 31 | 23- 0 1/2 | 7 3/4 | 5 3/4 |
| 32 | 23- 9 7/16 | 8 | 5 15/16 |
| 33 | 24- 6 3/8 | 8 1/4 | 6 1/8 |
| 34 | 25- 3 1/4 | 8 1/2 | 6 5/16 |
| 35 | 26- 0 3/16 | 8 3/4 | 6 1/2 |
| 36 | 26- 9 1/8 | 9 | 6 11/16 |
| 37 | 27- 6 1/16 | 9 1/4 | 6 7/8 |
| 38 | 28- 2 15/16 | 9 1/2 | 7 1/16 |
| 39 | 28-11 7/8 | 9 3/4 | 7 1/4 |
| 40 | 29- 8 13/16 | 10 | 7 7/16 |
| 41 | 30- 5 11/16 | 10 1/4 | 7 5/8 |
| 42 | 31- 2 5/8 | 10 1/2 | 7 13/16 |
| 43 | 31-11 9/16 | 10 3/4 | 8 |
| 44 | 32- 8 1/2 | 11 | 8 3/16 |
| 45 | 33- 5 3/8 | 11 1/4 | 8 3/8 |
| 46 | 34- 2 5/16 | 11 1/2 | 8 9/16 |
| 47 | 34-11 1/4 | 11 3/4 | 8 3/4 |
| 48 | 35- 8 1/8 | 12 | 8 15/16 |
| 49 | 36- 5 1/16 | | |
| 50 | 37- 2 | | |

# 6 in 12 – $26^1/_2$°

## *Cuts and Saw Settings*

**Tail Cut, Hip or Valley Rafters,** with Square Cut Common Rafters:
Framing Square: **6 13/16 & 24**; set saw at **39 1/4°**
(Compound Miter or Radial Arm Saws: set bevel at 39 1/4°, miter at 15 3/4°)
Explanation: Section 3(d), pages 12–13.

**Chamfer Cut, Hip or Valley Rafters:**
Set saw at **18 1/2°**. Explanation: Section 3(d), page 10.

**Fascia Board Miter Cut at Hip or Valley:**
Framing Square: **10 3/4 & 24**; set saw at **39 1/4°**
(Compound Miter or Radial Arm Saws: set bevel at 39 1/4°, miter at 24°)
Explanation: Section 5, pages 24–25.

**Frieze Block against Hip or Valley Rafter:**
Framing Square: **10 3/4 & 24**; set saw at **39 1/4°**
(Compound Miter or Radial Arm Saws: set bevel at 39 1/4°, miter at 24°)
Explanation: Section 6, page 26.

**Roof Sheathing at Hip or Valley:**
Framing Square: **12 & 13 7/16**; set saw at **18 1/2°**
(Compound Miter or Radial Arm Saws: set bevel at 18 1/2°, miter at 41 3/4°)
Distance "d" = **42 15/16"**. Explanation: Section 7, pages 26–27.

## *Calculating the Length of a Rafter*

**General Rule:** Span of building
less Thickness of ridge board (if any)
gives Length of rafter to plumb line of seat cut

*and* Span of building
less Thickness of ridge board (if any)
plus 2 × projection of overhang
gives Total length of rafter to tail cut

Explanation: Section 4, pages 13–19.

| *Jack Rafters* | | | | | |
|---|---|---|---|---|---|
| SPACED *Inches* | DIFFERENCE *Inches* | SPACED *Inches* | DIFFERENCE *Inches* | SPACED *Inches* | DIFFERENCE *Inches* |
| 1/4 | 1/4 | 15 | 16 3/4 | **32** | **35 3/4** |
| 1/2 | 9/16 | **16** | **17 7/8** | 33 | 36 7/8 |
| 3/4 | 13/16 | 17 | 19 | 34 | 38 |
| 1 | 1 1/8 | 18 | 20 1/8 | 35 | 39 1/8 |
| 2 | 2 1/4 | 19 | 21 1/4 | **36** | **40 1/4** |
| 3 | 3 3/8 | 20 | 22 3/8 | 37 | 41 3/8 |
| 4 | 4 1/2 | 21 | 23 1/2 | 38 | 42 1/2 |
| 5 | 5 9/16 | 22 | 24 5/8 | 39 | 43 5/8 |
| 6 | 6 11/16 | 23 | 25 11/16 | 40 | 44 3/4 |
| 7 | 7 13/16 | **24** | **26 13/16** | 41 | 45 13/16 |
| 8 | 8 15/16 | 25 | 27 15/16 | 42 | 46 15/16 |
| 9 | 10 1/16 | 26 | 29 1/16 | 43 | 48 1/16 |
| 10 | 11 3/16 | 27 | 30 3/16 | 44 | 49 3/16 |
| 11 | 12 5/16 | 28 | 31 5/16 | 45 | 50 5/16 |
| **12** | **13 7/16** | 29 | 32 7/16 | 46 | 51 7/16 |
| 13 | 14 9/16 | 30 | 33 9/16 | 47 | 52 9/16 |
| 14 | 15 5/8 | 31 | 34 11/16 | **48** | **53 11/16** |

## Common Rafters
## 6 in 12 – 26 1/2°

| SPAN Feet | LENGTH Feet Inches | SPAN Inches | LENGTH Inches |
|---|---|---|---|
| 1 | 0- 6 11/16 | 1/4 | 1/8 |
| 2 | 1- 1 7/16 | 1/2 | 1/4 |
| 3 | 1- 8 1/8 | 3/4 | 7/16 |
| 4 | 2- 2 13/16 | 1 | 9/16 |
| 5 | 2- 9 9/16 | 1 1/4 | 11/16 |
| 6 | 3- 4 1/4 | 1 1/2 | 13/16 |
| 7 | 3-10 15/16 | 1 3/4 | 1 |
| 8 | 4- 5 11/16 | 2 | 1 1/8 |
| 9 | 5- 0 3/8 | 2 1/4 | 1 1/4 |
| 10 | 5- 7 1/16 | 2 1/2 | 1 3/8 |
| 11 | 6- 1 13/16 | 2 3/4 | 1 9/16 |
| 12 | 6- 8 1/2 | 3 | 1 11/16 |
| 13 | 7- 3 3/16 | 3 1/4 | 1 13/16 |
| 14 | 7- 9 15/16 | 3 1/2 | 1 15/16 |
| 15 | 8- 4 5/8 | 3 3/4 | 2 1/8 |
| 16 | 8-11 5/16 | 4 | 2 1/4 |
| 17 | 9- 6 1/16 | 4 1/4 | 2 3/8 |
| 18 | 10- 0 3/4 | 4 1/2 | 2 1/2 |
| 19 | 10- 7 7/16 | 4 3/4 | 2 5/8 |
| 20 | 11- 2 3/16 | 5 | 2 13/16 |
| 21 | 11- 8 7/8 | 5 1/4 | 2 15/16 |
| 22 | 12- 3 9/16 | 5 1/2 | 3 1/16 |
| 23 | 12-10 5/16 | 5 3/4 | 3 3/16 |
| 24 | 13- 5 | 6 | 3 3/8 |
| 25 | 13-11 11/16 | 6 1/4 | 3 1/2 |
| 26 | 14- 6 7/16 | 6 1/2 | 3 5/8 |
| 27 | 15- 1 1/8 | 6 3/4 | 3 3/4 |
| 28 | 15- 7 13/16 | 7 | 3 15/16 |
| 29 | 16- 2 9/16 | 7 1/4 | 4 1/16 |
| 30 | 16- 9 1/4 | 7 1/2 | 4 3/16 |
| 31 | 17- 3 15/16 | 7 3/4 | 4 5/16 |
| 32 | 17-10 11/16 | 8 | 4 1/2 |
| 33 | 18- 5 3/8 | 8 1/4 | 4 5/8 |
| 34 | 19- 0 1/16 | 8 1/2 | 4 3/4 |
| 35 | 19- 6 13/16 | 8 3/4 | 4 7/8 |
| 36 | 20- 1 1/2 | 9 | 5 |
| 37 | 20- 8 3/16 | 9 1/4 | 5 3/16 |
| 38 | 21- 2 15/16 | 9 1/2 | 5 5/16 |
| 39 | 21- 9 5/8 | 9 3/4 | 5 7/16 |
| 40 | 22- 4 5/16 | 10 | 5 9/16 |
| 41 | 22-11 1/16 | 10 1/4 | 5 3/4 |
| 42 | 23- 5 3/4 | 10 1/2 | 5 7/8 |
| 43 | 24- 0 7/16 | 10 3/4 | 6 |
| 44 | 24- 7 3/16 | 11 | 6 1/8 |
| 45 | 25- 1 7/8 | 11 1/4 | 6 5/16 |
| 46 | 25- 8 9/16 | 11 1/2 | 6 7/16 |
| 47 | 26- 3 5/16 | 11 3/4 | 6 9/16 |
| 48 | 26-10 | 12 | 6 11/16 |
| 49 | 27- 4 11/16 | | |
| 50 | 27-11 7/16 | | |

## Hip or Valley Rafters
## 6 in 17 – 19 1/2°

| SPAN Feet | LENGTH Feet Inches | SPAN Inches | LENGTH Inches |
|---|---|---|---|
| 1 | 0 - 9 | 1/4 | 3/16 |
| 2 | 1 - 6 | 1/2 | 3/8 |
| 3 | 2 - 3 | 3/4 | 9/16 |
| 4 | 3 - 0 | 1 | 3/4 |
| 5 | 3 - 9 | 1 1/4 | 15/16 |
| 6 | 4 - 6 | 1 1/2 | 1 1/8 |
| 7 | 5 - 3 | 1 3/4 | 1 5/16 |
| 8 | 6 - 0 | 2 | 1 1/2 |
| 9 | 6 - 9 | 2 1/4 | 1 11/16 |
| 10 | 7 - 6 | 2 1/2 | 1 7/8 |
| 11 | 8 - 3 | 2 3/4 | 2 1/16 |
| 12 | 9 - 0 | 3 | 2 1/4 |
| 13 | 9 - 9 | 3 1/4 | 2 7/16 |
| 14 | 10 - 6 | 3 1/2 | 2 5/8 |
| 15 | 11 - 3 | 3 3/4 | 2 13/16 |
| 16 | 12 - 0 | 4 | 3 |
| 17 | 12 - 9 | 4 1/4 | 3 3/16 |
| 18 | 13 - 6 | 4 1/2 | 3 3/8 |
| 19 | 14 - 3 | 4 3/4 | 3 9/16 |
| 20 | 15 - 0 | 5 | 3 3/4 |
| 21 | 15 - 9 | 5 1/4 | 3 15/16 |
| 22 | 16 - 6 | 5 1/2 | 4 1/8 |
| 23 | 17 - 3 | 5 3/4 | 4 5/16 |
| 24 | 18 - 0 | 6 | 4 1/2 |
| 25 | 18 - 9 | 6 1/4 | 4 11/16 |
| 26 | 19 - 6 | 6 1/2 | 4 7/8 |
| 27 | 20 - 3 | 6 3/4 | 5 1/16 |
| 28 | 21 - 0 | 7 | 5 1/4 |
| 29 | 21 - 9 | 7 1/4 | 5 7/16 |
| 30 | 22 - 6 | 7 1/2 | 5 5/8 |
| 31 | 23 - 3 | 7 3/4 | 5 13/16 |
| 32 | 24 - 0 | 8 | 6 |
| 33 | 24 - 9 | 8 1/4 | 6 3/16 |
| 34 | 25 - 6 | 8 1/2 | 6 3/8 |
| 35 | 26 - 3 | 8 3/4 | 6 9/16 |
| 36 | 27 - 0 | 9 | 6 3/4 |
| 37 | 27 - 9 | 9 1/4 | 6 15/16 |
| 38 | 28 - 6 | 9 1/2 | 7 1/8 |
| 39 | 29 - 3 | 9 3/4 | 7 5/16 |
| 40 | 30 - 0 | 10 | 7 1/2 |
| 41 | 30 - 9 | 10 1/4 | 7 11/16 |
| 42 | 31 - 6 | 10 1/2 | 7 7/8 |
| 43 | 32 - 3 | 10 3/4 | 8 1/16 |
| 44 | 33 - 0 | 11 | 8 1/4 |
| 45 | 33 - 9 | 11 1/4 | 8 7/16 |
| 46 | 34 - 6 | 11 1/2 | 8 5/8 |
| 47 | 35 - 3 | 11 3/4 | 8 13/16 |
| 48 | 36 - 0 | 12 | 9 |
| 49 | 36 - 9 | | |
| 50 | 37 - 6 | | |

# 6 1/2 in 12 – 28 1/2°

## *Cuts and Saw Settings*

**Tail Cut, Hip or Valley Rafters,** with Square Cut Common Rafters:
Framing Square: **7 1/8 & 24**; set saw at **38 1/2°**
(Compound Miter or Radial Arm Saws: set bevel at 38 1/2°, miter at 16 1/2°)
Explanation: Section 3(d), pages 12–13.

**Chamfer Cut, Hip or Valley Rafters:**
Set saw at **19 3/4°**. Explanation: Section 3(d), page 10.

**Fascia Board Miter Cut at Hip or Valley:**
Framing Square: **11 7/16 & 24**; set saw at **38 1/2°**
(Compound Miter or Radial Arm Saws: set bevel at 38 1/2°, miter at 25 1/2°)
Explanation: Section 5, pages 24–25.

**Frieze Block against Hip or Valley Rafter:**
Framing Square: **11 7/16 & 24**; set saw at **38 1/2°**
(Compound Miter or Radial Arm Saws: set bevel at 38 1/2°, miter at 25 1/2°)
Explanation: Section 6, page 26.

**Roof Sheathing at Hip or Valley:**
Framing Square: **12 & 13 5/8**; set saw at **19 3/4°**
(Compound Miter or Radial Arm Saws: set bevel at 19 3/4°, miter at 41 1/4°)
Distance "d" = **42 3/16"**. Explanation: Section 7, pages 26–27.

## *Calculating the Length of a Rafter*

**General Rule:**

| | |
|---|---|
| | Span of building |
| less | Thickness of ridge board (if any) |
| gives | Length of rafter to plumb line of seat cut |

*and*

| | |
|---|---|
| | Span of building |
| less | Thickness of ridge board (if any) |
| plus | 2 × projection of overhang |
| gives | Total length of rafter to tail cut |

Explanation: Section 4, pages 13–19.

### *Jack Rafters*

| SPACED *Inches* | DIFFERENCE *Inches* | SPACED *Inches* | DIFFERENCE *Inches* | SPACED *Inches* | DIFFERENCE *Inches* |
|---|---|---|---|---|---|
| 1/4 | 5/16 | 15 | 17 1/16 | **32** | **36 3/8** |
| 1/2 | 9/16 | **16** | **18 3/16** | 33 | 37 1/2 |
| 3/4 | 7/8 | 17 | 19 5/16 | 34 | 38 11/16 |
| 1 | 1 1/8 | 18 | 20 1/2 | 35 | 39 13/16 |
| 2 | 2 1/4 | 19 | 21 5/8 | **36** | **40 15/16** |
| 3 | 3 7/16 | 20 | 22 3/4 | 37 | 42 1/16 |
| 4 | 4 9/16 | 21 | 23 7/8 | 38 | 43 3/16 |
| 5 | 5 11/16 | 22 | 25 | 39 | 44 3/8 |
| 6 | 6 13/16 | 23 | 26 3/16 | 40 | 45 1/2 |
| 7 | 7 15/16 | **24** | **27 5/16** | 41 | 46 5/8 |
| 8 | 9 1/8 | 25 | 28 7/16 | 42 | 47 3/4 |
| 9 | 10 1/4 | 26 | 29 9/16 | 43 | 48 7/8 |
| 10 | 11 3/8 | 27 | 30 11/16 | 44 | 50 1/16 |
| 11 | 12 1/2 | 28 | 31 7/8 | 45 | 51 3/16 |
| **12** | **13 5/8** | 29 | 33 | 46 | 52 5/16 |
| 13 | 14 13/16 | 30 | 34 1/8 | 47 | 53 7/16 |
| 14 | 15 15/16 | 31 | 35 1/4 | **48** | **54 9/16** |

## *Common Rafters*
## 6 1/2 in 12 – 28 1/2°

| SPAN Feet | LENGTH Feet Inches | SPAN Inches | LENGTH Inches |
|---|---|---|---|
| 1 | 0- 6 13/16 | 1/4 | 1/8 |
| 2 | 1- 1 5/8 | 1/2 | 5/16 |
| 3 | 1- 8 1/2 | 3/4 | 7/16 |
| 4 | 2- 3 5/16 | 1 | 9/16 |
| 5 | 2-10 1/8 | 1 1/4 | 11/16 |
| 6 | 3- 4 15/16 | 1 1/2 | 7/8 |
| 7 | 3-11 3/4 | 1 3/4 | 1 |
| 8 | 4- 6 9/16 | 2 | 1 1/8 |
| 9 | 5- 1 7/16 | 2 1/4 | 1 1/4 |
| 10 | 5- 8 1/4 | 2 1/2 | 1 7/16 |
| 11 | 6- 3 1/16 | 2 3/4 | 1 9/16 |
| 12 | 6- 9 7/8 | 3 | 1 11/16 |
| 13 | 7- 4 11/16 | 3 1/4 | 1 7/8 |
| 14 | 7-11 9/16 | 3 1/2 | 2 |
| 15 | 8- 6 3/8 | 3 3/4 | 2 1/8 |
| 16 | 9- 1 3/16 | 4 | 2 1/4 |
| 17 | 9- 8 | 4 1/4 | 2 7/16 |
| 18 | 10- 2 13/16 | 4 1/2 | 2 9/16 |
| 19 | 10- 9 5/8 | 4 3/4 | 2 11/16 |
| 20 | 11- 4 1/2 | 5 | 2 13/16 |
| 21 | 11-11 5/16 | 5 1/4 | 3 |
| 22 | 12- 6 1/8 | 5 1/2 | 3 1/8 |
| 23 | 13- 0 15/16 | 5 3/4 | 3 1/4 |
| 24 | 13- 7 3/4 | 6 | 3 7/16 |
| 25 | 14- 2 9/16 | 6 1/4 | 3 9/16 |
| 26 | 14- 9 7/16 | 6 1/2 | 3 11/16 |
| 27 | 15- 4 1/4 | 6 3/4 | 3 13/16 |
| 28 | 15-11 1/16 | 7 | 4 |
| 29 | 16- 5 7/8 | 7 1/4 | 4 1/8 |
| 30 | 17- 0 11/16 | 7 1/2 | 4 1/4 |
| 31 | 17- 7 9/16 | 7 3/4 | 4 7/16 |
| 32 | 18- 2 3/8 | 8 | 4 9/16 |
| 33 | 18- 9 3/16 | 8 1/4 | 4 11/16 |
| 34 | 19- 4 | 8 1/2 | 4 13/16 |
| 35 | 19-10 13/16 | 8 3/4 | 5 |
| 36 | 20- 5 5/8 | 9 | 5 1/8 |
| 37 | 21- 0 1/2 | 9 1/4 | 5 1/4 |
| 38 | 21- 7 5/16 | 9 1/2 | 5 3/8 |
| 39 | 22- 2 1/8 | 9 3/4 | 5 9/16 |
| 40 | 22- 8 15/16 | 10 | 5 11/16 |
| 41 | 23- 3 3/4 | 10 1/4 | 5 13/16 |
| 42 | 23-10 5/8 | 10 1/2 | 6 |
| 43 | 24- 5 7/16 | 10 3/4 | 6 1/8 |
| 44 | 25- 0 1/4 | 11 | 6 1/4 |
| 45 | 25- 7 1/16 | 11 1/4 | 6 3/8 |
| 46 | 26- 1 7/8 | 11 1/2 | 6 9/16 |
| 47 | 26- 8 11/16 | 11 3/4 | 6 11/16 |
| 48 | 27- 3 9/16 | 12 | 6 13/16 |
| 49 | 27-10 3/8 | | |
| 50 | 28- 5 3/16 | | |

## *Hip or Valley Rafters*
## 6 1/2 in 17 – 21°

| SPAN Feet | LENGTH Feet Inches | SPAN Inches | LENGTH Inches |
|---|---|---|---|
| 1 | 0- 9 1/16 | 1/4 | 3/16 |
| 2 | 1- 6 3/16 | 1/2 | 3/8 |
| 3 | 2- 3 1/4 | 3/4 | 9/16 |
| 4 | 3- 0 3/8 | 1 | 3/4 |
| 5 | 3- 9 7/16 | 1 1/4 | 15/16 |
| 6 | 4- 6 1/2 | 1 1/2 | 1 1/8 |
| 7 | 5- 3 5/8 | 1 3/4 | 1 5/16 |
| 8 | 6- 0 11/16 | 2 | 1 1/2 |
| 9 | 6- 9 3/4 | 2 1/4 | 1 11/16 |
| 10 | 7- 6 7/8 | 2 1/2 | 1 7/8 |
| 11 | 8- 3 15/16 | 2 3/4 | 2 1/16 |
| 12 | 9- 1 1/16 | 3 | 2 1/4 |
| 13 | 9-10 1/8 | 3 1/4 | 2 7/16 |
| 14 | 10- 7 3/16 | 3 1/2 | 2 5/8 |
| 15 | 11- 4 5/16 | 3 3/4 | 2 13/16 |
| 16 | 12- 1 3/8 | 4 | 3 |
| 17 | 12-10 7/16 | 4 1/4 | 3 3/16 |
| 18 | 13- 7 9/16 | 4 1/2 | 3 7/16 |
| 19 | 14- 4 5/8 | 4 3/4 | 3 5/8 |
| 20 | 15- 1 3/4 | 5 | 3 13/16 |
| 21 | 15-10 13/16 | 5 1/4 | 4 |
| 22 | 16- 7 7/8 | 5 1/2 | 4 3/16 |
| 23 | 17- 5 | 5 3/4 | 4 3/8 |
| 24 | 18- 2 1/16 | 6 | 4 9/16 |
| 25 | 18-11 3/16 | 6 1/4 | 4 3/4 |
| 26 | 19- 8 1/4 | 6 1/2 | 4 15/16 |
| 27 | 20- 5 5/16 | 6 3/4 | 5 1/8 |
| 28 | 21- 2 7/16 | 7 | 5 5/16 |
| 29 | 21-11 1/2 | 7 1/4 | 5 1/2 |
| 30 | 22- 8 9/16 | 7 1/2 | 5 11/16 |
| 31 | 23- 5 11/16 | 7 3/4 | 5 7/8 |
| 32 | 24- 2 3/4 | 8 | 6 1/16 |
| 33 | 24-11 7/8 | 8 1/4 | 6 1/4 |
| 34 | 25- 8 15/16 | 8 1/2 | 6 7/16 |
| 35 | 26- 6 | 8 3/4 | 6 5/8 |
| 36 | 27- 3 1/8 | 9 | 6 13/16 |
| 37 | 28- 0 3/16 | 9 1/4 | 7 |
| 38 | 28- 9 5/16 | 9 1/2 | 7 3/16 |
| 39 | 29- 6 3/8 | 9 3/4 | 7 3/8 |
| 40 | 30- 3 7/16 | 10 | 7 9/16 |
| 41 | 31- 0 9/16 | 10 1/4 | 7 3/4 |
| 42 | 31- 9 5/8 | 10 1/2 | 7 15/16 |
| 43 | 32- 6 11/16 | 10 3/4 | 8 1/8 |
| 44 | 33- 3 13/16 | 11 | 8 5/16 |
| 45 | 34- 0 7/8 | 11 1/4 | 8 1/2 |
| 46 | 34-10 | 11 1/2 | 8 11/16 |
| 47 | 35- 7 1/16 | 11 3/4 | 8 7/8 |
| 48 | 36- 4 1/8 | 12 | 9 1/16 |
| 49 | 37- 1 1/4 | | |
| 50 | 37-10 5/16 | | |

# 7 in 12 – 30 1/4°

## *Cuts and Saw Settings*

**Tail Cut, Hip or Valley Rafters,** with Square Cut Common Rafters:
Framing Square: **7 3/8 & 24**; set saw at **37 3/4°**
(Compound Miter or Radial Arm Saws: set bevel at 37 3/4°, miter at 17°)
Explanation: Section 3(d), pages 12–13.

**Chamfer Cut, Hip or Valley Rafters:**
Set saw at **20 3/4°**. Explanation: Section 3(d), page 10.

**Fascia Board Miter Cut at Hip or Valley:**
Framing Square: **12 1/16 & 24**; set saw at **37 3/4°**
(Compound Miter or Radial Arm Saws: set bevel at 37 3/4°, miter at 26 3/4°)
Explanation: Section 5, pages 24–25.

**Frieze Block against Hip or Valley Rafter:**
Framing Square: **12 1/16 & 24**; set saw at **37 3/4°**
(Compound Miter or Radial Arm Saws: set bevel at 37 3/4°, miter at 26 3/4°)
Explanation: Section 6, page 26.

**Roof Sheathing at Hip or Valley:**
Framing Square: **12 & 13 7/8**; set saw at **20 3/4°**
(Compound Miter or Radial Arm Saws: set bevel at 20 3/4°, miter at 40 3/4°)
Distance "d" = **41 7/16"**. Explanation: Section 7, pages 26–27.

## *Calculating the Length of a Rafter*

**General Rule:** Span of building
less Thickness of ridge board (if any)
gives Length of rafter to plumb line of seat cut

*and* Span of building
less Thickness of ridge board (if any)
plus 2 × projection of overhang
gives Total length of rafter to tail cut

Explanation: Section 4, pages 13–19.

| *Jack Rafters* | | | | | |
|---|---|---|---|---|---|
| SPACED *Inches* | DIFFERENCE *Inches* | SPACED *Inches* | DIFFERENCE *Inches* | SPACED *Inches* | DIFFERENCE *Inches* |
| 1/4 | 5/16 | 15 | 17 3/8 | **32** | **37 1/16** |
| 1/2 | 9/16 | **16** | **18 1/2** | 33 | 38 3/16 |
| 3/4 | 7/8 | 17 | 19 11/16 | 34 | 39 3/8 |
| 1 | 1 3/16 | 18 | 20 13/16 | 35 | 40 1/2 |
| 2 | 2 5/16 | 19 | 22 | **36** | **41 11/16** |
| 3 | 3 1/2 | 20 | 23 1/8 | 37 | 42 13/16 |
| 4 | 4 5/8 | 21 | 24 5/16 | 38 | 44 |
| 5 | 5 13/16 | 22 | 25 1/2 | 39 | 45 1/8 |
| 6 | 6 15/16 | 23 | 26 5/8 | 40 | 46 5/16 |
| 7 | 8 1/8 | **24** | **27 13/16** | 41 | 47 7/16 |
| 8 | 9 1/4 | 25 | 28 15/16 | 42 | 48 5/8 |
| 9 | 10 7/16 | 26 | 30 1/8 | 43 | 49 13/16 |
| 10 | 11 9/16 | 27 | 31 1/4 | 44 | 50 15/16 |
| 11 | 12 3/4 | 28 | 32 7/16 | 45 | 52 1/8 |
| **12** | **13 7/8** | 29 | 33 9/16 | 46 | 53 1/4 |
| 13 | 15 1/16 | 30 | 34 3/4 | 47 | 54 7/16 |
| 14 | 16 3/16 | 31 | 35 7/8 | **48** | **55 9/16** |

| *Common Rafters* 7 in 12 – 30 1/4° | | | |
|---|---|---|---|
| SPAN *Feet* | LENGTH *Feet Inches* | SPAN *Inches* | LENGTH *Inches* |
| 1 | 0- 6 15/16 | 1/4 | 1/8 |
| 2 | 1- 1 7/8 | 1/2 | 5/16 |
| 3 | 1- 8 13/16 | 3/4 | 7/16 |
| 4 | 2- 3 13/16 | 1 | 9/16 |
| 5 | 2-10 3/4 | 1 1/4 | 3/4 |
| 6 | 3- 5 11/16 | 1 1/2 | 7/8 |
| 7 | 4- 0 5/8 | 1 3/4 | 1 |
| 8 | 4- 7 9/16 | 2 | 1 3/16 |
| 9 | 5- 2 1/2 | 2 1/4 | 1 5/16 |
| 10 | 5- 9 7/16 | 2 1/2 | 1 7/16 |
| 11 | 6- 4 7/16 | 2 3/4 | 1 9/16 |
| 12 | 6-11 3/8 | 3 | 1 3/4 |
| 13 | 7- 6 5/16 | 3 1/4 | 1 7/8 |
| 14 | 8- 1 1/4 | 3 1/2 | 2 |
| 15 | 8- 8 3/16 | 3 3/4 | 2 3/16 |
| 16 | 9- 3 1/8 | 4 | 2 5/16 |
| 17 | 9-10 1/16 | 4 1/4 | 2 7/16 |
| 18 | 10- 5 1/16 | 4 1/2 | 2 5/8 |
| 19 | 11- 0 | 4 3/4 | 2 3/4 |
| 20 | 11- 6 15/16 | 5 | 2 7/8 |
| 21 | 12- 1 7/8 | 5 1/4 | 3 1/16 |
| 22 | 12- 8 13/16 | 5 1/2 | 3 3/16 |
| 23 | 13- 3 3/4 | 5 3/4 | 3 5/16 |
| 24 | 13-10 11/16 | 6 | 3 1/2 |
| 25 | 14- 5 5/8 | 6 1/4 | 3 5/8 |
| 26 | 15- 0 5/8 | 6 1/2 | 3 3/4 |
| 27 | 15- 7 9/16 | 6 3/4 | 3 15/16 |
| 28 | 16- 2 1/2 | 7 | 4 1/16 |
| 29 | 16- 9 7/16 | 7 1/4 | 4 3/16 |
| 30 | 17- 4 3/8 | 7 1/2 | 4 5/16 |
| 31 | 17-11 5/16 | 7 3/4 | 4 1/2 |
| 32 | 18- 6 1/4 | 8 | 4 5/8 |
| 33 | 19- 1 1/4 | 8 1/4 | 4 3/4 |
| 34 | 19- 8 3/16 | 8 1/2 | 4 15/16 |
| 35 | 20- 3 1/8 | 8 3/4 | 5 1/16 |
| 36 | 20-10 1/16 | 9 | 5 3/16 |
| 37 | 21- 5 | 9 1/4 | 5 3/8 |
| 38 | 21-11 15/16 | 9 1/2 | 5 1/2 |
| 39 | 22- 6 7/8 | 9 3/4 | 5 5/8 |
| 40 | 23- 1 7/8 | 10 | 5 13/16 |
| 41 | 23- 8 13/16 | 10 1/4 | 5 15/16 |
| 42 | 24- 3 3/4 | 10 1/2 | 6 1/16 |
| 43 | 24-10 11/16 | 10 3/4 | 6 1/4 |
| 44 | 25- 5 5/8 | 11 | 6 3/8 |
| 45 | 26- 0 9/16 | 11 1/4 | 6 1/2 |
| 46 | 26- 7 1/2 | 11 1/2 | 6 11/16 |
| 47 | 27- 2 1/2 | 11 3/4 | 6 13/16 |
| 48 | 27- 9 7/16 | 12 | 6 15/16 |
| 49 | 28- 4 3/8 | | |
| 50 | 28-11 5/16 | | |

| *Hip or Valley Rafters* 7 in 17 – 22 1/2° | | | |
|---|---|---|---|
| SPAN *Feet* | LENGTH *Feet Inches* | SPAN *Inches* | LENGTH *Inches* |
| 1 | 0- 9 3/16 | 1/4 | 3/16 |
| 2 | 1- 6 3/8 | 1/2 | 3/8 |
| 3 | 2- 3 9/16 | 3/4 | 9/16 |
| 4 | 3- 0 11/16 | 1 | 3/4 |
| 5 | 3- 9 7/8 | 1 1/4 | 15/16 |
| 6 | 4- 7 1/16 | 1 1/2 | 1 1/8 |
| 7 | 5- 4 1/4 | 1 3/4 | 1 5/16 |
| 8 | 6- 1 7/16 | 2 | 1 1/2 |
| 9 | 6-10 5/8 | 2 1/4 | 1 3/4 |
| 10 | 7- 7 13/16 | 2 1/2 | 1 15/16 |
| 11 | 8- 4 15/16 | 2 3/4 | 2 1/8 |
| 12 | 9- 2 1/8 | 3 | 2 5/16 |
| 13 | 9-11 5/16 | 3 1/4 | 2 1/2 |
| 14 | 10- 8 1/2 | 3 1/2 | 2 11/16 |
| 15 | 11- 5 11/16 | 3 3/4 | 2 7/8 |
| 16 | 12- 2 7/8 | 4 | 3 1/16 |
| 17 | 13- 0 1/16 | 4 1/4 | 3 1/4 |
| 18 | 13- 9 3/16 | 4 1/2 | 3 7/16 |
| 19 | 14- 6 3/8 | 4 3/4 | 3 5/8 |
| 20 | 15- 3 9/16 | 5 | 3 13/16 |
| 21 | 16- 0 3/4 | 5 1/4 | 4 |
| 22 | 16- 9 15/16 | 5 1/2 | 4 3/16 |
| 23 | 17- 7 1/8 | 5 3/4 | 4 3/8 |
| 24 | 18- 4 5/16 | 6 | 4 9/16 |
| 25 | 19- 1 1/2 | 6 1/4 | 4 3/4 |
| 26 | 19-10 5/8 | 6 1/2 | 5 |
| 27 | 20- 7 13/16 | 6 3/4 | 5 3/16 |
| 28 | 21- 5 | 7 | 5 3/8 |
| 29 | 22- 2 3/16 | 7 1/4 | 5 9/16 |
| 30 | 22-11 3/8 | 7 1/2 | 5 3/4 |
| 31 | 23- 8 9/16 | 7 3/4 | 5 15/16 |
| 32 | 24- 5 3/4 | 8 | 6 1/8 |
| 33 | 25- 2 7/8 | 8 1/4 | 6 5/16 |
| 34 | 26- 0 1/16 | 8 1/2 | 6 1/2 |
| 35 | 26- 9 1/4 | 8 3/4 | 6 11/16 |
| 36 | 27- 6 7/16 | 9 | 6 7/8 |
| 37 | 28- 3 5/8 | 9 1/4 | 7 1/16 |
| 38 | 29- 0 13/16 | 9 1/2 | 7 1/4 |
| 39 | 29-10 | 9 3/4 | 7 7/16 |
| 40 | 30- 7 1/8 | 10 | 7 5/8 |
| 41 | 31- 4 5/16 | 10 1/4 | 7 13/16 |
| 42 | 32- 1 1/2 | 10 1/2 | 8 1/16 |
| 43 | 32-10 11/16 | 10 3/4 | 8 1/4 |
| 44 | 33- 7 7/8 | 11 | 8 7/16 |
| 45 | 34- 5 1/16 | 11 1/4 | 8 5/8 |
| 46 | 35- 2 1/4 | 11 1/2 | 8 13/16 |
| 47 | 35-11 3/8 | 11 3/4 | 9 |
| 48 | 36- 8 9/16 | 12 | 9 3/16 |
| 49 | 37- 5 3/4 | | |
| 50 | 38- 2 15/16 | | |

# 7 1/2 in 12 – 32°

## *Cuts and Saw Settings*

**Tail Cut, Hip or Valley Rafters,** with Square Cut Common Rafters:
Framing Square: **7 5/8 & 24**; set saw at **36 3/4°**
(Compound Miter or Radial Arm Saws: set bevel at 36 3/4°, miter at 17 3/4°)
Explanation: Section 3(d), pages 12–13.

**Chamfer Cut, Hip or Valley Rafters:**
Set saw at **22°**. Explanation: Section 3(d), page 10.

**Fascia Board Miter Cut at Hip or Valley:**
Framing Square: **12 3/4 & 24**; set saw at **36 3/4°**
(Compound Miter or Radial Arm Saws: set bevel at 36 3/4°, miter at 28°)
Explanation: Section 5, pages 24–25.

**Frieze Block against Hip or Valley Rafter:**
Framing Square: **12 3/4 & 24**; set saw at **36 3/4°**
(Compound Miter or Radial Arm Saws: set bevel at 36 3/4°, miter at 28°)
Explanation: Section 6, page 26.

**Roof Sheathing at Hip or Valley:**
Framing Square: **12 & 14 1/8**; set saw at **22°**
(Compound Miter or Radial Arm Saws: set bevel at 22°, miter at 40 1/4°)
Distance "d" = **40 11/16"**. Explanation: Section 7, pages 26–27.

## *Calculating the Length of a Rafter*

**General Rule:** Span of building
less Thickness of ridge board (if any)
gives Length of rafter to plumb line of seat cut

*and* Span of building
less Thickness of ridge board (if any)
plus 2 × projection of overhang
gives Total length of rafter to tail cut

Explanation: Section 4, pages 13–19.

| *Jack Rafters* | | | | | |
|---|---|---|---|---|---|
| SPACED *Inches* | DIFFERENCE *Inches* | SPACED *Inches* | DIFFERENCE *Inches* | SPACED *Inches* | DIFFERENCE *Inches* |
| 1/4 | 5/16 | 15 | 17 11/16 | **32** | **37 3/4** |
| 1/2 | 9/16 | **16** | **18 7/8** | 33 | 38 15/16 |
| 3/4 | 7/8 | 17 | 20 1/16 | 34 | 40 1/8 |
| 1 | 1 3/16 | 18 | 21 1/4 | 35 | 41 1/4 |
| 2 | 2 3/8 | 19 | 22 3/8 | **36** | **42 7/16** |
| 3 | 3 9/16 | 20 | 23 9/16 | 37 | 43 5/8 |
| 4 | 4 11/16 | 21 | 24 3/4 | 38 | 44 13/16 |
| 5 | 5 7/8 | 22 | 25 15/16 | 39 | 46 |
| 6 | 7 1/16 | 23 | 27 1/8 | 40 | 47 3/16 |
| 7 | 8 1/4 | **24** | **28 5/16** | 41 | 48 3/8 |
| 8 | 9 7/16 | 25 | 29 1/2 | 42 | 49 1/2 |
| 9 | 10 5/8 | 26 | 30 11/16 | 43 | 50 11/16 |
| 10 | 11 13/16 | 27 | 31 13/16 | 44 | 51 7/8 |
| 11 | 13 | 28 | 33 | 45 | 53 1/16 |
| **12** | **14 1/8** | 29 | 34 3/16 | 46 | 54 1/4 |
| 13 | 15 5/16 | 30 | 35 3/8 | 47 | 55 7/16 |
| 14 | 16 1/2 | 31 | 36 9/16 | **48** | **56 5/8** |

| **Common Rafters 7 1/2 in 12 – 32°** | | | | |
|---|---|---|---|---|
| SPAN Feet | LENGTH Feet | LENGTH Inches | SPAN Inches | LENGTH Inches |
| 1 | 0- | 7 1/16 | 1/4 | 1/8 |
| 2 | 1- | 2 1/8 | 1/2 | 5/16 |
| 3 | 1- | 9 1/4 | 3/4 | 7/16 |
| 4 | 2- | 4 5/16 | 1 | 9/16 |
| 5 | 2- | 11 3/8 | 1 1/4 | 3/4 |
| 6 | 3- | 6 7/16 | 1 1/2 | 7/8 |
| 7 | 4- | 1 1/2 | 1 3/4 | 1 1/16 |
| 8 | 4- | 8 5/8 | 2 | 1 3/16 |
| 9 | 5- | 3 11/16 | 2 1/4 | 1 5/16 |
| 10 | 5- | 10 3/4 | 2 1/2 | 1 1/2 |
| 11 | 6- | 5 13/16 | 2 3/4 | 1 5/8 |
| 12 | 7- | 0 7/8 | 3 | 1 3/4 |
| 13 | 7- | 8 | 3 1/4 | 1 15/16 |
| 14 | 8- | 3 1/16 | 3 1/2 | 2 1/16 |
| 15 | 8- | 10 1/8 | 3 3/4 | 2 3/16 |
| 16 | 9- | 5 3/16 | 4 | 2 3/8 |
| 17 | 10- | 0 5/16 | 4 1/4 | 2 1/2 |
| 18 | 10- | 7 3/8 | 4 1/2 | 2 5/8 |
| 19 | 11- | 2 7/16 | 4 3/4 | 2 13/16 |
| 20 | 11- | 9 1/2 | 5 | 2 15/16 |
| 21 | 12- | 4 9/16 | 5 1/4 | 3 1/8 |
| 22 | 12- | 11 11/16 | 5 1/2 | 3 1/4 |
| 23 | 13- | 6 3/4 | 5 3/4 | 3 3/8 |
| 24 | 14- | 1 13/16 | 6 | 3 9/16 |
| 25 | 14- | 8 7/8 | 6 1/4 | 3 11/16 |
| 26 | 15- | 3 15/16 | 6 1/2 | 3 13/16 |
| 27 | 15- | 11 1/16 | 6 3/4 | 4 |
| 28 | 16- | 6 1/8 | 7 | 4 1/8 |
| 29 | 17- | 1 3/16 | 7 1/4 | 4 1/4 |
| 30 | 17- | 8 1/4 | 7 1/2 | 4 7/16 |
| 31 | 18- | 3 5/16 | 7 3/4 | 4 9/16 |
| 32 | 18- | 10 7/16 | 8 | 4 11/16 |
| 33 | 19- | 5 1/2 | 8 1/4 | 4 7/8 |
| 34 | 20- | 0 9/16 | 8 1/2 | 5 |
| 35 | 20- | 7 5/8 | 8 3/4 | 5 3/16 |
| 36 | 21- | 2 11/16 | 9 | 5 5/16 |
| 37 | 21- | 9 13/16 | 9 1/4 | 5 7/16 |
| 38 | 22- | 4 7/8 | 9 1/2 | 5 5/8 |
| 39 | 22- | 11 15/16 | 9 3/4 | 5 3/4 |
| 40 | 23- | 7 | 10 | 5 7/8 |
| 41 | 24- | 2 1/8 | 10 1/4 | 6 1/16 |
| 42 | 24- | 9 3/16 | 10 1/2 | 6 3/16 |
| 43 | 25- | 4 1/4 | 10 3/4 | 6 5/16 |
| 44 | 25- | 11 5/16 | 11 | 6 1/2 |
| 45 | 26- | 6 3/8 | 11 1/4 | 6 5/8 |
| 46 | 27- | 1 1/2 | 11 1/2 | 6 3/4 |
| 47 | 27- | 8 9/16 | 11 3/4 | 6 15/16 |
| 48 | 28- | 3 5/8 | 12 | 7 1/16 |
| 49 | 28- | 10 11/16 | | |
| 50 | 29- | 5 3/4 | | |

| **Hip or Valley Rafters 7 1/2 in 17 – 23 3/4°** | | | | |
|---|---|---|---|---|
| SPAN Feet | LENGTH Feet | LENGTH Inches | SPAN Inches | LENGTH Inches |
| 1 | 0- | 9 1/4 | 1/4 | 3/16 |
| 2 | 1- | 6 9/16 | 1/2 | 3/8 |
| 3 | 2- | 3 13/16 | 3/4 | 9/16 |
| 4 | 3- | 1 1/8 | 1 | 3/4 |
| 5 | 3- | 10 3/8 | 1 1/4 | 15/16 |
| 6 | 4- | 7 11/16 | 1 1/2 | 1 3/16 |
| 7 | 5- | 4 15/16 | 1 3/4 | 1 3/8 |
| 8 | 6- | 2 3/16 | 2 | 1 9/16 |
| 9 | 6- | 11 1/2 | 2 1/4 | 1 3/4 |
| 10 | 7- | 8 3/4 | 2 1/2 | 1 15/16 |
| 11 | 8- | 6 1/16 | 2 3/4 | 2 1/8 |
| 12 | 9- | 3 5/16 | 3 | 2 5/16 |
| 13 | 10- | 0 5/8 | 3 1/4 | 2 1/2 |
| 14 | 10- | 9 7/8 | 3 1/2 | 2 11/16 |
| 15 | 11- | 7 1/8 | 3 3/4 | 2 7/8 |
| 16 | 12- | 4 7/16 | 4 | 3 1/16 |
| 17 | 13- | 1 11/16 | 4 1/4 | 3 5/16 |
| 18 | 13- | 11 | 4 1/2 | 3 1/2 |
| 19 | 14- | 8 1/4 | 4 3/4 | 3 11/16 |
| 20 | 15- | 5 9/16 | 5 | 3 7/8 |
| 21 | 16- | 2 13/16 | 5 1/4 | 4 1/16 |
| 22 | 17- | 0 1/16 | 5 1/2 | 4 1/4 |
| 23 | 17- | 9 3/8 | 5 3/4 | 4 7/16 |
| 24 | 18- | 6 5/8 | 6 | 4 5/8 |
| 25 | 19- | 3 15/16 | 6 1/4 | 4 13/16 |
| 26 | 20- | 1 3/16 | 6 1/2 | 5 |
| 27 | 20- | 10 1/2 | 6 3/4 | 5 3/16 |
| 28 | 21- | 7 3/4 | 7 | 5 7/16 |
| 29 | 22- | 5 1/16 | 7 1/4 | 5 5/8 |
| 30 | 23- | 2 5/16 | 7 1/2 | 5 13/16 |
| 31 | 23- | 11 9/16 | 7 3/4 | 6 |
| 32 | 24- | 8 7/8 | 8 | 6 3/16 |
| 33 | 25- | 6 1/8 | 8 1/4 | 6 3/8 |
| 34 | 26- | 3 7/16 | 8 1/2 | 6 9/16 |
| 35 | 27- | 0 11/16 | 8 3/4 | 6 3/4 |
| 36 | 27- | 10 | 9 | 6 15/16 |
| 37 | 28- | 7 1/4 | 9 1/4 | 7 1/8 |
| 38 | 29- | 4 1/2 | 9 1/2 | 7 3/8 |
| 39 | 30- | 1 13/16 | 9 3/4 | 7 9/16 |
| 40 | 30- | 11 1/16 | 10 | 7 3/4 |
| 41 | 31- | 8 3/8 | 10 1/4 | 7 15/16 |
| 42 | 32- | 5 5/8 | 10 1/2 | 8 1/8 |
| 43 | 33- | 2 15/16 | 10 3/4 | 8 5/16 |
| 44 | 34- | 0 3/16 | 11 | 8 1/2 |
| 45 | 34- | 9 7/16 | 11 1/4 | 8 11/16 |
| 46 | 35- | 6 3/4 | 11 1/2 | 8 7/8 |
| 47 | 36- | 4 | 11 3/4 | 9 1/16 |
| 48 | 37- | 1 5/16 | 12 | 9 1/4 |
| 49 | 37- | 10 9/16 | | |
| 50 | 38- | 7 7/8 | | |

# 8 in 12 – 33 3/4°

## *Cuts and Saw Settings*

**Tail Cut, Hip or Valley Rafters,** with Square Cut Common Rafters:
Framing Square: **7 13/16 & 24**; set saw at **36°**
(Compound Miter or Radial Arm Saws: set bevel at 36°, miter at 18°)
Explanation: Section 3(d), pages 12–13.

**Chamfer Cut, Hip or Valley Rafters:**
Set saw at **23°**. Explanation: Section 3(d), page 10.

**Fascia Board Miter Cut at Hip or Valley:**
Framing Square: **13 5/16 & 24**; set saw at **36°**
(Compound Miter or Radial Arm Saws: set bevel at 36°, miter at 29°)
Explanation: Section 5, pages 24–25.

**Frieze Block against Hip or Valley Rafter:**
Framing Square: **13 5/16 & 24**; set saw at **36°**
(Compound Miter or Radial Arm Saws: set bevel at 36°, miter at 29°)
Explanation: Section 6, page 26.

**Roof Sheathing at Hip or Valley:**
Framing Square: **12 & 14 7/16**; set saw at **23°**
(Compound Miter or Radial Arm Saws: set bevel at 23°, miter at 39 3/4°)
Distance "d" = **39 15/16"**. Explanation: Section 7, pages 26–27.

## *Calculating the Length of a Rafter*

**General Rule:** Span of building
less Thickness of ridge board (if any)
gives Length of rafter to plumb line of seat cut

*and* Span of building
less Thickness of ridge board (if any)
plus 2 × projection of overhang
gives Total length of rafter to tail cut

Explanation: Section 4, pages 13–19.

*Jack Rafters*

| SPACED *Inches* | DIFFERENCE *Inches* | SPACED *Inches* | DIFFERENCE *Inches* | SPACED *Inches* | DIFFERENCE *Inches* |
|---|---|---|---|---|---|
| 1/4 | 5/16 | 15 | 18 | **32** | **38 7/16** |
| 1/2 | 5/8 | **16** | **19 1/4** | 33 | 39 11/16 |
| 3/4 | 7/8 | 17 | 20 7/16 | 34 | 40 7/8 |
| 1 | 1 3/16 | 18 | 21 5/8 | 35 | 42 1/16 |
| 2 | 2 3/8 | 19 | 22 13/16 | **36** | **43 1/4** |
| 3 | 3 5/8 | 20 | 24 1/16 | 37 | 44 7/16 |
| 4 | 4 13/16 | 21 | 25 1/4 | 38 | 45 11/16 |
| 5 | 6 | 22 | 26 7/16 | 39 | 46 7/8 |
| 6 | 7 3/16 | 23 | 27 5/8 | 40 | 48 1/16 |
| 7 | 8 7/16 | **24** | **28 7/8** | 41 | 49 1/4 |
| 8 | 9 5/8 | 25 | 30 1/16 | 42 | 50 1/2 |
| 9 | 10 13/16 | 26 | 31 1/4 | 43 | 51 11/16 |
| 10 | 12 | 27 | 32 7/16 | 44 | 52 7/8 |
| 11 | 13 1/4 | 28 | 33 5/8 | 45 | 54 1/16 |
| **12** | **14 7/16** | 29 | 34 7/8 | 46 | 55 5/16 |
| 13 | 15 5/8 | 30 | 36 1/16 | 47 | 56 1/2 |
| 14 | 16 13/16 | 31 | 37 1/4 | **48** | **57 11/16** |

### *Common Rafters*
### 8 in 12 – 33 3/4°

| SPAN Feet | LENGTH Feet Inches | SPAN Inches | LENGTH Inches |
|---|---|---|---|
| 1 | 0- 7 3/16 | 1/4 | 1/8 |
| 2 | 1- 2 7/16 | 1/2 | 5/16 |
| 3 | 1- 9 5/8 | 3/4 | 7/16 |
| 4 | 2- 4 7/8 | 1 | 5/8 |
| 5 | 3- 0 1/16 | 1 1/4 | 3/4 |
| 6 | 3- 7 1/4 | 1 1/2 | 7/8 |
| 7 | 4- 2 1/2 | 1 3/4 | 1 1/16 |
| 8 | 4- 9 11/16 | 2 | 1 3/16 |
| 9 | 5- 4 7/8 | 2 1/4 | 1 3/8 |
| 10 | 6- 0 1/8 | 2 1/2 | 1 1/2 |
| 11 | 6- 7 5/16 | 2 3/4 | 1 5/8 |
| 12 | 7- 2 9/16 | 3 | 1 13/16 |
| 13 | 7- 9 3/4 | 3 1/4 | 1 15/16 |
| 14 | 8- 4 15/16 | 3 1/2 | 2 1/8 |
| 15 | 9- 0 3/16 | 3 3/4 | 2 1/4 |
| 16 | 9- 7 3/8 | 4 | 2 3/8 |
| 17 | 10- 2 9/16 | 4 1/4 | 2 9/16 |
| 18 | 10- 9 13/16 | 4 1/2 | 2 11/16 |
| 19 | 11- 5 | 4 3/4 | 2 7/8 |
| 20 | 12- 0 1/4 | 5 | 3 |
| 21 | 12- 7 7/16 | 5 1/4 | 3 1/8 |
| 22 | 13- 2 5/8 | 5 1/2 | 3 5/16 |
| 23 | 13- 9 7/8 | 5 3/4 | 3 7/16 |
| 24 | 14- 5 1/16 | 6 | 3 5/8 |
| 25 | 15- 0 1/4 | 6 1/4 | 3 3/4 |
| 26 | 15- 7 1/2 | 6 1/2 | 3 7/8 |
| 27 | 16- 2 11/16 | 6 3/4 | 4 1/16 |
| 28 | 16- 9 15/16 | 7 | 4 3/16 |
| 29 | 17- 5 1/8 | 7 1/4 | 4 3/8 |
| 30 | 18- 0 5/16 | 7 1/2 | 4 1/2 |
| 31 | 18- 7 9/16 | 7 3/4 | 4 11/16 |
| 32 | 19- 2 3/4 | 8 | 4 13/16 |
| 33 | 19- 9 15/16 | 8 1/4 | 4 15/16 |
| 34 | 20- 5 3/16 | 8 1/2 | 5 1/8 |
| 35 | 21- 0 3/8 | 8 3/4 | 5 1/4 |
| 36 | 21- 7 5/8 | 9 | 5 7/16 |
| 37 | 22- 2 13/16 | 9 1/4 | 5 9/16 |
| 38 | 22-10 | 9 1/2 | 5 11/16 |
| 39 | 23- 5 1/4 | 9 3/4 | 5 7/8 |
| 40 | 24- 0 7/16 | 10 | 6 |
| 41 | 24- 7 5/8 | 10 1/4 | 6 3/16 |
| 42 | 25- 2 7/8 | 10 1/2 | 6 5/16 |
| 43 | 25-10 1/16 | 10 3/4 | 6 7/16 |
| 44 | 26- 5 5/16 | 11 | 6 5/8 |
| 45 | 27- 0 1/2 | 11 1/4 | 6 3/4 |
| 46 | 27- 7 11/16 | 11 1/2 | 6 15/16 |
| 47 | 28- 2 15/16 | 11 3/4 | 7 1/16 |
| 48 | 28-10 1/8 | 12 | 7 3/16 |
| 49 | 29- 5 3/8 | | |
| 50 | 30- 0 9/16 | | |

### *Hip or Valley Rafters*
### 8 in 17 – 25 1/4°

| SPAN Feet | LENGTH Feet Inches | SPAN Inches | LENGTH Inches |
|---|---|---|---|
| 1 | 0- 9 3/8 | 1/4 | 3/16 |
| 2 | 1- 6 3/4 | 1/2 | 3/8 |
| 3 | 2- 4 1/8 | 3/4 | 9/16 |
| 4 | 3- 1 1/2 | 1 | 13/16 |
| 5 | 3-10 7/8 | 1 1/4 | 1 |
| 6 | 4- 8 5/16 | 1 1/2 | 1 3/16 |
| 7 | 5- 5 11/16 | 1 3/4 | 1 3/8 |
| 8 | 6- 3 1/16 | 2 | 1 9/16 |
| 9 | 7- 0 7/16 | 2 1/4 | 1 3/4 |
| 10 | 7- 9 13/16 | 2 1/2 | 1 15/16 |
| 11 | 8- 7 3/16 | 2 3/4 | 2 1/8 |
| 12 | 9- 4 9/16 | 3 | 2 3/8 |
| 13 | 10- 1 15/16 | 3 1/4 | 2 9/16 |
| 14 | 10-11 5/16 | 3 1/2 | 2 3/4 |
| 15 | 11- 8 11/16 | 3 3/4 | 2 15/16 |
| 16 | 12- 6 1/16 | 4 | 3 1/8 |
| 17 | 13- 3 1/2 | 4 1/4 | 3 5/16 |
| 18 | 14- 0 7/8 | 4 1/2 | 3 1/2 |
| 19 | 14-10 1/4 | 4 3/4 | 3 11/16 |
| 20 | 15- 7 5/8 | 5 | 3 15/16 |
| 21 | 16- 5 | 5 1/4 | 4 1/8 |
| 22 | 17- 2 3/8 | 5 1/2 | 4 5/16 |
| 23 | 17-11 3/4 | 5 3/4 | 4 1/2 |
| 24 | 18- 9 1/8 | 6 | 4 11/16 |
| 25 | 19- 6 1/2 | 6 1/4 | 4 7/8 |
| 26 | 20- 3 7/8 | 6 1/2 | 5 1/16 |
| 27 | 21- 1 5/16 | 6 3/4 | 5 1/4 |
| 28 | 21-10 11/16 | 7 | 5 1/2 |
| 29 | 22- 8 1/16 | 7 1/4 | 5 11/16 |
| 30 | 23- 5 7/16 | 7 1/2 | 5 7/8 |
| 31 | 24- 2 13/16 | 7 3/4 | 6 1/16 |
| 32 | 25- 0 3/16 | 8 | 6 1/4 |
| 33 | 25- 9 9/16 | 8 1/4 | 6 7/16 |
| 34 | 26- 6 15/16 | 8 1/2 | 6 5/8 |
| 35 | 27- 4 5/16 | 8 3/4 | 6 13/16 |
| 36 | 28- 1 11/16 | 9 | 7 1/16 |
| 37 | 28-11 1/16 | 9 1/4 | 7 1/4 |
| 38 | 29- 8 1/2 | 9 1/2 | 7 7/16 |
| 39 | 30- 5 7/8 | 9 3/4 | 7 5/8 |
| 40 | 31- 3 1/4 | 10 | 7 13/16 |
| 41 | 32- 0 5/8 | 10 1/4 | 8 |
| 42 | 32-10 | 10 1/2 | 8 3/16 |
| 43 | 33- 7 3/8 | 10 3/4 | 8 3/8 |
| 44 | 34- 4 3/4 | 11 | 8 5/8 |
| 45 | 35- 2 1/8 | 11 1/4 | 8 13/16 |
| 46 | 35-11 1/2 | 11 1/2 | 9 |
| 47 | 36- 8 7/8 | 11 3/4 | 9 3/16 |
| 48 | 37- 6 1/4 | 12 | 9 3/8 |
| 49 | 38- 3 11/16 | | |
| 50 | 39- 1 1/16 | | |

# $8\frac{1}{2}$ in 12 – $35\frac{1}{4}°$

## *Cuts and Saw Settings*

**Tail Cut, Hip or Valley Rafters,** with Square Cut Common Rafters:
Framing Square: **8 & 24**; set saw at **35 1/4°**
(Compound Miter or Radial Arm Saws: set bevel at 35 1/4°, miter at 18 1/2°)
Explanation: Section 3(d), pages 12–13.

**Chamfer Cut, Hip or Valley Rafters:**
Set saw at **24°**. Explanation: Section 3(d), page 10.

**Fascia Board Miter Cut at Hip or Valley:**
Framing Square: **13 7/8 & 24**; set saw at **35 1/4°**
(Compound Miter or Radial Arm Saws: set bevel at 35 1/4°, miter at 30°)
Explanation: Section 5, pages 24–25.

**Frieze Block against Hip or Valley Rafter:**
Framing Square: **13 7/8 & 24**; set saw at **35 1/4°**
(Compound Miter or Radial Arm Saws: set bevel at 35 1/4°, miter at 30°)
Explanation: Section 6, page 26.

**Roof Sheathing at Hip or Valley:**
Framing Square: **12 & 14 11/16**; set saw at **24°**
(Compound Miter or Radial Arm Saws: set bevel at 24°, miter at 39 1/4°)
Distance "d" = **39 3/16"**. Explanation: Section 7, pages 26–27.

## *Calculating the Length of a Rafter*

**General Rule:**

| | | |
|---|---|---|
| | | Span of building |
| | less | Thickness of ridge board (if any) |
| | gives | Length of rafter to plumb line of seat cut |
| *and* | | Span of building |
| | less | Thickness of ridge board (if any) |
| | plus | 2 × projection of overhang |
| | gives | Total length of rafter to tail cut |

Explanation: Section 4, pages 13–19.

| *Jack Rafters* | | | | | |
|---|---|---|---|---|---|
| SPACED *Inches* | DIFFERENCE *Inches* | SPACED *Inches* | DIFFERENCE *Inches* | SPACED *Inches* | DIFFERENCE *Inches* |
| 1/4 | 5/16 | 15 | 18 3/8 | **32** | **39 3/16** |
| 1/2 | 5/8 | **16** | **19 5/8** | 33 | 40 7/16 |
| 3/4 | 15/16 | 17 | 20 13/16 | 34 | 41 11/16 |
| 1 | 1 1/4 | 18 | 22 1/16 | 35 | 42 7/8 |
| 2 | 2 7/16 | 19 | 23 5/16 | **36** | **44 1/8** |
| 3 | 3 11/16 | 20 | 24 1/2 | 37 | 45 5/16 |
| 4 | 4 7/8 | 21 | 25 3/4 | 38 | 46 9/16 |
| 5 | 6 1/8 | 22 | 26 15/16 | 39 | 47 13/16 |
| 6 | 7 3/8 | 23 | 28 3/16 | 40 | 49 |
| 7 | 8 9/16 | **24** | **29 7/16** | 41 | 50 1/4 |
| 8 | 9 13/16 | 25 | 30 5/8 | 42 | 51 1/2 |
| 9 | 11 | 26 | 31 7/8 | 43 | 52 11/16 |
| 10 | 12 1/4 | 27 | 33 1/16 | 44 | 53 15/16 |
| 11 | 13 1/2 | 28 | 34 5/16 | 45 | 55 1/8 |
| **12** | **14 11/16** | 29 | 35 9/16 | 46 | 56 3/8 |
| 13 | 15 15/16 | 30 | 36 3/4 | 47 | 57 5/8 |
| 14 | 17 3/16 | 31 | 38 | **48** | **58 13/16** |

## *Common Rafters*
## 8 1/2 in 12 – 35 1/4°

| SPAN *Feet* | LENGTH *Feet Inches* | SPAN *Inches* | LENGTH *Inches* |
|---|---|---|---|
| 1 | 0- 7 3/8 | 1/4 | 1/8 |
| 2 | 1- 2 11/16 | 1/2 | 5/16 |
| 3 | 1-10 1/16 | 3/4 | 7/16 |
| 4 | 2- 5 7/16 | 1 | 5/8 |
| 5 | 3- 0 3/4 | 1 1/4 | 3/4 |
| 6 | 3- 8 1/8 | 1 1/2 | 15/16 |
| 7 | 4- 3 1/2 | 1 3/4 | 1 1/16 |
| 8 | 4-10 13/16 | 2 | 1 1/4 |
| 9 | 5- 6 3/16 | 2 1/4 | 1 3/8 |
| 10 | 6- 1 1/2 | 2 1/2 | 1 9/16 |
| 11 | 6- 8 7/8 | 2 3/4 | 1 11/16 |
| 12 | 7- 4 1/4 | 3 | 1 13/16 |
| 13 | 7-11 9/16 | 3 1/4 | 2 |
| 14 | 8- 6 15/16 | 3 1/2 | 2 1/8 |
| 15 | 9- 2 5/16 | 3 3/4 | 2 5/16 |
| 16 | 9- 9 5/8 | 4 | 2 7/16 |
| 17 | 10- 5 | 4 1/4 | 2 5/8 |
| 18 | 11- 0 3/8 | 4 1/2 | 2 3/4 |
| 19 | 11- 7 11/16 | 4 3/4 | 2 15/16 |
| 20 | 12- 3 1/16 | 5 | 3 1/16 |
| 21 | 12-10 7/16 | 5 1/4 | 3 3/16 |
| 22 | 13- 5 3/4 | 5 1/2 | 3 3/8 |
| 23 | 14- 1 1/8 | 5 3/4 | 3 1/2 |
| 24 | 14- 8 7/16 | 6 | 3 11/16 |
| 25 | 15- 3 13/16 | 6 1/4 | 3 13/16 |
| 26 | 15-11 3/16 | 6 1/2 | 4 |
| 27 | 16- 6 1/2 | 6 3/4 | 4 1/8 |
| 28 | 17- 1 7/8 | 7 | 4 5/16 |
| 29 | 17- 9 1/4 | 7 1/4 | 4 7/16 |
| 30 | 18- 4 9/16 | 7 1/2 | 4 5/8 |
| 31 | 18-11 15/16 | 7 3/4 | 4 3/4 |
| 32 | 19- 7 5/16 | 8 | 4 7/8 |
| 33 | 20- 2 5/8 | 8 1/4 | 5 1/16 |
| 34 | 20-10 | 8 1/2 | 5 3/16 |
| 35 | 21- 5 3/8 | 8 3/4 | 5 3/8 |
| 36 | 22- 0 11/16 | 9 | 5 1/2 |
| 37 | 22- 8 1/16 | 9 1/4 | 5 11/16 |
| 38 | 23- 3 3/8 | 9 1/2 | 5 13/16 |
| 39 | 23-10 3/4 | 9 3/4 | 6 |
| 40 | 24- 6 1/8 | 10 | 6 1/8 |
| 41 | 25- 1 7/16 | 10 1/4 | 6 1/4 |
| 42 | 25- 8 13/16 | 10 1/2 | 6 7/16 |
| 43 | 26- 4 3/16 | 10 3/4 | 6 9/16 |
| 44 | 26-11 1/2 | 11 | 6 3/4 |
| 45 | 27- 6 7/8 | 11 1/4 | 6 7/8 |
| 46 | 28- 2 1/4 | 11 1/2 | 7 1/16 |
| 47 | 28- 9 9/16 | 11 3/4 | 7 3/16 |
| 48 | 29- 4 15/16 | 12 | 7 3/8 |
| 49 | 30- 0 5/16 | | |
| 50 | 30- 7 5/8 | | |

## *Hip or Valley Rafters*
## 8 1/2 in 17 – 26 1/2°

| SPAN *Feet* | LENGTH *Feet Inches* | SPAN *Inches* | LENGTH *Inches* |
|---|---|---|---|
| 1 | 0- 9 1/2 | 1/4 | 3/16 |
| 2 | 1- 7 | 1/2 | 3/8 |
| 3 | 2- 4 1/2 | 3/4 | 9/16 |
| 4 | 3- 1 15/16 | 1 | 13/16 |
| 5 | 3-11 7/16 | 1 1/4 | 1 |
| 6 | 4- 8 15/16 | 1 1/2 | 1 3/16 |
| 7 | 5- 6 7/16 | 1 3/4 | 1 3/8 |
| 8 | 6- 3 15/16 | 2 | 1 9/16 |
| 9 | 7- 1 7/16 | 2 1/4 | 1 3/4 |
| 10 | 7-10 7/8 | 2 1/2 | 2 |
| 11 | 8- 8 3/8 | 2 3/4 | 2 3/16 |
| 12 | 9- 5 7/8 | 3 | 2 3/8 |
| 13 | 10- 3 3/8 | 3 1/4 | 2 9/16 |
| 14 | 11- 0 7/8 | 3 1/2 | 2 3/4 |
| 15 | 11-10 3/8 | 3 3/4 | 2 15/16 |
| 16 | 12- 7 13/16 | 4 | 3 3/16 |
| 17 | 13- 5 5/16 | 4 1/4 | 3 3/8 |
| 18 | 14- 2 13/16 | 4 1/2 | 3 9/16 |
| 19 | 15- 0 5/16 | 4 3/4 | 3 3/4 |
| 20 | 15- 9 13/16 | 5 | 3 15/16 |
| 21 | 16- 7 5/16 | 5 1/4 | 4 1/8 |
| 22 | 17- 4 13/16 | 5 1/2 | 4 3/8 |
| 23 | 18- 2 1/4 | 5 3/4 | 4 9/16 |
| 24 | 18-11 3/4 | 6 | 4 3/4 |
| 25 | 19- 9 1/4 | 6 1/4 | 4 15/16 |
| 26 | 20- 6 3/4 | 6 1/2 | 5 1/8 |
| 27 | 21- 4 1/4 | 6 3/4 | 5 5/16 |
| 28 | 22- 1 3/4 | 7 | 5 9/16 |
| 29 | 22-11 3/16 | 7 1/4 | 5 3/4 |
| 30 | 23- 8 11/16 | 7 1/2 | 5 15/16 |
| 31 | 24- 6 3/16 | 7 3/4 | 6 1/8 |
| 32 | 25- 3 11/16 | 8 | 6 5/16 |
| 33 | 26- 1 3/16 | 8 1/4 | 6 1/2 |
| 34 | 26-10 11/16 | 8 1/2 | 6 3/4 |
| 35 | 27- 8 1/8 | 8 3/4 | 6 15/16 |
| 36 | 28- 5 5/8 | 9 | 7 1/8 |
| 37 | 29- 3 1/8 | 9 1/4 | 7 5/16 |
| 38 | 30- 0 5/8 | 9 1/2 | 7 1/2 |
| 39 | 30-10 1/8 | 9 3/4 | 7 11/16 |
| 40 | 31- 7 5/8 | 10 | 7 15/16 |
| 41 | 32- 5 1/8 | 10 1/4 | 8 1/8 |
| 42 | 33- 2 9/16 | 10 1/2 | 8 5/16 |
| 43 | 34- 0 1/16 | 10 3/4 | 8 1/2 |
| 44 | 34- 9 9/16 | 11 | 8 11/16 |
| 45 | 35- 7 1/16 | 11 1/4 | 8 7/8 |
| 46 | 36- 4 9/16 | 11 1/2 | 9 1/8 |
| 47 | 37- 2 1/16 | 11 3/4 | 9 5/16 |
| 48 | 37-11 1/2 | 12 | 9 1/2 |
| 49 | 38- 9 | | |
| 50 | 39- 6 1/2 | | |

# 9 in 12 – 36 3/4°

## *Cuts and Saw Settings*

**Tail Cut, Hip or Valley Rafters,** with Square Cut Common Rafters:
Framing Square: **8 1/8 & 24**; set saw at **34 1/2°**
(Compound Miter or Radial Arm Saws: set bevel at 34 1/2°, miter at 18 3/4°)
Explanation: Section 3(d), pages 12–13.

**Chamfer Cut, Hip or Valley Rafters:**
Set saw at **25°**. Explanation: Section 3(d), page 10.

**Fascia Board Miter Cut at Hip or Valley:**
Framing Square: **14 3/8 & 24**; set saw at **34 1/2°**
(Compound Miter or Radial Arm Saws: set bevel at 34 1/2°, miter at 31°)
Explanation: Section 5, pages 24–25.

**Frieze Block against Hip or Valley Rafter:**
Framing Square: **14 3/8 & 24**; set saw at **34 1/2°**
(Compound Miter or Radial Arm Saws: set bevel at 34 1/2°, miter at 31°)
Explanation: Section 6, page 26.

**Roof Sheathing at Hip or Valley:**
Framing Square: **12 & 15**; set saw at **25°**
(Compound Miter or Radial Arm Saws: set bevel at 25°, miter at 38 3/4°)
Distance "d" = **38 3/8"**. Explanation: Section 7, pages 26–27.

## *Calculating the Length of a Rafter*

| **General Rule:** | | Span of building |
|---|---|---|
| | less | Thickness of ridge board (if any) |
| | gives | Length of rafter to plumb line of seat cut |
| *and* | | Span of building |
| | less | Thickness of ridge board (if any) |
| | plus | 2 × projection of overhang |
| | gives | Total length of rafter to tail cut |

Explanation: Section 4, pages 13–19.

| *Jack Rafters* | | | | | |
|---|---|---|---|---|---|
| SPACED *Inches* | DIFFERENCE *Inches* | SPACED *Inches* | DIFFERENCE *Inches* | SPACED *Inches* | DIFFERENCE *Inches* |
| 1/4 | 5/16 | 15 | 18 3/4 | **32** | **40** |
| 1/2 | 5/8 | **16** | **20** | 33 | 41 1/4 |
| 3/4 | 15/16 | 17 | 21 1/4 | 34 | 42 1/2 |
| 1 | 1 1/4 | 18 | 22 1/2 | 35 | 43 3/4 |
| 2 | 2 1/2 | 19 | 23 3/4 | **36** | **45** |
| 3 | 3 3/4 | 20 | 25 | 37 | 46 1/4 |
| 4 | 5 | 21 | 26 1/4 | 38 | 47 1/2 |
| 5 | 6 1/4 | 22 | 27 1/2 | 39 | 48 3/4 |
| 6 | 7 1/2 | 23 | 28 3/4 | 40 | 50 |
| 7 | 8 3/4 | **24** | **30** | 41 | 51 1/4 |
| 8 | 10 | 25 | 31 1/4 | 42 | 52 1/2 |
| 9 | 11 1/4 | 26 | 32 1/2 | 43 | 53 3/4 |
| 10 | 12 1/2 | 27 | 33 3/4 | 44 | 55 |
| 11 | 13 3/4 | 28 | 35 | 45 | 56 1/4 |
| **12** | **15** | 29 | 36 1/4 | 46 | 57 1/2 |
| 13 | 16 1/4 | 30 | 37 1/2 | 47 | 58 3/4 |
| 14 | 17 1/2 | 31 | 38 3/4 | **48** | **60** |

| Common Rafters 9 in 12 – 36 3/4° | | | |
|---|---|---|---|
| SPAN Feet | LENGTH Feet Inches | SPAN Inches | LENGTH Inches |
| 1 | 0- 7 1/2 | 1/4 | 3/16 |
| 2 | 1- 3 | 1/2 | 5/16 |
| 3 | 1-10 1/2 | 3/4 | 1/2 |
| 4 | 2- 6 | 1 | 5/8 |
| 5 | 3- 1 1/2 | 1 1/4 | 13/16 |
| 6 | 3- 9 | 1 1/2 | 15/16 |
| 7 | 4- 4 1/2 | 1 3/4 | 1 1/8 |
| 8 | 5- 0 | 2 | 1 1/4 |
| 9 | 5- 7 1/2 | 2 1/4 | 1 7/16 |
| 10 | 6- 3 | 2 1/2 | 1 9/16 |
| 11 | 6-10 1/2 | 2 3/4 | 1 3/4 |
| 12 | 7- 6 | 3 | 1 7/8 |
| 13 | 8- 1 1/2 | 3 1/4 | 2 1/16 |
| 14 | 8- 9 | 3 1/2 | 2 3/16 |
| 15 | 9- 4 1/2 | 3 3/4 | 2 3/8 |
| 16 | 10- 0 | 4 | 2 1/2 |
| 17 | 10- 7 1/2 | 4 1/4 | 2 11/16 |
| 18 | 11- 3 | 4 1/2 | 2 13/16 |
| 19 | 11-10 1/2 | 4 3/4 | 3 |
| 20 | 12- 6 | 5 | 3 1/8 |
| 21 | 13- 1 1/2 | 5 1/4 | 3 5/16 |
| 22 | 13- 9 | 5 1/2 | 3 7/16 |
| 23 | 14- 4 1/2 | 5 3/4 | 3 5/8 |
| 24 | 15- 0 | 6 | 3 3/4 |
| 25 | 15- 7 1/2 | 6 1/4 | 3 15/16 |
| 26 | 16- 3 | 6 1/2 | 4 1/16 |
| 27 | 16-10 1/2 | 6 3/4 | 4 1/4 |
| 28 | 17- 6 | 7 | 4 3/8 |
| 29 | 18- 1 1/2 | 7 1/4 | 4 9/16 |
| 30 | 18- 9 | 7 1/2 | 4 11/16 |
| 31 | 19- 4 1/2 | 7 3/4 | 4 7/8 |
| 32 | 20- 0 | 8 | 5 |
| 33 | 20- 7 1/2 | 8 1/4 | 5 3/16 |
| 34 | 21- 3 | 8 1/2 | 5 5/16 |
| 35 | 21-10 1/2 | 8 3/4 | 5 1/2 |
| 36 | 22- 6 | 9 | 5 5/8 |
| 37 | 23- 1 1/2 | 9 1/4 | 5 13/16 |
| 38 | 23- 9 | 9 1/2 | 5 15/16 |
| 39 | 24- 4 1/2 | 9 3/4 | 6 1/8 |
| 40 | 25- 0 | 10 | 6 1/4 |
| 41 | 25- 7 1/2 | 10 1/4 | 6 7/16 |
| 42 | 26- 3 | 10 1/2 | 6 9/16 |
| 43 | 26-10 1/2 | 10 3/4 | 6 3/4 |
| 44 | 27- 6 | 11 | 6 7/8 |
| 45 | 28- 1 1/2 | 11 1/4 | 7 1/16 |
| 46 | 28- 9 | 11 1/2 | 7 3/16 |
| 47 | 29- 4 1/2 | 11 3/4 | 7 3/8 |
| 48 | 30- 0 | 12 | 7 1/2 |
| 49 | 30- 7 1/2 | | |
| 50 | 31- 3 | | |

| Hip or Valley Rafters 9 in 17 – 28° | | | |
|---|---|---|---|
| SPAN Feet | LENGTH Feet Inches | SPAN Inches | LENGTH Inches |
| 1 | 0 - 9 5/8 | 1/4 | 3/16 |
| 2 | 1 - 7 3/16 | 1/2 | 3/8 |
| 3 | 2 - 4 13/16 | 3/4 | 5/8 |
| 4 | 3 - 2 7/16 | 1 | 13/16 |
| 5 | 4 - 0 | 1 1/4 | 1 |
| 6 | 4 - 9 5/8 | 1 1/2 | 1 3/16 |
| 7 | 5 - 7 1/4 | 1 3/4 | 1 3/8 |
| 8 | 6 - 4 13/16 | 2 | 1 5/8 |
| 9 | 7 - 2 7/16 | 2 1/4 | 1 13/16 |
| 10 | 8 - 0 1/16 | 2 1/2 | 2 |
| 11 | 8 - 9 5/8 | 2 3/4 | 2 3/16 |
| 12 | 9 - 7 1/4 | 3 | 2 3/8 |
| 13 | 10 - 4 7/8 | 3 1/4 | 2 5/8 |
| 14 | 11 - 2 7/16 | 3 1/2 | 2 13/16 |
| 15 | 12 - 0 1/16 | 3 3/4 | 3 |
| 16 | 12 - 9 11/16 | 4 | 3 3/16 |
| 17 | 13 - 7 1/4 | 4 1/4 | 3 3/8 |
| 18 | 14 - 4 7/8 | 4 1/2 | 3 5/8 |
| 19 | 15 - 2 1/2 | 4 3/4 | 3 13/16 |
| 20 | 16 - 0 1/16 | 5 | 4 |
| 21 | 16 - 9 11/16 | 5 1/4 | 4 3/16 |
| 22 | 17 - 7 5/16 | 5 1/2 | 4 3/8 |
| 23 | 18 - 4 15/16 | 5 3/4 | 4 5/8 |
| 24 | 19 - 2 1/2 | 6 | 4 13/16 |
| 25 | 20 - 0 1/8 | 6 1/4 | 5 |
| 26 | 20 - 9 3/4 | 6 1/2 | 5 3/16 |
| 27 | 21 - 7 5/16 | 6 3/4 | 5 3/8 |
| 28 | 22 - 4 15/16 | 7 | 5 5/8 |
| 29 | 23 - 2 9/16 | 7 1/4 | 5 13/16 |
| 30 | 24 - 0 1/8 | 7 1/2 | 6 |
| 31 | 24 - 9 3/4 | 7 3/4 | 6 3/16 |
| 32 | 25 - 7 3/8 | 8 | 6 3/8 |
| 33 | 26 - 4 15/16 | 8 1/4 | 6 5/8 |
| 34 | 27 - 2 9/16 | 8 1/2 | 6 13/16 |
| 35 | 28 - 0 3/16 | 8 3/4 | 7 |
| 36 | 28 - 9 3/4 | 9 | 7 3/16 |
| 37 | 29 - 7 3/8 | 9 1/4 | 7 3/8 |
| 38 | 30 - 5 | 9 1/2 | 7 5/8 |
| 39 | 31 - 2 9/16 | 9 3/4 | 7 13/16 |
| 40 | 32 - 0 3/16 | 10 | 8 |
| 41 | 32 - 9 13/16 | 10 1/4 | 8 3/16 |
| 42 | 33 - 7 3/8 | 10 1/2 | 8 3/8 |
| 43 | 34 - 5 | 10 3/4 | 8 5/8 |
| 44 | 35 - 2 5/8 | 11 | 8 13/16 |
| 45 | 36 - 0 3/16 | 11 1/4 | 9 |
| 46 | 36 - 9 13/16 | 11 1/2 | 9 3/16 |
| 47 | 37 - 7 7/16 | 11 3/4 | 9 3/8 |
| 48 | 38 - 5 | 12 | 9 5/8 |
| 49 | 39 - 2 5/8 | | |
| 50 | 40 - 0 1/4 | | |

# 9 1/2 in 12 – 38 1/4°

## *Cuts and Saw Settings*

**Tail Cut, Hip or Valley Rafters,** with Square Cut Common Rafters:
Framing Square: **8 1/4 & 24**; set saw at **33 3/4°**
(Compound Miter or Radial Arm Saws: set bevel at 33 3/4°, miter at 19°)
Explanation: Section 3(d), pages 12–13.

**Chamfer Cut, Hip or Valley Rafters:**
Set saw at **26°**. Explanation: Section 3(d), page 10.

**Fascia Board Miter Cut at Hip or Valley:**
Framing Square: **14 7/8 & 24**; set saw at **33 3/4°**
(Compound Miter or Radial Arm Saws: set bevel at 33 3/4°, miter at 31 3/4°)
Explanation: Section 5, pages 24–25.

**Frieze Block against Hip or Valley Rafter:**
Framing Square: **14 7/8 & 24**; set saw at **33 3/4°**
(Compound Miter or Radial Arm Saws: set bevel at 33 3/4°, miter at 31 3/4°)
Explanation: Section 6, page 26.

**Roof Sheathing at Hip or Valley:**
Framing Square: **12 & 15 5/16**; set saw at **26°**
(Compound Miter or Radial Arm Saws: set bevel at 26°, miter at 38°)
Distance "d" = **37 5/8"**. Explanation: Section 7, pages 26–27.

## *Calculating the Length of a Rafter*

**General Rule:**

| | | |
|---|---|---|
| | | Span of building |
| | less | Thickness of ridge board (if any) |
| | gives | Length of rafter to plumb line of seat cut |
| *and* | | Span of building |
| | less | Thickness of ridge board (if any) |
| | plus | 2 × projection of overhang |
| | gives | Total length of rafter to tail cut |

Explanation: Section 4, pages 13–19.

| *Jack Rafters* | | | | | |
|---|---|---|---|---|---|
| SPACED *Inches* | DIFFERENCE *Inches* | SPACED *Inches* | DIFFERENCE *Inches* | SPACED *Inches* | DIFFERENCE *Inches* |
| 1/4 | 5/16 | 15 | 19 1/8 | **32** | **40 13/16** |
| 1/2 | 5/8 | **16** | **20 7/16** | 33 | 42 1/16 |
| 3/4 | 15/16 | 17 | 21 11/16 | 34 | 43 3/8 |
| 1 | 1 1/4 | 18 | 22 15/16 | 35 | 44 5/8 |
| 2 | 2 9/16 | 19 | 24 1/4 | **36** | **45 15/16** |
| 3 | 3 13/16 | 20 | 25 1/2 | 37 | 47 3/16 |
| 4 | 5 1/8 | 21 | 26 13/16 | 38 | 48 7/16 |
| 5 | 6 3/8 | 22 | 28 1/16 | 39 | 49 3/4 |
| 6 | 7 5/8 | 23 | 29 5/16 | 40 | 51 |
| 7 | 8 15/16 | **24** | **30 5/8** | 41 | 52 5/16 |
| 8 | 10 3/16 | 25 | 31 7/8 | 42 | 53 9/16 |
| 9 | 11 1/2 | 26 | 33 3/16 | 43 | 54 13/16 |
| 10 | 12 3/4 | 27 | 34 7/16 | 44 | 56 1/8 |
| 11 | 14 | 28 | 35 11/16 | 45 | 57 3/8 |
| **12** | **15 5/16** | 29 | 37 | 46 | 58 11/16 |
| 13 | 16 9/16 | 30 | 38 1/4 | 47 | 59 15/16 |
| 14 | 17 7/8 | 31 | 39 9/16 | **48** | **61 1/4** |

| **Common Rafters 9 1/2 in 12 – 38 1/4°** | | | |
|---|---|---|---|
| **SPAN** *Feet* | **LENGTH** *Feet Inches* | **SPAN** *Inches* | **LENGTH** *Inches* |
| 1 | 0- 7 5/8 | 1/4 | 3/16 |
| 2 | 1- 3 5/16 | 1/2 | 5/16 |
| 3 | 1-10 15/16 | 3/4 | 1/2 |
| 4 | 2- 6 5/8 | 1 | 5/8 |
| 5 | 3- 2 1/4 | 1 1/4 | 13/16 |
| 6 | 3- 9 15/16 | 1 1/2 | 15/16 |
| 7 | 4- 5 9/16 | 1 3/4 | 1 1/8 |
| 8 | 5- 1 1/4 | 2 | 1 1/4 |
| 9 | 5- 8 7/8 | 2 1/4 | 1 7/16 |
| 10 | 6- 4 1/2 | 2 1/2 | 1 5/8 |
| 11 | 7- 0 3/16 | 2 3/4 | 1 3/4 |
| 12 | 7- 7 13/16 | 3 | 1 15/16 |
| 13 | 8- 3 1/2 | 3 1/4 | 2 1/16 |
| 14 | 8-11 1/8 | 3 1/2 | 2 1/4 |
| 15 | 9- 6 13/16 | 3 3/4 | 2 3/8 |
| 16 | 10- 2 7/16 | 4 | 2 9/16 |
| 17 | 10-10 1/8 | 4 1/4 | 2 11/16 |
| 18 | 11- 5 3/4 | 4 1/2 | 2 7/8 |
| 19 | 12- 1 3/8 | 4 3/4 | 3 |
| 20 | 12- 9 1/16 | 5 | 3 3/16 |
| 21 | 13- 4 11/16 | 5 1/4 | 3 3/8 |
| 22 | 14- 0 3/8 | 5 1/2 | 3 1/2 |
| 23 | 14- 8 | 5 3/4 | 3 11/16 |
| 24 | 15- 3 11/16 | 6 | 3 13/16 |
| 25 | 15-11 5/16 | 6 1/4 | 4 |
| 26 | 16- 6 15/16 | 6 1/2 | 4 1/8 |
| 27 | 17- 2 5/8 | 6 3/4 | 4 5/16 |
| 28 | 17-10 1/4 | 7 | 4 7/16 |
| 29 | 18- 5 15/16 | 7 1/4 | 4 5/8 |
| 30 | 19- 1 9/16 | 7 1/2 | 4 13/16 |
| 31 | 19- 9 1/4 | 7 3/4 | 4 15/16 |
| 32 | 20- 4 7/8 | 8 | 5 1/8 |
| 33 | 21- 0 9/16 | 8 1/4 | 5 1/4 |
| 34 | 21- 8 3/16 | 8 1/2 | 5 7/16 |
| 35 | 22- 3 13/16 | 8 3/4 | 5 9/16 |
| 36 | 22-11 1/2 | 9 | 5 3/4 |
| 37 | 23- 7 1/8 | 9 1/4 | 5 7/8 |
| 38 | 24- 2 13/16 | 9 1/2 | 6 1/16 |
| 39 | 24-10 7/16 | 9 3/4 | 6 3/16 |
| 40 | 25- 6 1/8 | 10 | 6 3/8 |
| 41 | 26- 1 3/4 | 10 1/4 | 6 9/16 |
| 42 | 26- 9 7/16 | 10 1/2 | 6 11/16 |
| 43 | 27- 5 1/16 | 10 3/4 | 6 7/8 |
| 44 | 28- 0 11/16 | 11 | 7 |
| 45 | 28- 8 3/8 | 11 1/4 | 7 3/16 |
| 46 | 29- 4 | 11 1/2 | 7 5/16 |
| 47 | 29-11 11/16 | 11 3/4 | 7 1/2 |
| 48 | 30- 7 5/16 | 12 | 7 5/8 |
| 49 | 31- 3 | | |
| 50 | 31-10 5/8 | | |

| **Hip or Valley Rafters 9 1/2 in 17 – 29 1/4°** | | | |
|---|---|---|---|
| **SPAN** *Feet* | **LENGTH** *Feet Inches* | **SPAN** *Inches* | **LENGTH** *Inches* |
| 1 | 0- 9 3/4 | 1/4 | 3/16 |
| 2 | 1- 7 7/16 | 1/2 | 3/8 |
| 3 | 2- 5 3/16 | 3/4 | 5/8 |
| 4 | 3- 2 7/8 | 1 | 13/16 |
| 5 | 4- 0 5/8 | 1 1/4 | 1 |
| 6 | 4-10 3/8 | 1 1/2 | 1 3/16 |
| 7 | 5- 8 1/16 | 1 3/4 | 1 7/16 |
| 8 | 6- 5 13/16 | 2 | 1 5/8 |
| 9 | 7- 3 1/2 | 2 1/4 | 1 13/16 |
| 10 | 8- 1 1/4 | 2 1/2 | 2 |
| 11 | 8-10 15/16 | 2 3/4 | 2 1/4 |
| 12 | 9- 8 11/16 | 3 | 2 7/16 |
| 13 | 10- 6 7/16 | 3 1/4 | 2 5/8 |
| 14 | 11- 4 1/8 | 3 1/2 | 2 13/16 |
| 15 | 12- 1 7/8 | 3 3/4 | 3 1/16 |
| 16 | 12-11 9/16 | 4 | 3 1/4 |
| 17 | 13- 9 5/16 | 4 1/4 | 3 7/16 |
| 18 | 14- 7 1/16 | 4 1/2 | 3 5/8 |
| 19 | 15- 4 3/4 | 4 3/4 | 3 7/8 |
| 20 | 16- 2 1/2 | 5 | 4 1/16 |
| 21 | 17- 0 3/16 | 5 1/4 | 4 1/4 |
| 22 | 17- 9 15/16 | 5 1/2 | 4 7/16 |
| 23 | 18- 7 11/16 | 5 3/4 | 4 11/16 |
| 24 | 19- 5 3/8 | 6 | 4 7/8 |
| 25 | 20- 3 1/8 | 6 1/4 | 5 1/16 |
| 26 | 21- 0 13/16 | 6 1/2 | 5 1/4 |
| 27 | 21-10 9/16 | 6 3/4 | 5 1/2 |
| 28 | 22- 8 1/4 | 7 | 5 11/16 |
| 29 | 23- 6 | 7 1/4 | 5 7/8 |
| 30 | 24- 3 3/4 | 7 1/2 | 6 1/16 |
| 31 | 25- 1 7/16 | 7 3/4 | 6 1/4 |
| 32 | 25-11 3/16 | 8 | 6 1/2 |
| 33 | 26- 8 7/8 | 8 1/4 | 6 11/16 |
| 34 | 27- 6 5/8 | 8 1/2 | 6 7/8 |
| 35 | 28- 4 3/8 | 8 3/4 | 7 1/16 |
| 36 | 29- 2 1/16 | 9 | 7 5/16 |
| 37 | 29-11 13/16 | 9 1/4 | 7 1/2 |
| 38 | 30- 9 1/2 | 9 1/2 | 7 11/16 |
| 39 | 31- 7 1/4 | 9 3/4 | 7 7/8 |
| 40 | 32- 5 | 10 | 8 1/8 |
| 41 | 33- 2 11/16 | 10 1/4 | 8 5/16 |
| 42 | 34- 0 7/16 | 10 1/2 | 8 1/2 |
| 43 | 34-10 1/8 | 10 3/4 | 8 11/16 |
| 44 | 35- 7 7/8 | 11 | 8 15/16 |
| 45 | 36- 5 5/8 | 11 1/4 | 9 1/8 |
| 46 | 37- 3 5/16 | 11 1/2 | 9 5/16 |
| 47 | 38- 1 1/16 | 11 3/4 | 9 1/2 |
| 48 | 38-10 3/4 | 12 | 9 3/4 |
| 49 | 39- 8 1/2 | | |
| 50 | 40- 6 3/16 | | |

## *Cuts and Saw Settings*

**Tail Cut, Hip or Valley Rafters,** with Square Cut Common Rafters:
Framing Square: **8 3/8 & 24**; set saw at **33°**
(Compound Miter or Radial Arm Saws: set bevel at 33°, miter at 19 1/4°)
Explanation: Section 3(d), pages 12–13.

**Chamfer Cut, Hip or Valley Rafters:**
Set saw at **27°**. Explanation: Section 3(d), page 10.

**Fascia Board Miter Cut at Hip or Valley:**
Framing Square: **15 3/8 & 24**; set saw at **33°**
(Compound Miter or Radial Arm Saws: set bevel at 33°, miter at 32 3/4°)
Explanation: Section 5, pages 24–25.

**Frieze Block against Hip or Valley Rafter:**
Framing Square: **15 3/8 & 24**; set saw at **33°**
(Compound Miter or Radial Arm Saws: set bevel at 33°, miter at 32 3/4°)
Explanation: Section 6, page 26.

**Roof Sheathing at Hip or Valley:**
Framing Square: **12 & 15 5/8**; set saw at **27°**
(Compound Miter or Radial Arm Saws: set bevel at 27°, miter at 37 1/2°)
Distance "d" = **36 7/8"**. Explanation: Section 7, pages 26–27.

## *Calculating the Length of a Rafter*

**General Rule:** Span of building
less Thickness of ridge board (if any)
gives Length of rafter to plumb line of seat cut

*and* Span of building
less Thickness of ridge board (if any)
plus 2 × projection of overhang
gives Total length of rafter to tail cut

Explanation: Section 4, pages 13–19.

| *Jack Rafters* | | | | | |
|---|---|---|---|---|---|
| SPACED *Inches* | DIFFERENCE *Inches* | SPACED *Inches* | DIFFERENCE *Inches* | SPACED *Inches* | DIFFERENCE *Inches* |
| 1/4 | 5/16 | 15 | 19 1/2 | **32** | **41 5/8** |
| 1/2 | 5/8 | **16** | **20 13/16** | 33 | 42 15/16 |
| 3/4 | 1 | 17 | 22 1/8 | 34 | 44 1/4 |
| 1 | 1 5/16 | 18 | 23 7/16 | 35 | 45 9/16 |
| 2 | 2 5/8 | 19 | 24 3/4 | **36** | **46 7/8** |
| 3 | 3 7/8 | 20 | 26 1/16 | 37 | 48 3/16 |
| 4 | 5 3/16 | 21 | 27 5/16 | 38 | 49 7/16 |
| 5 | 6 1/2 | 22 | 28 5/8 | 39 | 50 3/4 |
| 6 | 7 13/16 | 23 | 29 15/16 | 40 | 52 1/16 |
| 7 | 9 1/8 | **24** | **31 1/4** | 41 | 53 3/8 |
| 8 | 10 7/16 | 25 | 32 9/16 | 42 | 54 11/16 |
| 9 | 11 11/16 | 26 | 33 7/8 | 43 | 56 |
| 10 | 13 | 27 | 35 1/8 | 44 | 57 1/4 |
| 11 | 14 5/16 | 28 | 36 7/16 | 45 | 58 9/16 |
| **12** | **15 5/8** | 29 | 37 3/4 | 46 | 59 7/8 |
| 13 | 16 15/16 | 30 | 39 1/16 | 47 | 61 3/16 |
| 14 | 18 1/4 | 31 | 40 3/8 | **48** | **62 1/2** |

| ***Common Rafters* 10 in 12 – 39 3/4°** | | | |
|---|---|---|---|
| **SPAN** *Feet* | **LENGTH** *Feet Inches* | **SPAN** *Inches* | **LENGTH** *Inches* |
| 1 | 0- 7 13/16 | 1/4 | 3/16 |
| 2 | 1- 3 5/8 | 1/2 | 5/16 |
| 3 | 1-11 7/16 | 3/4 | 1/2 |
| 4 | 2- 7 1/4 | 1 | 5/8 |
| 5 | 3- 3 1/16 | 1 1/4 | 13/16 |
| 6 | 3-10 7/8 | 1 1/2 | 1 |
| 7 | 4- 6 11/16 | 1 3/4 | 1 1/8 |
| 8 | 5- 2 1/2 | 2 | 1 5/16 |
| 9 | 5-10 5/16 | 2 1/4 | 1 7/16 |
| 10 | 6- 6 1/8 | 2 1/2 | 1 5/8 |
| 11 | 7- 1 15/16 | 2 3/4 | 1 13/16 |
| 12 | 7- 9 3/4 | 3 | 1 15/16 |
| 13 | 8- 5 9/16 | 3 1/4 | 2 1/8 |
| 14 | 9- 1 5/16 | 3 1/2 | 2 1/4 |
| 15 | 9- 9 1/8 | 3 3/4 | 2 7/16 |
| 16 | 10- 4 15/16 | 4 | 2 5/8 |
| 17 | 11- 0 3/4 | 4 1/4 | 2 3/4 |
| 18 | 11- 8 9/16 | 4 1/2 | 2 15/16 |
| 19 | 12- 4 3/8 | 4 3/4 | 3 1/16 |
| 20 | 13- 0 3/16 | 5 | 3 1/4 |
| 21 | 13- 8 | 5 1/4 | 3 7/16 |
| 22 | 14- 3 13/16 | 5 1/2 | 3 9/16 |
| 23 | 14-11 5/8 | 5 3/4 | 3 3/4 |
| 24 | 15- 7 7/16 | 6 | 3 7/8 |
| 25 | 16- 3 1/4 | 6 1/4 | 4 1/16 |
| 26 | 16-11 1/16 | 6 1/2 | 4 1/4 |
| 27 | 17- 6 7/8 | 6 3/4 | 4 3/8 |
| 28 | 18- 2 11/16 | 7 | 4 9/16 |
| 29 | 18-10 1/2 | 7 1/4 | 4 11/16 |
| 30 | 19- 6 5/16 | 7 1/2 | 4 7/8 |
| 31 | 20- 2 1/8 | 7 3/4 | 5 1/16 |
| 32 | 20- 9 15/16 | 8 | 5 3/16 |
| 33 | 21- 5 3/4 | 8 1/4 | 5 3/8 |
| 34 | 22- 1 9/16 | 8 1/2 | 5 9/16 |
| 35 | 22- 9 3/8 | 8 3/4 | 5 11/16 |
| 36 | 23- 5 3/16 | 9 | 5 7/8 |
| 37 | 24- 1 | 9 1/4 | 6 |
| 38 | 24- 8 13/16 | 9 1/2 | 6 3/16 |
| 39 | 25- 4 5/8 | 9 3/4 | 6 3/8 |
| 40 | 26- 0 7/16 | 10 | 6 1/2 |
| 41 | 26- 8 1/4 | 10 1/4 | 6 11/16 |
| 42 | 27- 4 | 10 1/2 | 6 13/16 |
| 43 | 27-11 13/16 | 10 3/4 | 7 |
| 44 | 28- 7 5/8 | 11 | 7 3/16 |
| 45 | 29- 3 7/16 | 11 1/4 | 7 5/16 |
| 46 | 29-11 1/4 | 11 1/2 | 7 1/2 |
| 47 | 30- 7 1/16 | 11 3/4 | 7 5/8 |
| 48 | 31- 2 7/8 | 12 | 7 13/16 |
| 49 | 31-10 11/16 | | |
| 50 | 32- 6 1/2 | | |

| ***Hip or Valley Rafters* 10 in 17 – 30 1/2°** | | | |
|---|---|---|---|
| **SPAN** *Feet* | **LENGTH** *Feet Inches* | **SPAN** *Inches* | **LENGTH** *Inches* |
| 1 | 0- 9 7/8 | 1/4 | 3/16 |
| 2 | 1- 7 11/16 | 1/2 | 7/16 |
| 3 | 2- 5 9/16 | 3/4 | 5/8 |
| 4 | 3- 3 3/8 | 1 | 13/16 |
| 5 | 4- 1 1/4 | 1 1/4 | 1 |
| 6 | 4-11 1/16 | 1 1/2 | 1 1/4 |
| 7 | 5- 8 15/16 | 1 3/4 | 1 7/16 |
| 8 | 6- 6 13/16 | 2 | 1 5/8 |
| 9 | 7- 4 5/8 | 2 1/4 | 1 7/8 |
| 10 | 8- 2 1/2 | 2 1/2 | 2 1/16 |
| 11 | 9- 0 5/16 | 2 3/4 | 2 1/4 |
| 12 | 9-10 3/16 | 3 | 2 7/16 |
| 13 | 10- 8 1/16 | 3 1/4 | 2 11/16 |
| 14 | 11- 5 7/8 | 3 1/2 | 2 7/8 |
| 15 | 12- 3 3/4 | 3 3/4 | 3 1/16 |
| 16 | 13- 1 9/16 | 4 | 3 5/16 |
| 17 | 13-11 7/16 | 4 1/4 | 3 1/2 |
| 18 | 14- 9 1/4 | 4 1/2 | 3 11/16 |
| 19 | 15- 7 1/8 | 4 3/4 | 3 7/8 |
| 20 | 16- 5 | 5 | 4 1/8 |
| 21 | 17- 2 13/16 | 5 1/4 | 4 5/16 |
| 22 | 18- 0 11/16 | 5 1/2 | 4 1/2 |
| 23 | 18-10 1/2 | 5 3/4 | 4 3/4 |
| 24 | 19- 8 3/8 | 6 | 4 15/16 |
| 25 | 20- 6 1/4 | 6 1/4 | 5 1/8 |
| 26 | 21- 4 1/16 | 6 1/2 | 5 5/16 |
| 27 | 22- 1 15/16 | 6 3/4 | 5 9/16 |
| 28 | 22-11 3/4 | 7 | 5 3/4 |
| 29 | 23- 9 5/8 | 7 1/4 | 5 15/16 |
| 30 | 24- 7 7/16 | 7 1/2 | 6 1/8 |
| 31 | 25- 5 5/16 | 7 3/4 | 6 3/8 |
| 32 | 26- 3 3/16 | 8 | 6 9/16 |
| 33 | 27- 1 | 8 1/4 | 6 3/4 |
| 34 | 27-10 7/8 | 8 1/2 | 7 |
| 35 | 28- 8 11/16 | 8 3/4 | 7 3/16 |
| 36 | 29- 6 9/16 | 9 | 7 3/8 |
| 37 | 30- 4 7/16 | 9 1/4 | 7 9/16 |
| 38 | 31- 2 1/4 | 9 1/2 | 7 13/16 |
| 39 | 32- 0 1/8 | 9 3/4 | 8 |
| 40 | 32- 9 15/16 | 10 | 8 3/16 |
| 41 | 33- 7 13/16 | 10 1/4 | 8 7/16 |
| 42 | 34- 5 5/8 | 10 1/2 | 8 5/8 |
| 43 | 35- 3 1/2 | 10 3/4 | 8 13/16 |
| 44 | 36- 1 3/8 | 11 | 9 |
| 45 | 36-11 3/16 | 11 1/4 | 9 1/4 |
| 46 | 37- 9 1/16 | 11 1/2 | 9 7/16 |
| 47 | 38- 6 7/8 | 11 3/4 | 9 5/8 |
| 48 | 39- 4 3/4 | 12 | 9 7/8 |
| 49 | 40- 2 5/8 | | |
| 50 | 41- 0 7/16 | | |

# 10½ in 12 – 41¼°

## *Cuts and Saw Settings*

**Tail Cut, Hip or Valley Rafters,** with Square Cut Common Rafters:
Framing Square: **8 7/16 & 24**; set saw at **32 1/4°**
(Compound Miter or Radial Arm Saws: set bevel at 32 1/4°, miter at 19 1/4°)
Explanation: Section 3(d), pages 12–13.

**Chamfer Cut, Hip or Valley Rafters:**
Set saw at **27 3/4°**. Explanation: Section 3(d), page 10.

**Fascia Board Miter Cut at Hip or Valley:**
Framing Square: **15 13/16 & 24**; set saw at **32 1/4°**
(Compound Miter or Radial Arm Saws: set bevel at 32 1/4°, miter at 33 1/4°)
Explanation: Section 5, pages 24–25.

**Frieze Block against Hip or Valley Rafter:**
Framing Square: **15 13/16 & 24**; set saw at **32 1/4°**
(Compound Miter or Radial Arm Saws: set bevel at 32 1/4°, miter at 33 1/4°)
Explanation: Section 6, page 26.

**Roof Sheathing at Hip or Valley:**
Framing Square: **12 & 15 15/16**; set saw at **27 3/4°**
(Compound Miter or Radial Arm Saws: set bevel at 27 3/4°, miter at 37°)
Distance "d" = **36 1/8"**. Explanation: Section 7, pages 26–27.

## *Calculating the Length of a Rafter*

**General Rule:** Span of building
less Thickness of ridge board (if any)
gives Length of rafter to plumb line of seat cut

*and* Span of building
less Thickness of ridge board (if any)
plus 2 × projection of overhang
gives Total length of rafter to tail cut

Explanation: Section 4, pages 13–19.

| *Jack Rafters* | | | | | |
|---|---|---|---|---|---|
| SPACED *Inches* | DIFFERENCE *Inches* | SPACED *Inches* | DIFFERENCE *Inches* | SPACED *Inches* | DIFFERENCE *Inches* |
| 1/4 | 5/16 | 15 | 19 15/16 | **32** | **42 1/2** |
| 1/2 | 11/16 | **16** | **21 1/4** | 33 | 43 7/8 |
| 3/4 | 1 | 17 | 22 9/16 | 34 | 45 3/16 |
| 1 | 1 5/16 | 18 | 23 15/16 | 35 | 46 1/2 |
| 2 | 2 11/16 | 19 | 25 1/4 | **36** | **47 13/16** |
| 3 | 4 | 20 | 26 9/16 | 37 | 49 3/16 |
| 4 | 5 5/16 | 21 | 27 7/8 | 38 | 50 1/2 |
| 5 | 6 5/8 | 22 | 29 1/4 | 39 | 51 13/16 |
| 6 | 8 | 23 | 30 9/16 | 40 | 53 1/8 |
| 7 | 9 5/16 | **24** | **31 7/8** | 41 | 54 1/2 |
| 8 | 10 5/8 | 25 | 33 1/4 | 42 | 55 13/16 |
| 9 | 11 15/16 | 26 | 34 9/16 | 43 | 57 1/8 |
| 10 | 13 5/16 | 27 | 35 7/8 | 44 | 58 7/16 |
| 11 | 14 5/8 | 28 | 37 3/16 | 45 | 59 13/16 |
| **12** | **15 15/16** | 29 | 38 9/16 | 46 | 61 1/8 |
| 13 | 17 1/4 | 30 | 39 7/8 | 47 | 62 7/16 |
| 14 | 18 5/8 | 31 | 41 3/16 | **48** | **63 3/4** |

## Common Rafters
## 10 1/2 in 12 – 41 1/4°

| SPAN Feet | LENGTH Feet Inches | SPAN Inches | LENGTH Inches |
|---|---|---|---|
| 1 | 0- 8 | 1/4 | 3/16 |
| 2 | 1- 3 15/16 | 1/2 | 5/16 |
| 3 | 1-11 15/16 | 3/4 | 1/2 |
| 4 | 2- 7 7/8 | 1 | 11/16 |
| 5 | 3- 3 7/8 | 1 1/4 | 13/16 |
| 6 | 3-11 13/16 | 1 1/2 | 1 |
| 7 | 4- 7 13/16 | 1 3/4 | 1 3/16 |
| 8 | 5- 3 3/4 | 2 | 1 5/16 |
| 9 | 5-11 3/4 | 2 1/4 | 1 1/2 |
| 10 | 6- 7 3/4 | 2 1/2 | 1 11/16 |
| 11 | 7- 3 11/16 | 2 3/4 | 1 13/16 |
| 12 | 7-11 11/16 | 3 | 2 |
| 13 | 8- 7 5/8 | 3 1/4 | 2 3/16 |
| 14 | 9- 3 5/8 | 3 1/2 | 2 5/16 |
| 15 | 9-11 9/16 | 3 3/4 | 2 1/2 |
| 16 | 10- 7 9/16 | 4 | 2 11/16 |
| 17 | 11- 3 9/16 | 4 1/4 | 2 13/16 |
| 18 | 11-11 1/2 | 4 1/2 | 3 |
| 19 | 12- 7 1/2 | 4 3/4 | 3 1/8 |
| 20 | 13- 3 7/16 | 5 | 3 5/16 |
| 21 | 13-11 7/16 | 5 1/4 | 3 1/2 |
| 22 | 14- 7 3/8 | 5 1/2 | 3 5/8 |
| 23 | 15- 3 3/8 | 5 3/4 | 3 13/16 |
| 24 | 15-11 5/16 | 6 | 4 |
| 25 | 16- 7 5/16 | 6 1/4 | 4 1/8 |
| 26 | 17- 3 5/16 | 6 1/2 | 4 5/16 |
| 27 | 17-11 1/4 | 6 3/4 | 4 1/2 |
| 28 | 18- 7 1/4 | 7 | 4 5/8 |
| 29 | 19- 3 3/16 | 7 1/4 | 4 13/16 |
| 30 | 19-11 3/16 | 7 1/2 | 5 |
| 31 | 20- 7 1/8 | 7 3/4 | 5 1/8 |
| 32 | 21- 3 1/8 | 8 | 5 5/16 |
| 33 | 21-11 1/8 | 8 1/4 | 5 1/2 |
| 34 | 22- 7 1/16 | 8 1/2 | 5 5/8 |
| 35 | 23- 3 1/16 | 8 3/4 | 5 13/16 |
| 36 | 23-11 | 9 | 6 |
| 37 | 24- 7 | 9 1/4 | 6 1/8 |
| 38 | 25- 2 15/16 | 9 1/2 | 6 5/16 |
| 39 | 25-10 15/16 | 9 3/4 | 6 1/2 |
| 40 | 26- 6 7/8 | 10 | 6 5/8 |
| 41 | 27- 2 7/8 | 10 1/4 | 6 13/16 |
| 42 | 27-10 7/8 | 10 1/2 | 7 |
| 43 | 28- 6 13/16 | 10 3/4 | 7 1/8 |
| 44 | 29- 2 13/16 | 11 | 7 5/16 |
| 45 | 29-10 3/4 | 11 1/4 | 7 1/2 |
| 46 | 30- 6 3/4 | 11 1/2 | 7 5/8 |
| 47 | 31- 2 11/16 | 11 3/4 | 7 13/16 |
| 48 | 31-10 11/16 | 12 | 8 |
| 49 | 32- 6 11/16 | | |
| 50 | 33- 2 5/8 | | |

## Hip or Valley Rafters
## 10 1/2 in 17 – 31 3/4°

| SPAN Feet | LENGTH Feet Inches | SPAN Inches | LENGTH Inches |
|---|---|---|---|
| 1 | 0-10 | 1/4 | 3/16 |
| 2 | 1- 7 15/16 | 1/2 | 7/16 |
| 3 | 2- 5 15/16 | 3/4 | 5/8 |
| 4 | 3- 3 15/16 | 1 | 13/16 |
| 5 | 4- 1 7/8 | 1 1/4 | 1 1/16 |
| 6 | 4-11 7/8 | 1 1/2 | 1 1/4 |
| 7 | 5- 9 7/8 | 1 3/4 | 1 7/16 |
| 8 | 6- 7 13/16 | 2 | 1 11/16 |
| 9 | 7- 5 13/16 | 2 1/4 | 1 7/8 |
| 10 | 8- 3 3/4 | 2 1/2 | 2 1/16 |
| 11 | 9- 1 3/4 | 2 3/4 | 2 5/16 |
| 12 | 9-11 3/4 | 3 | 2 1/2 |
| 13 | 10- 9 11/16 | 3 1/4 | 2 11/16 |
| 14 | 11- 7 11/16 | 3 1/2 | 2 15/16 |
| 15 | 12- 5 11/16 | 3 3/4 | 3 1/8 |
| 16 | 13- 3 5/8 | 4 | 3 5/16 |
| 17 | 14- 1 5/8 | 4 1/4 | 3 9/16 |
| 18 | 14-11 5/8 | 4 1/2 | 3 3/4 |
| 19 | 15- 9 9/16 | 4 3/4 | 3 15/16 |
| 20 | 16- 7 9/16 | 5 | 4 3/16 |
| 21 | 17- 5 9/16 | 5 1/4 | 4 3/8 |
| 22 | 18- 3 1/2 | 5 1/2 | 4 9/16 |
| 23 | 19- 1 1/2 | 5 3/4 | 4 3/4 |
| 24 | 19-11 1/2 | 6 | 5 |
| 25 | 20- 9 7/16 | 6 1/4 | 5 3/16 |
| 26 | 21- 7 7/16 | 6 1/2 | 5 3/8 |
| 27 | 22- 5 7/16 | 6 3/4 | 5 5/8 |
| 28 | 23- 3 3/8 | 7 | 5 13/16 |
| 29 | 24- 1 3/8 | 7 1/4 | 6 |
| 30 | 24-11 5/16 | 7 1/2 | 6 1/4 |
| 31 | 25- 9 5/16 | 7 3/4 | 6 7/16 |
| 32 | 26- 7 5/16 | 8 | 6 5/8 |
| 33 | 27- 5 1/4 | 8 1/4 | 6 7/8 |
| 34 | 28- 3 1/4 | 8 1/2 | 7 1/16 |
| 35 | 29- 1 1/4 | 8 3/4 | 7 1/4 |
| 36 | 29-11 3/16 | 9 | 7 1/2 |
| 37 | 30- 9 3/16 | 9 1/4 | 7 11/16 |
| 38 | 31- 7 3/16 | 9 1/2 | 7 7/8 |
| 39 | 32- 5 1/8 | 9 3/4 | 8 1/8 |
| 40 | 33- 3 1/8 | 10 | 8 5/16 |
| 41 | 34- 1 1/8 | 10 1/4 | 8 1/2 |
| 42 | 34-11 1/16 | 10 1/2 | 8 3/4 |
| 43 | 35- 9 1/16 | 10 3/4 | 8 15/16 |
| 44 | 36- 7 1/16 | 11 | 9 1/8 |
| 45 | 37- 5 | 11 1/4 | 9 3/8 |
| 46 | 38- 3 | 11 1/2 | 9 9/16 |
| 47 | 39- 1 | 11 3/4 | 9 3/4 |
| 48 | 39-10 15/16 | 12 | 10 |
| 49 | 40- 8 15/16 | | |
| 50 | 41- 6 7/8 | | |

# 11 in 12 – $42^1/_2$°

## *Cuts and Saw Settings*

**Tail Cut, Hip or Valley Rafters,** with Square Cut Common Rafters:
Framing Square: **8 7/16 & 24**; set saw at **31 1/2°**
(Compound Miter or Radial Arm Saws: set bevel at 31 1/2°, miter at 19 1/2°)
Explanation: Section 3(d), pages 12–13.

**Chamfer Cut, Hip or Valley Rafters:**
Set saw at **28 1/2°**. Explanation: Section 3(d), page 10.

**Fascia Board Miter Cut at Hip or Valley:**
Framing Square: **8 1/8 & 12**; set saw at **31 1/2°**
(Compound Miter or Radial Arm Saws: set bevel at 31 1/2°, miter at 34°)
Explanation: Section 5, pages 24–25.

**Frieze Block against Hip or Valley Rafter:**
Framing Square: **8 1/8 & 12**; set saw at **31 1/2°**
(Compound Miter or Radial Arm Saws: set bevel at 31 1/2°, miter at 34°)
Explanation: Section 6, page 26.

**Roof Sheathing at Hip or Valley:**
Framing Square: **12 & 16 1/4**; set saw at **28 1/2°**
(Compound Miter or Radial Arm Saws: set bevel at 28 1/2°, miter at 36 1/2°)
Distance "d" = **35 3/8"**. Explanation: Section 7, pages 26–27.

## *Calculating the Length of a Rafter*

**General Rule:** Span of building
less Thickness of ridge board (if any)
gives Length of rafter to plumb line of seat cut

*and* Span of building
less Thickness of ridge board (if any)
plus 2 × projection of overhang
gives Total length of rafter to tail cut

Explanation: Section 4, pages 13–19.

| *Jack Rafters* | | | | | |
|---|---|---|---|---|---|
| SPACED *Inches* | DIFFERENCE *Inches* | SPACED *Inches* | DIFFERENCE *Inches* | SPACED *Inches* | DIFFERENCE *Inches* |
| 1/4 | 5/16 | 15 | 20 3/8 | **32** | **43 7/16** |
| 1/2 | 11/16 | **16** | **21 11/16** | 33 | 44 3/4 |
| 3/4 | 1 | 17 | 23 1/16 | 34 | 46 1/8 |
| 1 | 1 3/8 | 18 | 24 7/16 | 35 | 47 1/2 |
| 2 | 2 11/16 | 19 | 25 3/4 | **36** | **48 13/16** |
| 3 | 4 1/16 | 20 | 27 1/8 | 37 | 50 3/16 |
| 4 | 5 7/16 | 21 | 28 1/2 | 38 | 51 9/16 |
| 5 | 6 13/16 | 22 | 29 7/8 | 39 | 52 7/8 |
| 6 | 8 1/8 | 23 | 31 3/16 | 40 | 54 1/4 |
| 7 | 9 1/2 | **24** | **32 9/16** | 41 | 55 5/8 |
| 8 | 10 7/8 | 25 | 33 15/16 | 42 | 57 |
| 9 | 12 3/16 | 26 | 35 1/4 | 43 | 58 5/16 |
| 10 | 13 9/16 | 27 | 36 5/8 | 44 | 59 11/16 |
| 11 | 14 15/16 | 28 | 38 | 45 | 61 1/16 |
| **12** | **16 1/4** | 29 | 39 5/16 | 46 | 62 3/8 |
| 13 | 17 5/8 | 30 | 40 11/16 | 47 | 63 3/4 |
| 14 | 19 | 31 | 42 1/16 | **48** | **65 1/8** |

| **Common Rafters 11 in 12 – 42 1/2°** | | | | |
|---|---|---|---|---|
| SPAN Feet | LENGTH Feet Inches | | SPAN Inches | LENGTH Inches |
| 1 | 0- 8 1/8 | | 1/4 | 3/16 |
| 2 | 1- 4 1/4 | | 1/2 | 5/16 |
| 3 | 2- 0 7/16 | | 3/4 | 1/2 |
| 4 | 2- 8 9/16 | | 1 | 11/16 |
| 5 | 3- 4 11/16 | | 1 1/4 | 7/8 |
| 6 | 4- 0 13/16 | | 1 1/2 | 1 |
| 7 | 4- 9 | | 1 3/4 | 1 3/16 |
| 8 | 5- 5 1/8 | | 2 | 1 3/8 |
| 9 | 6- 1 1/4 | | 2 1/4 | 1 1/2 |
| 10 | 6- 9 3/8 | | 2 1/2 | 1 11/16 |
| 11 | 7- 5 9/16 | | 2 3/4 | 1 7/8 |
| 12 | 8- 1 11/16 | | 3 | 2 1/16 |
| 13 | 8- 9 13/16 | | 3 1/4 | 2 3/16 |
| 14 | 9- 5 15/16 | | 3 1/2 | 2 3/8 |
| 15 | 10- 2 1/16 | | 3 3/4 | 2 9/16 |
| 16 | 10-10 1/4 | | 4 | 2 11/16 |
| 17 | 11- 6 3/8 | | 4 1/4 | 2 7/8 |
| 18 | 12- 2 1/2 | | 4 1/2 | 3 1/16 |
| 19 | 12-10 5/8 | | 4 3/4 | 3 1/4 |
| 20 | 13- 6 13/16 | | 5 | 3 3/8 |
| 21 | 14- 2 15/16 | | 5 1/4 | 3 9/16 |
| 22 | 14-11 1/16 | | 5 1/2 | 3 3/4 |
| 23 | 15- 7 3/16 | | 5 3/4 | 3 7/8 |
| 24 | 16- 3 3/8 | | 6 | 4 1/16 |
| 25 | 16-11 1/2 | | 6 1/4 | 4 1/4 |
| 26 | 17- 7 5/8 | | 6 1/2 | 4 7/16 |
| 27 | 18- 3 3/4 | | 6 3/4 | 4 9/16 |
| 28 | 18-11 7/8 | | 7 | 4 3/4 |
| 29 | 19- 8 1/16 | | 7 1/4 | 4 15/16 |
| 30 | 20- 4 3/16 | | 7 1/2 | 5 1/16 |
| 31 | 21- 0 5/16 | | 7 3/4 | 5 1/4 |
| 32 | 21- 8 7/16 | | 8 | 5 7/16 |
| 33 | 22- 4 5/8 | | 8 1/4 | 5 5/8 |
| 34 | 23- 0 3/4 | | 8 1/2 | 5 3/4 |
| 35 | 23- 8 7/8 | | 8 3/4 | 5 15/16 |
| 36 | 24- 5 | | 9 | 6 1/8 |
| 37 | 25- 1 3/16 | | 9 1/4 | 6 1/4 |
| 38 | 25- 9 5/16 | | 9 1/2 | 6 7/16 |
| 39 | 26- 5 7/16 | | 9 3/4 | 6 5/8 |
| 40 | 27- 1 9/16 | | 10 | 6 13/16 |
| 41 | 27- 9 11/16 | | 10 1/4 | 6 15/16 |
| 42 | 28- 5 7/8 | | 10 1/2 | 7 1/8 |
| 43 | 29- 2 | | 10 3/4 | 7 5/16 |
| 44 | 29-10 1/8 | | 11 | 7 7/16 |
| 45 | 30- 6 1/4 | | 11 1/4 | 7 5/8 |
| 46 | 31- 2 7/16 | | 11 1/2 | 7 13/16 |
| 47 | 31-10 9/16 | | 11 3/4 | 8 |
| 48 | 32- 6 11/16 | | 12 | 8 1/8 |
| 49 | 33- 2 13/16 | | | |
| 50 | 33-11 | | | |

| **Hip or Valley Rafters 11 in 17 – 33°** | | | | |
|---|---|---|---|---|
| SPAN Feet | LENGTH Feet Inches | | SPAN Inches | LENGTH Inches |
| 1 | 0-10 1/8 | | 1/4 | 3/16 |
| 2 | 1- 8 1/4 | | 1/2 | 7/16 |
| 3 | 2- 6 5/16 | | 3/4 | 5/8 |
| 4 | 3- 4 7/16 | | 1 | 13/16 |
| 5 | 4- 2 9/16 | | 1 1/4 | 1 1/16 |
| 6 | 5- 0 11/16 | | 1 1/2 | 1 1/4 |
| 7 | 5-10 13/16 | | 1 3/4 | 1 1/2 |
| 8 | 6- 8 7/8 | | 2 | 1 11/16 |
| 9 | 7- 7 | | 2 1/4 | 1 7/8 |
| 10 | 8- 5 1/8 | | 2 1/2 | 2 1/8 |
| 11 | 9- 3 1/4 | | 2 3/4 | 2 5/16 |
| 12 | 10- 1 5/16 | | 3 | 2 1/2 |
| 13 | 10-11 7/16 | | 3 1/4 | 2 3/4 |
| 14 | 11- 9 9/16 | | 3 1/2 | 2 15/16 |
| 15 | 12- 7 11/16 | | 3 3/4 | 3 3/16 |
| 16 | 13- 5 13/16 | | 4 | 3 3/8 |
| 17 | 14- 3 7/8 | | 4 1/4 | 3 9/16 |
| 18 | 15- 2 | | 4 1/2 | 3 13/16 |
| 19 | 16- 0 1/8 | | 4 3/4 | 4 |
| 20 | 16-10 1/4 | | 5 | 4 3/16 |
| 21 | 17- 8 3/8 | | 5 1/4 | 4 7/16 |
| 22 | 18- 6 7/16 | | 5 1/2 | 4 5/8 |
| 23 | 19- 4 9/16 | | 5 3/4 | 4 7/8 |
| 24 | 20- 2 11/16 | | 6 | 5 1/16 |
| 25 | 21- 0 13/16 | | 6 1/4 | 5 1/4 |
| 26 | 21-10 15/16 | | 6 1/2 | 5 1/2 |
| 27 | 22- 9 | | 6 3/4 | 5 11/16 |
| 28 | 23- 7 1/8 | | 7 | 5 7/8 |
| 29 | 24- 5 1/4 | | 7 1/4 | 6 1/8 |
| 30 | 25- 3 3/8 | | 7 1/2 | 6 5/16 |
| 31 | 26- 1 7/16 | | 7 3/4 | 6 1/2 |
| 32 | 26-11 9/16 | | 8 | 6 3/4 |
| 33 | 27- 9 11/16 | | 8 1/4 | 6 15/16 |
| 34 | 28- 7 13/16 | | 8 1/2 | 7 3/16 |
| 35 | 29- 5 15/16 | | 8 3/4 | 7 3/8 |
| 36 | 30- 4 | | 9 | 7 9/16 |
| 37 | 31- 2 1/8 | | 9 1/4 | 7 13/16 |
| 38 | 32- 0 1/4 | | 9 1/2 | 8 |
| 39 | 32-10 3/8 | | 9 3/4 | 8 3/16 |
| 40 | 33- 8 1/2 | | 10 | 8 7/16 |
| 41 | 34- 6 9/16 | | 10 1/4 | 8 5/8 |
| 42 | 35- 4 11/16 | | 10 1/2 | 8 7/8 |
| 43 | 36- 2 13/16 | | 10 3/4 | 9 1/16 |
| 44 | 37- 0 15/16 | | 11 | 9 1/4 |
| 45 | 37-11 1/16 | | 11 1/4 | 9 1/2 |
| 46 | 38- 9 1/8 | | 11 1/2 | 9 11/16 |
| 47 | 39- 7 1/4 | | 11 3/4 | 9 7/8 |
| 48 | 40- 5 3/8 | | 12 | 10 1/8 |
| 49 | 41- 3 1/2 | | | |
| 50 | 42- 1 9/16 | | | |

# 11 1/2 in 12 – 43 3/4°

## *Cuts and Saw Settings*

**Tail Cut, Hip or Valley Rafters,** with Square Cut Common Rafters:
Framing Square: **8 1/2 & 24**; set saw at **30 3/4°**
(Compound Miter or Radial Arm Saws: set bevel at 30 3/4°, miter at 19 1/2°)
Explanation: Section 3(d), pages 12–13.

**Chamfer Cut, Hip or Valley Rafters:**
Set saw at **29 1/4°**. Explanation: Section 3(d), page 10.

**Fascia Board Miter Cut at Hip or Valley:**
Framing Square: **8 5/16 & 12**; set saw at **30 3/4°**
(Compound Miter or Radial Arm Saws: set bevel at 30 3/4°, miter at 34 3/4°)
Explanation: Section 5, pages 24–25.

**Frieze Block against Hip or Valley Rafter:**
Framing Square: **8 5/16 & 12**; set saw at **30 3/4°**
(Compound Miter or Radial Arm Saws: set bevel at 30 3/4°, miter at 34 3/4°)
Explanation: Section 6, page 26.

**Roof Sheathing at Hip or Valley:**
Framing Square: **12 & 16 5/8**; set saw at **29 1/4°**
(Compound Miter or Radial Arm Saws: set bevel at 29 1/4°, miter at 35 3/4°)
Distance "d" = **34 5/8"**. Explanation: Section 7, pages 26–27.

## *Calculating the Length of a Rafter*

**General Rule:**

| | | |
|---|---|---|
| | | Span of building |
| | less | Thickness of ridge board (if any) |
| | gives | Length of rafter to plumb line of seat cut |
| *and* | | Span of building |
| | less | Thickness of ridge board (if any) |
| | plus | 2 × projection of overhang |
| | gives | Total length of rafter to tail cut |

Explanation: Section 4, pages 13–19.

### *Jack Rafters*

| SPACED *Inches* | DIFFERENCE *Inches* | SPACED *Inches* | DIFFERENCE *Inches* | SPACED *Inches* | DIFFERENCE *Inches* |
|---|---|---|---|---|---|
| 1/4 | 3/8 | 15 | 20 3/4 | **32** | **44 5/16** |
| 1/2 | 11/16 | **16** | **22 3/16** | 33 | 45 11/16 |
| 3/4 | 1 1/16 | 17 | 23 9/16 | 34 | 47 1/16 |
| 1 | 1 3/8 | 18 | 24 15/16 | 35 | 48 1/2 |
| 2 | 2 3/4 | 19 | 26 5/16 | **36** | **49 7/8** |
| 3 | 4 1/8 | 20 | 27 11/16 | 37 | 51 1/4 |
| 4 | 5 9/16 | 21 | 29 1/16 | 38 | 52 5/8 |
| 5 | 6 15/16 | 22 | 30 1/2 | 39 | 54 |
| 6 | 8 5/16 | 23 | 31 7/8 | 40 | 55 3/8 |
| 7 | 9 11/16 | **24** | **33 1/4** | 41 | 56 13/16 |
| 8 | 11 1/16 | 25 | 34 5/8 | 42 | 58 3/16 |
| 9 | 12 7/16 | 26 | 36 | 43 | 59 9/16 |
| 10 | 13 7/8 | 27 | 37 3/8 | 44 | 60 15/16 |
| 11 | 15 1/4 | 28 | 38 13/16 | 45 | 62 5/16 |
| **12** | **16 5/8** | 29 | 40 3/16 | 46 | 63 11/16 |
| 13 | 18 | 30 | 41 9/16 | 47 | 65 1/8 |
| 14 | 19 3/8 | 31 | 42 15/16 | **48** | **66 1/2** |

| **Common Rafters<br>11 1/2 in 12 – 43 3/4°** | | | |
|---|---|---|---|
| **SPAN** *Feet* | **LENGTH** *Feet Inches* | **SPAN** *Inches* | **LENGTH** *Inches* |
| 1 | 0- 8 5/16 | 1/4 | 3/16 |
| 2 | 1- 4 5/8 | 1/2 | 3/8 |
| 3 | 2- 0 15/16 | 3/4 | 1/2 |
| 4 | 2- 9 1/4 | 1 | 11/16 |
| 5 | 3- 5 9/16 | 1 1/4 | 7/8 |
| 6 | 4- 1 7/8 | 1 1/2 | 1 1/16 |
| 7 | 4-10 3/16 | 1 3/4 | 1 3/16 |
| 8 | 5- 6 1/2 | 2 | 1 3/8 |
| 9 | 6- 2 13/16 | 2 1/4 | 1 9/16 |
| 10 | 6-11 1/8 | 2 1/2 | 1 3/4 |
| 11 | 7- 7 7/16 | 2 3/4 | 1 7/8 |
| 12 | 8- 3 3/4 | 3 | 2 1/16 |
| 13 | 9- 0 1/16 | 3 1/4 | 2 1/4 |
| 14 | 9- 8 3/8 | 3 1/2 | 2 7/16 |
| 15 | 10- 4 5/8 | 3 3/4 | 2 5/8 |
| 16 | 11- 0 15/16 | 4 | 2 3/4 |
| 17 | 11- 9 1/4 | 4 1/4 | 2 15/16 |
| 18 | 12- 5 9/16 | 4 1/2 | 3 1/8 |
| 19 | 13- 1 7/8 | 4 3/4 | 3 5/16 |
| 20 | 13-10 3/16 | 5 | 3 7/16 |
| 21 | 14- 6 1/2 | 5 1/4 | 3 5/8 |
| 22 | 15- 2 13/16 | 5 1/2 | 3 13/16 |
| 23 | 15-11 1/8 | 5 3/4 | 4 |
| 24 | 16- 7 7/16 | 6 | 4 1/8 |
| 25 | 17- 3 3/4 | 6 1/4 | 4 5/16 |
| 26 | 18- 0 1/16 | 6 1/2 | 4 1/2 |
| 27 | 18- 8 3/8 | 6 3/4 | 4 11/16 |
| 28 | 19- 4 11/16 | 7 | 4 7/8 |
| 29 | 20- 1 | 7 1/4 | 5 |
| 30 | 20- 9 5/16 | 7 1/2 | 5 3/16 |
| 31 | 21- 5 5/8 | 7 3/4 | 5 3/8 |
| 32 | 22- 1 15/16 | 8 | 5 9/16 |
| 33 | 22-10 1/4 | 8 1/4 | 5 11/16 |
| 34 | 23- 6 9/16 | 8 1/2 | 5 7/8 |
| 35 | 24- 2 7/8 | 8 3/4 | 6 1/16 |
| 36 | 24-11 3/16 | 9 | 6 1/4 |
| 37 | 25- 7 1/2 | 9 1/4 | 6 3/8 |
| 38 | 26- 3 13/16 | 9 1/2 | 6 9/16 |
| 39 | 27- 0 1/8 | 9 3/4 | 6 3/4 |
| 40 | 27- 8 7/16 | 10 | 6 15/16 |
| 41 | 28- 4 3/4 | 10 1/4 | 7 1/8 |
| 42 | 29- 1 1/16 | 10 1/2 | 7 1/4 |
| 43 | 29- 9 3/8 | 10 3/4 | 7 7/16 |
| 44 | 30- 5 11/16 | 11 | 7 5/8 |
| 45 | 31- 1 15/16 | 11 1/4 | 7 13/16 |
| 46 | 31-10 1/4 | 11 1/2 | 7 15/16 |
| 47 | 32- 6 9/16 | 11 3/4 | 8 1/8 |
| 48 | 33- 2 7/8 | 12 | 8 5/16 |
| 49 | 33-11 3/16 | | |
| 50 | 34- 7 1/2 | | |

| **Hip or Valley Rafters<br>11 1/2 in 17 – 34°** | | | |
|---|---|---|---|
| **SPAN** *Feet* | **LENGTH** *Feet Inches* | **SPAN** *Inches* | **LENGTH** *Inches* |
| 1 | 0-10 1/4 | 1/4 | 3/16 |
| 2 | 1- 8 1/2 | 1/2 | 7/16 |
| 3 | 2- 6 3/4 | 3/4 | 5/8 |
| 4 | 3- 5 | 1 | 7/8 |
| 5 | 4- 3 1/4 | 1 1/4 | 1 1/16 |
| 6 | 5- 1 1/2 | 1 1/2 | 1 5/16 |
| 7 | 5-11 3/4 | 1 3/4 | 1 1/2 |
| 8 | 6-10 | 2 | 1 11/16 |
| 9 | 7- 8 1/4 | 2 1/4 | 1 15/16 |
| 10 | 8- 6 1/2 | 2 1/2 | 2 1/8 |
| 11 | 9- 4 3/4 | 2 3/4 | 2 3/8 |
| 12 | 10- 3 | 3 | 2 9/16 |
| 13 | 11- 1 1/4 | 3 1/4 | 2 3/4 |
| 14 | 11-11 1/2 | 3 1/2 | 3 |
| 15 | 12- 9 3/4 | 3 3/4 | 3 3/16 |
| 16 | 13- 8 | 4 | 3 7/16 |
| 17 | 14- 6 1/4 | 4 1/4 | 3 5/8 |
| 18 | 15- 4 1/2 | 4 1/2 | 3 7/8 |
| 19 | 16- 2 3/4 | 4 3/4 | 4 1/16 |
| 20 | 17- 1 | 5 | 4 1/4 |
| 21 | 17-11 1/4 | 5 1/4 | 4 1/2 |
| 22 | 18- 9 1/2 | 5 1/2 | 4 11/16 |
| 23 | 19- 7 3/4 | 5 3/4 | 4 15/16 |
| 24 | 20- 6 | 6 | 5 1/8 |
| 25 | 21- 4 1/4 | 6 1/4 | 5 5/16 |
| 26 | 22- 2 1/2 | 6 1/2 | 5 9/16 |
| 27 | 23- 0 3/4 | 6 3/4 | 5 3/4 |
| 28 | 23-11 | 7 | 6 |
| 29 | 24- 9 1/4 | 7 1/4 | 6 3/16 |
| 30 | 25- 7 1/2 | 7 1/2 | 6 7/16 |
| 31 | 26- 5 3/4 | 7 3/4 | 6 5/8 |
| 32 | 27- 4 | 8 | 6 13/16 |
| 33 | 28- 2 1/4 | 8 1/4 | 7 1/16 |
| 34 | 29- 0 1/2 | 8 1/2 | 7 1/4 |
| 35 | 29-10 3/4 | 8 3/4 | 7 1/2 |
| 36 | 30- 9 | 9 | 7 11/16 |
| 37 | 31- 7 1/4 | 9 1/4 | 7 7/8 |
| 38 | 32- 5 1/2 | 9 1/2 | 8 1/8 |
| 39 | 33- 3 3/4 | 9 3/4 | 8 5/16 |
| 40 | 34- 2 | 10 | 8 9/16 |
| 41 | 35- 0 1/4 | 10 1/4 | 8 3/4 |
| 42 | 35-10 1/2 | 10 1/2 | 9 |
| 43 | 36- 8 3/4 | 10 3/4 | 9 3/16 |
| 44 | 37- 7 | 11 | 9 3/8 |
| 45 | 38- 5 1/4 | 11 1/4 | 9 5/8 |
| 46 | 39- 3 1/2 | 11 1/2 | 9 13/16 |
| 47 | 40- 1 3/4 | 11 3/4 | 10 1/16 |
| 48 | 41- 0 | 12 | 10 1/4 |
| 49 | 41-10 1/4 | | |
| 50 | 42- 8 1/2 | | |

# 12 in 12 – 45°

## *Cuts and Saw Settings*

**Tail Cut, Hip or Valley Rafters,** with Square Cut Common Rafters:
Framing Square: **8 1/2 & 24**; set saw at **30**°
(Compound Miter or Radial Arm Saws: set bevel at 30°, miter at 19 1/2°)
Explanation: Section 3(d), pages 12–13.

**Chamfer Cut, Hip or Valley Rafters:**
Set saw at **30°**. Explanation: Section 3(d), page 10.

**Fascia Board Miter Cut at Hip or Valley:**
Framing Square: **8 1/2 & 12**; set saw at **30**°
(Compound Miter or Radial Arm Saws: set bevel at 30°, miter at 35 1/4°)
Explanation: Section 5, pages 24–25.

**Frieze Block against Hip or Valley Rafter:**
Framing Square: **8 1/2 & 12**; set saw at **30**°
(Compound Miter or Radial Arm Saws: set bevel at 30°, miter at 35 1/4°)
Explanation: Section 6, page 26.

**Roof Sheathing at Hip or Valley:**
Framing Square: **12 & 17**; set saw at **30**°
(Compound Miter or Radial Arm Saws: set bevel at 30°, miter at 35 1/4°)
Distance "d" = **33 15/16"**. Explanation: Section 7, pages 26–27.

## *Calculating the Length of a Rafter*

**General Rule:** Span of building
less Thickness of ridge board (if any)
gives Length of rafter to plumb line of seat cut

*and* Span of building
less Thickness of ridge board (if any)
plus 2 × projection of overhang
gives Total length of rafter to tail cut

Explanation: Section 4, pages 13–19.

| *Jack Rafters* | | | | | |
|---|---|---|---|---|---|
| SPACED *Inches* | DIFFERENCE *Inches* | SPACED *Inches* | DIFFERENCE *Inches* | SPACED *Inches* | DIFFERENCE *Inches* |
| 1/4 | 3/8 | 15 | 21 3/16 | **32** | **45 1/4** |
| 1/2 | 11/16 | **16** | **22 5/8** | 33 | 46 11/16 |
| 3/4 | 1 1/16 | 17 | 24 1/16 | 34 | 48 1/16 |
| 1 | 1 7/16 | 18 | 25 7/16 | 35 | 49 1/2 |
| 2 | 2 13/16 | 19 | 26 7/8 | **36** | **50 15/16** |
| 3 | 4 1/4 | 20 | 28 5/16 | 37 | 52 5/16 |
| 4 | 5 11/16 | 21 | 29 11/16 | 38 | 53 3/4 |
| 5 | 7 1/16 | 22 | 31 1/8 | 39 | 55 1/8 |
| 6 | 8 1/2 | 23 | 32 1/2 | 40 | 56 9/16 |
| 7 | 9 7/8 | **24** | **33 15/16** | 41 | 58 |
| 8 | 11 5/16 | 25 | 35 3/8 | 42 | 59 3/8 |
| 9 | 12 3/4 | 26 | 36 3/4 | 43 | 60 13/16 |
| 10 | 14 1/8 | 27 | 38 3/16 | 44 | 62 1/4 |
| 11 | 15 9/16 | 28 | 39 5/8 | 45 | 63 5/8 |
| **12** | **17** | 29 | 41 | 46 | 65 1/16 |
| 13 | 18 3/8 | 30 | 42 7/16 | 47 | 66 7/16 |
| 14 | 19 13/16 | 31 | 43 13/16 | **48** | **67 7/8** |

## Common Rafters
## 12 in 12 – 45°

| SPAN *Feet* | LENGTH *Feet Inches* | SPAN *Inches* | LENGTH *Inches* |
|---|---|---|---|
| 1 | 0- 8 1/2 | 1/4 | 3/16 |
| 2 | 1- 5 | 1/2 | 3/8 |
| 3 | 2- 1 7/16 | 3/4 | 1/2 |
| 4 | 2- 9 15/16 | 1 | 11/16 |
| 5 | 3- 6 7/16 | 1 1/4 | 7/8 |
| 6 | 4- 2 15/16 | 1 1/2 | 1 1/16 |
| 7 | 4-11 3/8 | 1 3/4 | 1 1/4 |
| 8 | 5- 7 7/8 | 2 | 1 7/16 |
| 9 | 6- 4 3/8 | 2 1/4 | 1 9/16 |
| 10 | 7- 0 7/8 | 2 1/2 | 1 3/4 |
| 11 | 7- 9 5/16 | 2 3/4 | 1 15/16 |
| 12 | 8- 5 13/16 | 3 | 2 1/8 |
| 13 | 9- 2 5/16 | 3 1/4 | 2 5/16 |
| 14 | 9-10 13/16 | 3 1/2 | 2 1/2 |
| 15 | 10- 7 1/4 | 3 3/4 | 2 5/8 |
| 16 | 11- 3 3/4 | 4 | 2 13/16 |
| 17 | 12- 0 1/4 | 4 1/4 | 3 |
| 18 | 12- 8 3/4 | 4 1/2 | 3 3/16 |
| 19 | 13- 5 1/4 | 4 3/4 | 3 3/8 |
| 20 | 14- 1 11/16 | 5 | 3 9/16 |
| 21 | 14-10 3/16 | 5 1/4 | 3 11/16 |
| 22 | 15- 6 11/16 | 5 1/2 | 3 7/8 |
| 23 | 16- 3 3/16 | 5 3/4 | 4 1/16 |
| 24 | 16-11 5/8 | 6 | 4 1/4 |
| 25 | 17- 8 1/8 | 6 1/4 | 4 7/16 |
| 26 | 18- 4 5/8 | 6 1/2 | 4 5/8 |
| 27 | 19- 1 1/8 | 6 3/4 | 4 3/4 |
| 28 | 19- 9 9/16 | 7 | 4 15/16 |
| 29 | 20- 6 1/16 | 7 1/4 | 5 1/8 |
| 30 | 21- 2 9/16 | 7 1/2 | 5 5/16 |
| 31 | 21-11 1/16 | 7 3/4 | 5 1/2 |
| 32 | 22- 7 1/2 | 8 | 5 11/16 |
| 33 | 23- 4 | 8 1/4 | 5 13/16 |
| 34 | 24- 0 1/2 | 8 1/2 | 6 |
| 35 | 24- 9 | 8 3/4 | 6 3/16 |
| 36 | 25- 5 1/2 | 9 | 6 3/8 |
| 37 | 26- 1 15/16 | 9 1/4 | 6 9/16 |
| 38 | 26-10 7/16 | 9 1/2 | 6 11/16 |
| 39 | 27- 6 15/16 | 9 3/4 | 6 7/8 |
| 40 | 28- 3 7/16 | 10 | 7 1/16 |
| 41 | 28-11 7/8 | 10 1/4 | 7 1/4 |
| 42 | 29- 8 3/8 | 10 1/2 | 7 7/16 |
| 43 | 30- 4 7/8 | 10 3/4 | 7 5/8 |
| 44 | 31- 1 3/8 | 11 | 7 3/4 |
| 45 | 31- 9 13/16 | 11 1/4 | 7 15/16 |
| 46 | 32- 6 5/16 | 11 1/2 | 8 1/8 |
| 47 | 33- 2 13/16 | 11 3/4 | 8 5/16 |
| 48 | 33-11 5/16 | 12 | 8 1/2 |
| 49 | 34- 7 3/4 | | |
| 50 | 35- 4 1/4 | | |

## Hip or Valley Rafters
## 12 in 17 – 35 1/4°

| SPAN *Feet* | LENGTH *Feet Inches* | SPAN *Inches* | LENGTH *Inches* |
|---|---|---|---|
| 1 | 0-10 3/8 | 1/4 | 3/16 |
| 2 | 1- 8 13/16 | 1/2 | 7/16 |
| 3 | 2- 7 3/16 | 3/4 | 5/8 |
| 4 | 3- 5 9/16 | 1 | 7/8 |
| 5 | 4- 3 15/16 | 1 1/4 | 1 1/16 |
| 6 | 5- 2 3/8 | 1 1/2 | 1 5/16 |
| 7 | 6- 0 3/4 | 1 3/4 | 1 1/2 |
| 8 | 6-11 1/8 | 2 | 1 3/4 |
| 9 | 7- 9 1/2 | 2 1/4 | 1 15/16 |
| 10 | 8- 7 15/16 | 2 1/2 | 2 3/16 |
| 11 | 9- 6 5/16 | 2 3/4 | 2 3/8 |
| 12 | 10- 4 11/16 | 3 | 2 5/8 |
| 13 | 11- 3 1/8 | 3 1/4 | 2 13/16 |
| 14 | 12- 1 1/2 | 3 1/2 | 3 |
| 15 | 12-11 7/8 | 3 3/4 | 3 1/4 |
| 16 | 13-10 1/4 | 4 | 3 7/16 |
| 17 | 14- 8 11/16 | 4 1/4 | 3 11/16 |
| 18 | 15- 7 1/16 | 4 1/2 | 3 7/8 |
| 19 | 16- 5 7/16 | 4 3/4 | 4 1/8 |
| 20 | 17- 3 7/8 | 5 | 4 5/16 |
| 21 | 18- 2 1/4 | 5 1/4 | 4 9/16 |
| 22 | 19- 0 5/8 | 5 1/2 | 4 3/4 |
| 23 | 19-11 | 5 3/4 | 5 |
| 24 | 20- 9 7/16 | 6 | 5 3/16 |
| 25 | 21- 7 13/16 | 6 1/4 | 5 7/16 |
| 26 | 22- 6 3/16 | 6 1/2 | 5 5/8 |
| 27 | 23- 4 9/16 | 6 3/4 | 5 7/8 |
| 28 | 24- 3 | 7 | 6 1/16 |
| 29 | 25- 1 3/8 | 7 1/4 | 6 1/4 |
| 30 | 25-11 3/4 | 7 1/2 | 6 1/2 |
| 31 | 26-10 3/16 | 7 3/4 | 6 11/16 |
| 32 | 27- 8 9/16 | 8 | 6 15/16 |
| 33 | 28- 6 15/16 | 8 1/4 | 7 1/8 |
| 34 | 29- 5 5/16 | 8 1/2 | 7 3/8 |
| 35 | 30- 3 3/4 | 8 3/4 | 7 9/16 |
| 36 | 31- 2 1/8 | 9 | 7 13/16 |
| 37 | 32- 0 1/2 | 9 1/4 | 8 |
| 38 | 32-10 15/16 | 9 1/2 | 8 1/4 |
| 39 | 33- 9 5/16 | 9 3/4 | 8 7/16 |
| 40 | 34- 7 11/16 | 10 | 8 11/16 |
| 41 | 35- 6 1/16 | 10 1/4 | 8 7/8 |
| 42 | 36- 4 1/2 | 10 1/2 | 9 1/16 |
| 43 | 37- 2 7/8 | 10 3/4 | 9 5/16 |
| 44 | 38- 1 1/4 | 11 | 9 1/2 |
| 45 | 38-11 5/8 | 11 1/4 | 9 3/4 |
| 46 | 39-10 1/16 | 11 1/2 | 9 15/16 |
| 47 | 40- 8 7/16 | 11 3/4 | 10 3/16 |
| 48 | 41- 6 13/16 | 12 | 10 3/8 |
| 49 | 42- 5 1/4 | | |
| 50 | 43- 3 5/8 | | |

# 12 1/2 in 12 – 46 1/4°

## *Cuts and Saw Settings*

**Tail Cut, Hip or Valley Rafters,** with Square Cut Common Rafters:
Framing Square: **8 1/2 & 24**; set saw at **29 1/4°**
(Compound Miter or Radial Arm Saws: set bevel at 29 1/4°, miter at 19 1/2°)
Explanation: Section 3(d), pages 12–13.

**Chamfer Cut, Hip or Valley Rafters:**
Set saw at **30 3/4°**. Explanation: Section 3(d), page 10.

**Fascia Board Miter Cut at Hip or Valley:**
Framing Square: **8 11/16 & 12**; set saw at **29 1/4°**
(Compound Miter or Radial Arm Saws: set bevel at 29 1/4°, miter at 35 3/4°)
Explanation: Section 5, pages 24–25.

**Frieze Block against Hip or Valley Rafter:**
Framing Square: **8 11/16 & 12**; set saw at **29 1/4°**
(Compound Miter or Radial Arm Saws: set bevel at 29 1/4°, miter at 35 3/4°)
Explanation: Section 6, page 26.

**Roof Sheathing at Hip or Valley:**
Framing Square: **12 & 17 5/16**; set saw at **30 3/4°**
(Compound Miter or Radial Arm Saws: set bevel at 30 3/4°, miter at 34 3/4°)
Distance "d" = **33 1/4"**. Explanation: Section 7, pages 26–27.

## *Calculating the Length of a Rafter*

**General Rule:** Span of building
less Thickness of ridge board (if any)
gives Length of rafter to plumb line of seat cut

*and* Span of building
less Thickness of ridge board (if any)
plus 2 × projection of overhang
gives Total length of rafter to tail cut

Explanation: Section 4, pages 13–19.

| *Jack Rafters* | | | | | |
|---|---|---|---|---|---|
| SPACED *Inches* | DIFFERENCE *Inches* | SPACED *Inches* | DIFFERENCE *Inches* | SPACED *Inches* | DIFFERENCE *Inches* |
| 1/4 | 3/8 | 15 | 21 11/16 | **32** | **46 3/16** |
| 1/2 | 3/4 | **16** | **23 1/8** | 33 | 47 5/8 |
| 3/4 | 1 1/16 | 17 | 24 9/16 | 34 | 49 1/8 |
| 1 | 1 7/16 | 18 | 26 | 35 | 50 9/16 |
| 2 | 2 7/8 | 19 | 27 7/16 | **36** | **52** |
| 3 | 4 5/16 | 20 | 28 7/8 | 37 | 53 7/16 |
| 4 | 5 3/4 | 21 | 30 5/16 | 38 | 54 7/8 |
| 5 | 7 1/4 | 22 | 31 3/4 | 39 | 56 5/16 |
| 6 | 8 11/16 | 23 | 33 3/16 | 40 | 57 3/4 |
| 7 | 10 1/8 | **24** | **34 5/8** | 41 | 59 3/16 |
| 8 | 11 9/16 | 25 | 36 1/8 | 42 | 60 5/8 |
| 9 | 13 | 26 | 37 9/16 | 43 | 62 1/16 |
| 10 | 14 7/16 | 27 | 39 | 44 | 63 9/16 |
| 11 | 15 7/8 | 28 | 40 7/16 | 45 | 65 |
| **12** | **17 5/16** | 29 | 41 7/8 | 46 | 66 7/16 |
| 13 | 18 3/4 | 30 | 43 5/16 | 47 | 67 7/8 |
| 14 | 20 3/16 | 31 | 44 3/4 | **48** | **69 5/16** |

## *Common Rafters*
## 12 1/2 in 12 – 46 1/4°

| SPAN *Feet* | LENGTH *Feet Inches* | SPAN *Inches* | LENGTH *Inches* |
|---|---|---|---|
| 1 | 0-8 11/16 | 1/4 | 3/16 |
| 2 | 1-5 5/16 | 1/2 | 3/8 |
| 3 | 2-2 | 3/4 | 9/16 |
| 4 | 2-10 5/8 | 1 | 3/4 |
| 5 | 3-7 5/16 | 1 1/4 | 7/8 |
| 6 | 4-4 | 1 1/2 | 1 1/16 |
| 7 | 5-0 5/8 | 1 3/4 | 1 1/4 |
| 8 | 5-9 5/16 | 2 | 1 7/16 |
| 9 | 6-6 | 2 1/4 | 1 5/8 |
| 10 | 7-2 5/8 | 2 1/2 | 1 13/16 |
| 11 | 7-11 5/16 | 2 3/4 | 2 |
| 12 | 8-7 15/16 | 3 | 2 3/16 |
| 13 | 9-4 5/8 | 3 1/4 | 2 3/8 |
| 14 | 10-1 5/16 | 3 1/2 | 2 1/2 |
| 15 | 10-9 15/16 | 3 3/4 | 2 11/16 |
| 16 | 11-6 5/8 | 4 | 2 7/8 |
| 17 | 12-3 5/16 | 4 1/4 | 3 1/16 |
| 18 | 12-11 15/16 | 4 1/2 | 3 1/4 |
| 19 | 13-8 5/8 | 4 3/4 | 3 7/16 |
| 20 | 14-5 1/4 | 5 | 3 5/8 |
| 21 | 15-1 15/16 | 5 1/4 | 3 13/16 |
| 22 | 15-10 5/8 | 5 1/2 | 4 |
| 23 | 16-7 1/4 | 5 3/4 | 4 1/8 |
| 24 | 17-3 15/16 | 6 | 4 5/16 |
| 25 | 18-0 5/8 | 6 1/4 | 4 1/2 |
| 26 | 18-9 1/4 | 6 1/2 | 4 11/16 |
| 27 | 19-5 15/16 | 6 3/4 | 4 7/8 |
| 28 | 20-2 9/16 | 7 | 5 1/16 |
| 29 | 20-11 1/4 | 7 1/4 | 5 1/4 |
| 30 | 21-7 15/16 | 7 1/2 | 5 7/16 |
| 31 | 22-4 9/16 | 7 3/4 | 5 5/8 |
| 32 | 23-1 1/4 | 8 | 5 3/4 |
| 33 | 23-9 15/16 | 8 1/4 | 5 15/16 |
| 34 | 24-6 9/16 | 8 1/2 | 6 1/8 |
| 35 | 25-3 1/4 | 8 3/4 | 6 5/16 |
| 36 | 25-11 7/8 | 9 | 6 1/2 |
| 37 | 26-8 9/16 | 9 1/4 | 6 11/16 |
| 38 | 27-5 1/4 | 9 1/2 | 6 7/8 |
| 39 | 28-1 7/8 | 9 3/4 | 7 1/16 |
| 40 | 28-10 9/16 | 10 | 7 1/4 |
| 41 | 29-7 3/16 | 10 1/4 | 7 3/8 |
| 42 | 30-3 7/8 | 10 1/2 | 7 9/16 |
| 43 | 31-0 9/16 | 10 3/4 | 7 3/4 |
| 44 | 31-9 3/16 | 11 | 7 15/16 |
| 45 | 32-5 7/8 | 11 1/4 | 8 1/8 |
| 46 | 33-2 9/16 | 11 1/2 | 8 5/16 |
| 47 | 33-11 3/16 | 11 3/4 | 8 1/2 |
| 48 | 34-7 7/8 | 12 | 8 11/16 |
| 49 | 35-4 1/2 | | |
| 50 | 36-1 3/16 | | |

## *Hip or Valley Rafters*
## 12 1/2 in 17 – 36 1/2°

| SPAN *Feet* | LENGTH *Feet Inches* | SPAN *Inches* | LENGTH *Inches* |
|---|---|---|---|
| 1 | 0-10 9/16 | 1/4 | 1/4 |
| 2 | 1-9 1/16 | 1/2 | 7/16 |
| 3 | 2-7 5/8 | 3/4 | 11/16 |
| 4 | 3-6 1/8 | 1 | 7/8 |
| 5 | 4-4 11/16 | 1 1/4 | 1 1/8 |
| 6 | 5-3 1/4 | 1 1/2 | 1 5/16 |
| 7 | 6-1 3/4 | 1 3/4 | 1 9/16 |
| 8 | 7-0 5/16 | 2 | 1 3/4 |
| 9 | 7-10 7/8 | 2 1/4 | 2 |
| 10 | 8-9 3/8 | 2 1/2 | 2 3/16 |
| 11 | 9-7 15/16 | 2 3/4 | 2 7/16 |
| 12 | 10-6 7/16 | 3 | 2 5/8 |
| 13 | 11-5 | 3 1/4 | 2 7/8 |
| 14 | 12-3 9/16 | 3 1/2 | 3 1/16 |
| 15 | 13-2 1/16 | 3 3/4 | 3 5/16 |
| 16 | 14-0 5/8 | 4 | 3 1/2 |
| 17 | 14-11 3/16 | 4 1/4 | 3 3/4 |
| 18 | 15-9 11/16 | 4 1/2 | 3 15/16 |
| 19 | 16-8 1/4 | 4 3/4 | 4 3/16 |
| 20 | 17-6 3/4 | 5 | 4 3/8 |
| 21 | 18-5 5/16 | 5 1/4 | 4 5/8 |
| 22 | 19-3 7/8 | 5 1/2 | 4 13/16 |
| 23 | 20-2 3/8 | 5 3/4 | 5 1/16 |
| 24 | 21-0 15/16 | 6 | 5 1/4 |
| 25 | 21-11 7/16 | 6 1/4 | 5 1/2 |
| 26 | 22-10 | 6 1/2 | 5 11/16 |
| 27 | 23-8 9/16 | 6 3/4 | 5 15/16 |
| 28 | 24-7 1/16 | 7 | 6 1/8 |
| 29 | 25-5 5/8 | 7 1/4 | 6 3/8 |
| 30 | 26-4 3/16 | 7 1/2 | 6 9/16 |
| 31 | 27-2 11/16 | 7 3/4 | 6 13/16 |
| 32 | 28-1 1/4 | 8 | 7 |
| 33 | 28-11 3/4 | 8 1/4 | 7 1/4 |
| 34 | 29-10 5/16 | 8 1/2 | 7 7/16 |
| 35 | 30-8 7/8 | 8 3/4 | 7 11/16 |
| 36 | 31-7 3/8 | 9 | 7 7/8 |
| 37 | 32-5 15/16 | 9 1/4 | 8 1/8 |
| 38 | 33-4 7/16 | 9 1/2 | 8 5/16 |
| 39 | 34-3 | 9 3/4 | 8 9/16 |
| 40 | 35-1 9/16 | 10 | 8 13/16 |
| 41 | 36-0 1/16 | 10 1/4 | 9 |
| 42 | 36-10 5/8 | 10 1/2 | 9 1/4 |
| 43 | 37-9 3/16 | 10 3/4 | 9 7/16 |
| 44 | 38-7 11/16 | 11 | 9 11/16 |
| 45 | 39-6 1/4 | 11 1/4 | 9 7/8 |
| 46 | 40-4 3/4 | 11 1/2 | 10 1/8 |
| 47 | 41-3 5/16 | 11 3/4 | 10 5/16 |
| 48 | 42-1 7/8 | 12 | 10 9/16 |
| 49 | 43-0 3/8 | | |
| 50 | 43-10 15/16 | | |

# 13 in 12 – 47 1/4°

## *Cuts and Saw Settings*

**Tail Cut, Hip or Valley Rafters,** with Square Cut Common Rafters:
Framing Square: **8 7/16 & 24**; set saw at **28 3/4°**
(Compound Miter or Radial Arm Saws: set bevel at 28 3/4°, miter at 19 1/2°)
Explanation: Section 3(d), pages 12–13.

**Chamfer Cut, Hip or Valley Rafters:**
Set saw at **31 1/4°**. Explanation: Section 3(d), page 10.

**Fascia Board Miter Cut at Hip or Valley:**
Framing Square: **8 13/16 & 12**; set saw at **28 3/4°**
(Compound Miter or Radial Arm Saws: set bevel at 28 3/4°, miter at 36 1/4°)
Explanation: Section 5, pages 24–25.

**Frieze Block against Hip or Valley Rafter:**
Framing Square: **8 13/16 & 12**; set saw at **28 3/4°**
(Compound Miter or Radial Arm Saws: set bevel at 28 3/4°, miter at 36 1/4°)
Explanation: Section 6, page 26.

**Roof Sheathing at Hip or Valley:**
Framing Square: **12 & 17 11/16**; set saw at **31 1/4°**
(Compound Miter or Radial Arm Saws: set bevel at 31 1/4°, miter at 34 1/4°)
Distance "d" = **32 9/16"**. Explanation: Section 7, pages 26–27.

## *Calculating the Length of a Rafter*

**General Rule:**

| | | |
|---|---|---|
| | | Span of building |
| | less | Thickness of ridge board (if any) |
| | gives | Length of rafter to plumb line of seat cut |
| *and* | | Span of building |
| | less | Thickness of ridge board (if any) |
| | plus | 2 × projection of overhang |
| | gives | Total length of rafter to tail cut |

Explanation: Section 4, pages 13–19.

| *Jack Rafters* | | | | | |
|---|---|---|---|---|---|
| SPACED *Inches* | DIFFERENCE *Inches* | SPACED *Inches* | DIFFERENCE *Inches* | SPACED *Inches* | DIFFERENCE *Inches* |
| 1/4 | 3/8 | 15 | 22 1/8 | **32** | **47 3/16** |
| 1/2 | 3/4 | **16** | **23 9/16** | 33 | 48 5/8 |
| 3/4 | 1 1/8 | 17 | 25 1/16 | 34 | 50 1/8 |
| 1 | 1 1/2 | 18 | 26 9/16 | 35 | 51 5/8 |
| 2 | 2 15/16 | 19 | 28 | **36** | **53 1/16** |
| 3 | 4 7/16 | 20 | 29 1/2 | 37 | 54 9/16 |
| 4 | 5 7/8 | 21 | 30 15/16 | 38 | 56 |
| 5 | 7 3/8 | 22 | 32 7/16 | 39 | 57 1/2 |
| 6 | 8 7/8 | 23 | 33 15/16 | 40 | 59 |
| 7 | 10 5/16 | **24** | **35 3/8** | 41 | 60 7/16 |
| 8 | 11 13/16 | 25 | 36 7/8 | 42 | 61 15/16 |
| 9 | 13 1/4 | 26 | 38 5/16 | 43 | 63 3/8 |
| 10 | 14 3/4 | 27 | 39 13/16 | 44 | 64 7/8 |
| 11 | 16 3/16 | 28 | 41 1/4 | 45 | 66 3/8 |
| **12** | **17 11/16** | 29 | 42 3/4 | 46 | 67 13/16 |
| 13 | 19 3/16 | 30 | 44 1/4 | 47 | 69 5/16 |
| 14 | 20 5/8 | 31 | 45 11/16 | **48** | **70 3/4** |

## *Common Rafters*
## 13 in 12 – 47 1/4°

| SPAN *Feet* | LENGTH *Feet Inches* | SPAN *Inches* | LENGTH *Inches* |
|---|---|---|---|
| 1 | 0- 8 7/8 | 1/4 | 3/16 |
| 2 | 1- 5 11/16 | 1/2 | 3/8 |
| 3 | 2- 2 9/16 | 3/4 | 9/16 |
| 4 | 2-11 3/8 | 1 | 3/4 |
| 5 | 3- 8 1/4 | 1 1/4 | 15/16 |
| 6 | 4- 5 1/16 | 1 1/2 | 1 1/8 |
| 7 | 5- 1 15/16 | 1 3/4 | 1 5/16 |
| 8 | 5-10 3/4 | 2 | 1 1/2 |
| 9 | 6- 7 5/8 | 2 1/4 | 1 11/16 |
| 10 | 7- 4 7/16 | 2 1/2 | 1 13/16 |
| 11 | 8- 1 5/16 | 2 3/4 | 2 |
| 12 | 8-10 1/8 | 3 | 2 3/16 |
| 13 | 9- 7 | 3 1/4 | 2 3/8 |
| 14 | 10- 3 13/16 | 3 1/2 | 2 9/16 |
| 15 | 11- 0 11/16 | 3 3/4 | 2 3/4 |
| 16 | 11- 9 9/16 | 4 | 2 15/16 |
| 17 | 12- 6 3/8 | 4 1/4 | 3 1/8 |
| 18 | 13- 3 1/4 | 4 1/2 | 3 5/16 |
| 19 | 14- 0 1/16 | 4 3/4 | 3 1/2 |
| 20 | 14- 8 15/16 | 5 | 3 11/16 |
| 21 | 15- 5 3/4 | 5 1/4 | 3 7/8 |
| 22 | 16- 2 5/8 | 5 1/2 | 4 1/16 |
| 23 | 16-11 7/16 | 5 3/4 | 4 1/4 |
| 24 | 17- 8 5/16 | 6 | 4 7/16 |
| 25 | 18- 5 1/8 | 6 1/4 | 4 5/8 |
| 26 | 19- 2 | 6 1/2 | 4 13/16 |
| 27 | 19-10 13/16 | 6 3/4 | 5 |
| 28 | 20- 7 11/16 | 7 | 5 3/16 |
| 29 | 21- 4 1/2 | 7 1/4 | 5 3/8 |
| 30 | 22- 1 3/8 | 7 1/2 | 5 1/2 |
| 31 | 22-10 1/4 | 7 3/4 | 5 11/16 |
| 32 | 23- 7 1/16 | 8 | 5 7/8 |
| 33 | 24- 3 15/16 | 8 1/4 | 6 1/16 |
| 34 | 25- 0 3/4 | 8 1/2 | 6 1/4 |
| 35 | 25- 9 5/8 | 8 3/4 | 6 7/16 |
| 36 | 26- 6 7/16 | 9 | 6 5/8 |
| 37 | 27- 3 5/16 | 9 1/4 | 6 13/16 |
| 38 | 28- 0 1/8 | 9 1/2 | 7 |
| 39 | 28- 9 | 9 3/4 | 7 3/16 |
| 40 | 29- 5 13/16 | 10 | 7 3/8 |
| 41 | 30- 2 11/16 | 10 1/4 | 7 9/16 |
| 42 | 30-11 1/2 | 10 1/2 | 7 3/4 |
| 43 | 31- 8 3/8 | 10 3/4 | 7 15/16 |
| 44 | 32- 5 1/4 | 11 | 8 1/8 |
| 45 | 33- 2 1/16 | 11 1/4 | 8 5/16 |
| 46 | 33-10 15/16 | 11 1/2 | 8 1/2 |
| 47 | 34- 7 3/4 | 11 3/4 | 8 11/16 |
| 48 | 35- 4 5/8 | 12 | 8 7/8 |
| 49 | 36- 1 7/16 | | |
| 50 | 36-10 5/16 | | |

## *Hip or Valley Rafters*
## 13 in 17 – 37 1/2°

| SPAN *Feet* | LENGTH *Feet Inches* | SPAN *Inches* | LENGTH *Inches* |
|---|---|---|---|
| 1 | 0-10 11/16 | 1/4 | 1/4 |
| 2 | 1- 9 3/8 | 1/2 | 7/16 |
| 3 | 2- 8 1/16 | 3/4 | 11/16 |
| 4 | 3- 6 3/4 | 1 | 7/8 |
| 5 | 4- 5 7/16 | 1 1/4 | 1 1/8 |
| 6 | 5- 4 1/8 | 1 1/2 | 1 5/16 |
| 7 | 6- 2 13/16 | 1 3/4 | 1 9/16 |
| 8 | 7- 1 1/2 | 2 | 1 13/16 |
| 9 | 8- 0 3/16 | 2 1/4 | 2 |
| 10 | 8-10 7/8 | 2 1/2 | 2 1/4 |
| 11 | 9- 9 9/16 | 2 3/4 | 2 7/16 |
| 12 | 10- 8 1/4 | 3 | 2 11/16 |
| 13 | 11- 6 15/16 | 3 1/4 | 2 7/8 |
| 14 | 12- 5 5/8 | 3 1/2 | 3 1/8 |
| 15 | 13- 4 5/16 | 3 3/4 | 3 5/16 |
| 16 | 14- 3 | 4 | 3 9/16 |
| 17 | 15- 1 11/16 | 4 1/4 | 3 13/16 |
| 18 | 16- 0 3/8 | 4 1/2 | 4 |
| 19 | 16-11 1/16 | 4 3/4 | 4 1/4 |
| 20 | 17- 9 3/4 | 5 | 4 7/16 |
| 21 | 18- 8 7/16 | 5 1/4 | 4 11/16 |
| 22 | 19- 7 1/8 | 5 1/2 | 4 7/8 |
| 23 | 20- 5 13/16 | 5 3/4 | 5 1/8 |
| 24 | 21- 4 1/2 | 6 | 5 3/8 |
| 25 | 22- 3 1/4 | 6 1/4 | 5 9/16 |
| 26 | 23- 1 15/16 | 6 1/2 | 5 13/16 |
| 27 | 24- 0 5/8 | 6 3/4 | 6 |
| 28 | 24-11 5/16 | 7 | 6 1/4 |
| 29 | 25-10 | 7 1/4 | 6 7/16 |
| 30 | 26- 8 11/16 | 7 1/2 | 6 11/16 |
| 31 | 27- 7 3/8 | 7 3/4 | 6 7/8 |
| 32 | 28- 6 1/16 | 8 | 7 1/8 |
| 33 | 29- 4 3/4 | 8 1/4 | 7 3/8 |
| 34 | 30- 3 7/16 | 8 1/2 | 7 9/16 |
| 35 | 31- 2 1/8 | 8 3/4 | 7 13/16 |
| 36 | 32- 0 13/16 | 9 | 8 |
| 37 | 32-11 1/2 | 9 1/4 | 8 1/4 |
| 38 | 33-10 3/16 | 9 1/2 | 8 7/16 |
| 39 | 34- 8 7/8 | 9 3/4 | 8 11/16 |
| 40 | 35- 7 9/16 | 10 | 8 15/16 |
| 41 | 36- 6 1/4 | 10 1/4 | 9 1/8 |
| 42 | 37- 4 15/16 | 10 1/2 | 9 3/8 |
| 43 | 38- 3 5/8 | 10 3/4 | 9 9/16 |
| 44 | 39- 2 5/16 | 11 | 9 13/16 |
| 45 | 40- 1 | 11 1/4 | 10 |
| 46 | 40-11 11/16 | 11 1/2 | 10 1/4 |
| 47 | 41-10 3/8 | 11 3/4 | 10 7/16 |
| 48 | 42- 9 1/16 | 12 | 10 11/16 |
| 49 | 43- 7 3/4 | | |
| 50 | 44- 6 7/16 | | |

# $13^1/_2$ in 12 – $48^1/_4$°

## *Cuts and Saw Settings*

**Tail Cut, Hip or Valley Rafters,** with Square Cut Common Rafters:
Framing Square: **8 7/16 & 24**; set saw at **28°**
(Compound Miter or Radial Arm Saws: set bevel at 28°, miter at 19 1/4°)
Explanation: Section 3(d), pages 12–13.

**Chamfer Cut, Hip or Valley Rafters:**
Set saw at **32°**. Explanation: Section 3(d), page 10.

**Fascia Board Miter Cut at Hip or Valley:**
Framing Square: **9 & 12**; set saw at **28°**
(Compound Miter or Radial Arm Saws: set bevel at 28°, miter at 36 3/4°)
Explanation: Section 5, pages 24–25.

**Frieze Block against Hip or Valley Rafter:**
Framing Square: **9 & 12**; set saw at **28°**
(Compound Miter or Radial Arm Saws: set bevel at 28°, miter at 36 3/4°)
Explanation: Section 6, page 26.

**Roof Sheathing at Hip or Valley:**
Framing Square: **12 & 18 1/16**; set saw at **32°**
(Compound Miter or Radial Arm Saws: set bevel at 32°, miter at 33 1/2°)
Distance "d" = **31 7/8"**. Explanation: Section 7, pages 26–27.

## *Calculating the Length of a Rafter*

**General Rule:**

| | | |
|---|---|---|
| | | Span of building |
| | less | Thickness of ridge board (if any) |
| | gives | Length of rafter to plumb line of seat cut |
| *and* | | Span of building |
| | less | Thickness of ridge board (if any) |
| | plus | 2 × projection of overhang |
| | gives | Total length of rafter to tail cut |

Explanation: Section 4, pages 13–19.

| *Jack Rafters* | | | | | |
|---|---|---|---|---|---|
| SPACED *Inches* | DIFFERENCE *Inches* | SPACED *Inches* | DIFFERENCE *Inches* | SPACED *Inches* | DIFFERENCE *Inches* |
| 1/4 | 3/8 | 15 | 22 9/16 | **32** | **48 3/16** |
| 1/2 | 3/4 | **16** | **24 1/16** | 33 | 49 11/16 |
| 3/4 | 1 1/8 | 17 | 25 9/16 | 34 | 51 3/16 |
| 1 | 1 1/2 | 18 | 27 1/16 | 35 | 52 11/16 |
| 2 | 3 | 19 | 28 5/8 | **36** | **54 3/16** |
| 3 | 4 1/2 | 20 | 30 1/8 | 37 | 55 11/16 |
| 4 | 6 | 21 | 31 5/8 | 38 | 57 3/16 |
| 5 | 7 1/2 | 22 | 33 1/8 | 39 | 58 11/16 |
| 6 | 9 | 23 | 34 5/8 | 40 | 60 3/16 |
| 7 | 10 9/16 | **24** | **36 1/8** | 41 | 61 11/16 |
| 8 | 12 1/16 | 25 | 37 5/8 | 42 | 63 3/16 |
| 9 | 13 9/16 | 26 | 39 1/8 | 43 | 64 3/4 |
| 10 | 15 1/16 | 27 | 40 5/8 | 44 | 66 1/4 |
| 11 | 16 9/16 | 28 | 42 1/8 | 45 | 67 3/4 |
| **12** | **18 1/16** | 29 | 43 5/8 | 46 | 69 1/4 |
| 13 | 19 9/16 | 30 | 45 1/8 | 47 | 70 3/4 |
| 14 | 21 1/16 | 31 | 46 11/16 | **48** | **72 1/4** |

| Common Rafters 13 1/2 in 12 – 48 1/4° | | | |
|---|---|---|---|
| SPAN *Feet* | LENGTH *Feet Inches* | SPAN *Inches* | LENGTH *Inches* |
| 1 | 0- 9 | 1/4 | 3/16 |
| 2 | 1- 6 1/16 | 1/2 | 3/8 |
| 3 | 2- 3 1/16 | 3/4 | 9/16 |
| 4 | 3- 0 1/8 | 1 | 3/4 |
| 5 | 3- 9 1/8 | 1 1/4 | 15/16 |
| 6 | 4- 6 3/16 | 1 1/2 | 1 1/8 |
| 7 | 5- 3 3/16 | 1 3/4 | 1 5/16 |
| 8 | 6- 0 1/4 | 2 | 1 1/2 |
| 9 | 6- 9 1/4 | 2 1/4 | 1 11/16 |
| 10 | 7- 6 5/16 | 2 1/2 | 1 7/8 |
| 11 | 8- 3 5/16 | 2 3/4 | 2 1/16 |
| 12 | 9- 0 3/8 | 3 | 2 1/4 |
| 13 | 9- 9 3/8 | 3 1/4 | 2 7/16 |
| 14 | 10- 6 7/16 | 3 1/2 | 2 5/8 |
| 15 | 11- 3 7/16 | 3 3/4 | 2 13/16 |
| 16 | 12- 0 1/2 | 4 | 3 |
| 17 | 12- 9 1/2 | 4 1/4 | 3 3/16 |
| 18 | 13- 6 9/16 | 4 1/2 | 3 3/8 |
| 19 | 14- 3 9/16 | 4 3/4 | 3 9/16 |
| 20 | 15- 0 5/8 | 5 | 3 3/4 |
| 21 | 15- 9 5/8 | 5 1/4 | 3 15/16 |
| 22 | 16- 6 11/16 | 5 1/2 | 4 1/8 |
| 23 | 17- 3 11/16 | 5 3/4 | 4 5/16 |
| 24 | 18- 0 3/4 | 6 | 4 1/2 |
| 25 | 18- 9 3/4 | 6 1/4 | 4 11/16 |
| 26 | 19- 6 13/16 | 6 1/2 | 4 7/8 |
| 27 | 20- 3 13/16 | 6 3/4 | 5 1/16 |
| 28 | 21- 0 7/8 | 7 | 5 1/4 |
| 29 | 21- 9 7/8 | 7 1/4 | 5 7/16 |
| 30 | 22- 6 15/16 | 7 1/2 | 5 5/8 |
| 31 | 23- 3 15/16 | 7 3/4 | 5 13/16 |
| 32 | 24- 1 | 8 | 6 |
| 33 | 24-10 | 8 1/4 | 6 3/16 |
| 34 | 25- 7 1/16 | 8 1/2 | 6 3/8 |
| 35 | 26- 4 1/16 | 8 3/4 | 6 9/16 |
| 36 | 27- 1 1/8 | 9 | 6 3/4 |
| 37 | 27-10 1/8 | 9 1/4 | 6 15/16 |
| 38 | 28- 7 3/16 | 9 1/2 | 7 1/8 |
| 39 | 29- 4 3/16 | 9 3/4 | 7 5/16 |
| 40 | 30- 1 1/4 | 10 | 7 1/2 |
| 41 | 30-10 1/4 | 10 1/4 | 7 11/16 |
| 42 | 31- 7 5/16 | 10 1/2 | 7 7/8 |
| 43 | 32- 4 5/16 | 10 3/4 | 8 1/16 |
| 44 | 33- 1 3/8 | 11 | 8 1/4 |
| 45 | 33-10 3/8 | 11 1/4 | 8 7/16 |
| 46 | 34- 7 7/16 | 11 1/2 | 8 5/8 |
| 47 | 35- 4 7/16 | 11 3/4 | 8 13/16 |
| 48 | 36- 1 1/2 | 12 | 9 |
| 49 | 36-10 1/2 | | |
| 50 | 37- 7 9/16 | | |

| Hip or Valley Rafters 13 1/2 in 17 – 38 1/2° | | | |
|---|---|---|---|
| SPAN *Feet* | LENGTH *Feet Inches* | SPAN *Inches* | LENGTH *Inches* |
| 1 | 0-10 13/16 | 1/4 | 1/4 |
| 2 | 1- 9 11/16 | 1/2 | 7/16 |
| 3 | 2- 8 1/2 | 3/4 | 11/16 |
| 4 | 3- 7 3/8 | 1 | 7/8 |
| 5 | 4- 6 3/16 | 1 1/4 | 1 1/8 |
| 6 | 5- 5 1/16 | 1 1/2 | 1 3/8 |
| 7 | 6- 3 7/8 | 1 3/4 | 1 9/16 |
| 8 | 7- 2 3/4 | 2 | 1 13/16 |
| 9 | 8- 1 9/16 | 2 1/4 | 2 1/16 |
| 10 | 9- 0 7/16 | 2 1/2 | 2 1/4 |
| 11 | 9-11 1/4 | 2 3/4 | 2 1/2 |
| 12 | 10-10 1/8 | 3 | 2 11/16 |
| 13 | 11- 8 15/16 | 3 1/4 | 2 15/16 |
| 14 | 12- 7 13/16 | 3 1/2 | 3 3/16 |
| 15 | 13- 6 5/8 | 3 3/4 | 3 3/8 |
| 16 | 14- 5 1/2 | 4 | 3 5/8 |
| 17 | 15- 4 5/16 | 4 1/4 | 3 13/16 |
| 18 | 16- 3 3/16 | 4 1/2 | 4 1/16 |
| 19 | 17- 2 | 4 3/4 | 4 5/16 |
| 20 | 18- 0 7/8 | 5 | 4 1/2 |
| 21 | 18-11 11/16 | 5 1/4 | 4 3/4 |
| 22 | 19-10 9/16 | 5 1/2 | 5 |
| 23 | 20- 9 3/8 | 5 3/4 | 5 3/16 |
| 24 | 21- 8 1/4 | 6 | 5 7/16 |
| 25 | 22- 7 1/16 | 6 1/4 | 5 5/8 |
| 26 | 23- 5 15/16 | 6 1/2 | 5 7/8 |
| 27 | 24- 4 3/4 | 6 3/4 | 6 1/8 |
| 28 | 25- 3 9/16 | 7 | 6 5/16 |
| 29 | 26- 2 7/16 | 7 1/4 | 6 9/16 |
| 30 | 27- 1 1/4 | 7 1/2 | 6 3/4 |
| 31 | 28- 0 1/8 | 7 3/4 | 7 |
| 32 | 28-10 15/16 | 8 | 7 1/4 |
| 33 | 29- 9 13/16 | 8 1/4 | 7 7/16 |
| 34 | 30- 8 5/8 | 8 1/2 | 7 11/16 |
| 35 | 31- 7 1/2 | 8 3/4 | 7 7/8 |
| 36 | 32- 6 5/16 | 9 | 8 1/8 |
| 37 | 33- 5 3/16 | 9 1/4 | 8 3/8 |
| 38 | 34- 4 | 9 1/2 | 8 9/16 |
| 39 | 35- 2 7/8 | 9 3/4 | 8 13/16 |
| 40 | 36- 1 11/16 | 10 | 9 1/16 |
| 41 | 37- 0 9/16 | 10 1/4 | 9 1/4 |
| 42 | 37-11 3/8 | 10 1/2 | 9 1/2 |
| 43 | 38-10 1/4 | 10 3/4 | 9 11/16 |
| 44 | 39- 9 1/16 | 11 | 9 15/16 |
| 45 | 40- 7 15/16 | 11 1/4 | 10 3/16 |
| 46 | 41- 6 3/4 | 11 1/2 | 10 3/8 |
| 47 | 42- 5 5/8 | 11 3/4 | 10 5/8 |
| 48 | 43- 4 7/16 | 12 | 10 13/16 |
| 49 | 44- 3 5/16 | | |
| 50 | 45- 2 1/8 | | |

# 14 in 12 – 49 1/2°

## *Cuts and Saw Settings*

**Tail Cut, Hip or Valley Rafters,** with Square Cut Common Rafters:
Framing Square: **8 3/8 & 24**; set saw at **27 1/2°**
(Compound Miter or Radial Arm Saws: set bevel at 27 1/2°, miter at 19 1/4°)
Explanation: Section 3(d), pages 12–13.

**Chamfer Cut, Hip or Valley Rafters:**
Set saw at **32 1/2°**. Explanation: Section 3(d), page 10.

**Fascia Board Miter Cut at Hip or Valley:**
Framing Square: **9 1/8 & 12**; set saw at **27 1/2°**
(Compound Miter or Radial Arm Saws: set bevel at 27 1/2°, miter at 37 1/4°)
Explanation: Section 5, pages 24–25.

**Frieze Block against Hip or Valley Rafter:**
Framing Square: **9 1/8 & 12**; set saw at **27 1/2°**
(Compound Miter or Radial Arm Saws: set bevel at 27 1/2°, miter at 37 1/4°)
Explanation: Section 6, page 26.

**Roof Sheathing at Hip or Valley:**
Framing Square: **12 & 18 7/16**; set saw at **32 1/2°**
(Compound Miter or Radial Arm Saws: set bevel at 32 1/2°, miter at 33°)
Distance "d" = **31 1/4"**. Explanation: Section 7, pages 26–27.

## *Calculating the Length of a Rafter*

**General Rule:** Span of building
less Thickness of ridge board (if any)
gives Length of rafter to plumb line of seat cut

*and* Span of building
less Thickness of ridge board (if any)
plus 2 × projection of overhang
gives Total length of rafter to tail cut

Explanation: Section 4, pages 13–19.

| *Jack Rafters* | | | | | |
|---|---|---|---|---|---|
| SPACED *Inches* | DIFFERENCE *Inches* | SPACED *Inches* | DIFFERENCE *Inches* | SPACED *Inches* | DIFFERENCE *Inches* |
| 1/4 | 3/8 | 15 | 23 1/16 | **32** | **49 3/16** |
| 1/2 | 3/4 | **16** | **24 9/16** | 33 | 50 11/16 |
| 3/4 | 1 1/8 | 17 | 26 1/8 | 34 | 52 1/4 |
| 1 | 1 9/16 | 18 | 27 11/16 | 35 | 53 3/4 |
| 2 | 3 1/16 | 19 | 29 3/16 | **36** | **55 5/16** |
| 3 | 4 5/8 | 20 | 30 3/4 | 37 | 56 7/8 |
| 4 | 6 1/8 | 21 | 32 1/4 | 38 | 58 3/8 |
| 5 | 7 11/16 | 22 | 33 13/16 | 39 | 59 15/16 |
| 6 | 9 1/4 | 23 | 35 5/16 | 40 | 61 7/16 |
| 7 | 10 3/4 | **24** | **36 7/8** | 41 | 63 |
| 8 | 12 5/16 | 25 | 38 7/16 | 42 | 64 9/16 |
| 9 | 13 13/16 | 26 | 39 15/16 | 43 | 66 1/16 |
| 10 | 15 3/8 | 27 | 41 1/2 | 44 | 67 5/8 |
| 11 | 16 7/8 | 28 | 43 | 45 | 69 1/8 |
| **12** | **18 7/16** | 29 | 44 9/16 | 46 | 70 11/16 |
| 13 | 20 | 30 | 46 1/8 | 47 | 72 1/4 |
| 14 | 21 1/2 | 31 | 47 5/8 | **48** | **73 3/4** |

**Common Rafters**
**14 in 12 – 49 1/2°**

| SPAN *Feet* | LENGTH *Feet Inches* | SPAN *Inches* | LENGTH *Inches* |
|---|---|---|---|
| 1 | 0 - 9 1/4 | 1/4 | 3/16 |
| 2 | 1 - 6 7/16 | 1/2 | 3/8 |
| 3 | 2 - 3 11/16 | 3/4 | 9/16 |
| 4 | 3 - 0 7/8 | 1 | 3/4 |
| 5 | 3 -10 1/8 | 1 1/4 | 15/16 |
| 6 | 4 - 7 5/16 | 1 1/2 | 1 1/8 |
| 7 | 5 - 4 9/16 | 1 3/4 | 1 3/8 |
| 8 | 6 - 1 3/4 | 2 | 1 9/16 |
| 9 | 6 -11 | 2 1/4 | 1 3/4 |
| 10 | 7 - 8 3/16 | 2 1/2 | 1 15/16 |
| 11 | 8 - 5 7/16 | 2 3/4 | 2 1/8 |
| 12 | 9 - 2 5/8 | 3 | 2 5/16 |
| 13 | 9 -11 7/8 | 3 1/4 | 2 1/2 |
| 14 | 10 - 9 1/16 | 3 1/2 | 2 11/16 |
| 15 | 11 - 6 5/16 | 3 3/4 | 2 7/8 |
| 16 | 12 - 3 1/2 | 4 | 3 1/16 |
| 17 | 13 - 0 3/4 | 4 1/4 | 3 1/4 |
| 18 | 13 - 9 15/16 | 4 1/2 | 3 7/16 |
| 19 | 14 - 7 3/16 | 4 3/4 | 3 5/8 |
| 20 | 15 - 4 3/8 | 5 | 3 13/16 |
| 21 | 16 - 1 5/8 | 5 1/4 | 4 1/16 |
| 22 | 16 -10 13/16 | 5 1/2 | 4 1/4 |
| 23 | 17 - 8 1/16 | 5 3/4 | 4 7/16 |
| 24 | 18 - 5 1/4 | 6 | 4 5/8 |
| 25 | 19 - 2 1/2 | 6 1/4 | 4 13/16 |
| 26 | 19 -11 11/16 | 6 1/2 | 5 |
| 27 | 20 - 8 15/16 | 6 3/4 | 5 3/16 |
| 28 | 21 - 6 1/8 | 7 | 5 3/8 |
| 29 | 22 - 3 3/8 | 7 1/4 | 5 9/16 |
| 30 | 23 - 0 9/16 | 7 1/2 | 5 3/4 |
| 31 | 23 - 9 13/16 | 7 3/4 | 5 15/16 |
| 32 | 24 - 7 | 8 | 6 1/8 |
| 33 | 25 - 4 1/4 | 8 1/4 | 6 5/16 |
| 34 | 26 - 1 7/16 | 8 1/2 | 6 1/2 |
| 35 | 26 -10 11/16 | 8 3/4 | 6 3/4 |
| 36 | 27 - 7 7/8 | 9 | 6 15/16 |
| 37 | 28 - 5 1/8 | 9 1/4 | 7 1/8 |
| 38 | 29 - 2 5/16 | 9 1/2 | 7 5/16 |
| 39 | 29 -11 9/16 | 9 3/4 | 7 1/2 |
| 40 | 30 - 8 13/16 | 10 | 7 11/16 |
| 41 | 31 - 6 | 10 1/4 | 7 7/8 |
| 42 | 32 - 3 1/4 | 10 1/2 | 8 1/16 |
| 43 | 33 - 0 7/16 | 10 3/4 | 8 1/4 |
| 44 | 33 - 9 11/16 | 11 | 8 7/16 |
| 45 | 34 - 6 7/8 | 11 1/4 | 8 5/8 |
| 46 | 35 - 4 1/8 | 11 1/2 | 8 13/16 |
| 47 | 36 - 1 5/16 | 11 3/4 | 9 |
| 48 | 36 -10 9/16 | 12 | 9 1/4 |
| 49 | 37 - 7 3/4 | | |
| 50 | 38 - 5 | | |

**Hip or Valley Rafters**
**14 in 17 – 39 1/2°**

| SPAN *Feet* | LENGTH *Feet Inches* | SPAN *Inches* | LENGTH *Inches* |
|---|---|---|---|
| 1 | 0 - 11 | 1/4 | 1/4 |
| 2 | 1 - 10 | 1/2 | 7/16 |
| 3 | 2 - 9 | 3/4 | 11/16 |
| 4 | 3 - 8 | 1 | 15/16 |
| 5 | 4 - 7 | 1 1/4 | 1 1/8 |
| 6 | 5 - 6 | 1 1/2 | 1 3/8 |
| 7 | 6 - 5 | 1 3/4 | 1 5/8 |
| 8 | 7 - 4 | 2 | 1 13/16 |
| 9 | 8 - 3 | 2 1/4 | 2 1/16 |
| 10 | 9 - 2 | 2 1/2 | 2 5/16 |
| 11 | 10 - 1 | 2 3/4 | 2 1/2 |
| 12 | 11 - 0 | 3 | 2 3/4 |
| 13 | 11 - 11 | 3 1/4 | 3 |
| 14 | 12 - 10 | 3 1/2 | 3 3/16 |
| 15 | 13 - 9 | 3 3/4 | 3 7/16 |
| 16 | 14 - 8 | 4 | 3 11/16 |
| 17 | 15 - 7 | 4 1/4 | 3 7/8 |
| 18 | 16 - 6 | 4 1/2 | 4 1/8 |
| 19 | 17 - 5 | 4 3/4 | 4 3/8 |
| 20 | 18 - 4 | 5 | 4 9/16 |
| 21 | 19 - 3 | 5 1/4 | 4 13/16 |
| 22 | 20 - 2 | 5 1/2 | 5 1/16 |
| 23 | 21 - 1 | 5 3/4 | 5 1/4 |
| 24 | 22 - 0 | 6 | 5 1/2 |
| 25 | 22 - 11 | 6 1/4 | 5 3/4 |
| 26 | 23 - 10 | 6 1/2 | 5 15/16 |
| 27 | 24 - 9 | 6 3/4 | 6 3/16 |
| 28 | 25 - 8 | 7 | 6 7/16 |
| 29 | 26 - 7 | 7 1/4 | 6 5/8 |
| 30 | 27 - 6 | 7 1/2 | 6 7/8 |
| 31 | 28 - 5 | 7 3/4 | 7 1/8 |
| 32 | 29 - 4 | 8 | 7 5/16 |
| 33 | 30 - 3 | 8 1/4 | 7 9/16 |
| 34 | 31 - 2 | 8 1/2 | 7 13/16 |
| 35 | 32 - 1 | 8 3/4 | 8 |
| 36 | 33 - 0 | 9 | 8 1/4 |
| 37 | 33 - 11 | 9 1/4 | 8 1/2 |
| 38 | 34 - 10 | 9 1/2 | 8 11/16 |
| 39 | 35 - 9 | 9 3/4 | 8 15/16 |
| 40 | 36 - 8 | 10 | 9 3/16 |
| 41 | 37 - 7 | 10 1/4 | 9 3/8 |
| 42 | 38 - 6 | 10 1/2 | 9 5/8 |
| 43 | 39 - 5 | 10 3/4 | 9 7/8 |
| 44 | 40 - 4 | 11 | 10 1/16 |
| 45 | 41 - 3 | 11 1/4 | 10 5/16 |
| 46 | 42 - 2 | 11 1/2 | 10 9/16 |
| 47 | 43 - 1 | 11 3/4 | 10 3/4 |
| 48 | 44 - 0 | 12 | 11 |
| 49 | 44 - 11 | | |
| 50 | 45 - 10 | | |

# $14^1/_2$ in 12 – $50^1/_2$°

## *Cuts and Saw Settings*

**Tail Cut, Hip or Valley Rafters,** with Square Cut Common Rafters:
Framing Square: **8 5/16 & 24**; set saw at **26 3/4°**
(Compound Miter or Radial Arm Saws: set bevel at 26 3/4°, miter at 19 1/4°)
Explanation: Section 3(d), pages 12–13.

**Chamfer Cut, Hip or Valley Rafters:**
Set saw at **33°**. Explanation: Section 3(d), page 10.

**Fascia Board Miter Cut at Hip or Valley:**
Framing Square: **9 1/4 & 12**; set saw at **26 3/4°**
(Compound Miter or Radial Arm Saws: set bevel at 26 3/4°, miter at 37 1/2°)
Explanation: Section 5, pages 24–25.

**Frieze Block against Hip or Valley Rafter:**
Framing Square: **9 1/4 & 12**; set saw at **26 3/4°**
(Compound Miter or Radial Arm Saws: set bevel at 26 3/4°, miter at 37 1/2°)
Explanation: Section 6, page 26.

**Roof Sheathing at Hip or Valley:**
Framing Square: **12 & 18 13/16**; set saw at **33°**
(Compound Miter or Radial Arm Saws: set bevel at 33°, miter at 32 1/2°)
Distance "d" = **30 5/8"**. Explanation: Section 7, pages 26–27.

## *Calculating the Length of a Rafter*

**General Rule:**

| | | |
|---|---|---|
| | | Span of building |
| | less | Thickness of ridge board (if any) |
| | gives | Length of rafter to plumb line of seat cut |
| *and* | | Span of building |
| | less | Thickness of ridge board (if any) |
| | plus | 2 × projection of overhang |
| | gives | Total length of rafter to tail cut |

Explanation: Section 4, pages 13–19.

| *Jack Rafters* | | | | | |
|---|---|---|---|---|---|
| SPACED *Inches* | DIFFERENCE *Inches* | SPACED *Inches* | DIFFERENCE *Inches* | SPACED *Inches* | DIFFERENCE *Inches* |
| 1/4 | 3/8 | 15 | 23 1/2 | **32** | **50 3/16** |
| 1/2 | 13/16 | **16** | **25 1/8** | 33 | 51 3/4 |
| 3/4 | 1 3/16 | 17 | 26 11/16 | 34 | 53 5/16 |
| 1 | 1 9/16 | 18 | 28 1/4 | 35 | 54 7/8 |
| 2 | 3 1/8 | 19 | 29 13/16 | **36** | **56 7/16** |
| 3 | 4 11/16 | 20 | 31 3/8 | 37 | 58 1/16 |
| 4 | 6 1/4 | 21 | 32 15/16 | 38 | 59 5/8 |
| 5 | 7 13/16 | 22 | 34 1/2 | 39 | 61 3/16 |
| 6 | 9 7/16 | 23 | 36 1/16 | 40 | 62 3/4 |
| 7 | 11 | **24** | **37 5/8** | 41 | 64 5/16 |
| 8 | 12 9/16 | 25 | 39 3/16 | 42 | 65 7/8 |
| 9 | 14 1/8 | 26 | 40 3/4 | 43 | 67 7/16 |
| 10 | 15 11/16 | 27 | 42 3/8 | 44 | 69 |
| 11 | 17 1/4 | 28 | 43 15/16 | 45 | 70 9/16 |
| **12** | **18 13/16** | 29 | 45 1/2 | 46 | 72 1/8 |
| 13 | 20 3/8 | 30 | 47 1/16 | 47 | 73 11/16 |
| 14 | 21 15/16 | 31 | 48 5/8 | **48** | **75 5/16** |

## Common Rafters
### 14 1/2 in 12 – 50 1/2°

| SPAN Feet | LENGTH Feet Inches | SPAN Inches | LENGTH Inches |
|---|---|---|---|
| 1 | 0- 9 7/16 | 1/4 | 3/16 |
| 2 | 1- 6 13/16 | 1/2 | 3/8 |
| 3 | 2- 4 1/4 | 3/4 | 9/16 |
| 4 | 3- 1 5/8 | 1 | 13/16 |
| 5 | 3-11 1/16 | 1 1/4 | 1 |
| 6 | 4- 8 7/16 | 1 1/2 | 1 3/16 |
| 7 | 5- 5 7/8 | 1 3/4 | 1 3/8 |
| 8 | 6- 3 5/16 | 2 | 1 9/16 |
| 9 | 7- 0 11/16 | 2 1/4 | 1 3/4 |
| 10 | 7-10 1/8 | 2 1/2 | 1 15/16 |
| 11 | 8- 7 1/2 | 2 3/4 | 2 3/16 |
| 12 | 9- 4 15/16 | 3 | 2 3/8 |
| 13 | 10- 2 5/16 | 3 1/4 | 2 9/16 |
| 14 | 10-11 3/4 | 3 1/2 | 2 3/4 |
| 15 | 11- 9 3/16 | 3 3/4 | 2 15/16 |
| 16 | 12- 6 9/16 | 4 | 3 1/8 |
| 17 | 13- 4 | 4 1/4 | 3 5/16 |
| 18 | 14- 1 3/8 | 4 1/2 | 3 1/2 |
| 19 | 14-10 13/16 | 4 3/4 | 3 3/4 |
| 20 | 15- 8 3/16 | 5 | 3 15/16 |
| 21 | 16- 5 5/8 | 5 1/4 | 4 1/8 |
| 22 | 17- 3 1/16 | 5 1/2 | 4 5/16 |
| 23 | 18- 0 7/16 | 5 3/4 | 4 1/2 |
| 24 | 18- 9 7/8 | 6 | 4 11/16 |
| 25 | 19- 7 1/4 | 6 1/4 | 4 7/8 |
| 26 | 20- 4 11/16 | 6 1/2 | 5 1/8 |
| 27 | 21- 2 1/16 | 6 3/4 | 5 5/16 |
| 28 | 21-11 1/2 | 7 | 5 1/2 |
| 29 | 22- 8 15/16 | 7 1/4 | 5 11/16 |
| 30 | 23- 6 5/16 | 7 1/2 | 5 7/8 |
| 31 | 24- 3 3/4 | 7 3/4 | 6 1/16 |
| 32 | 25- 1 1/8 | 8 | 6 1/4 |
| 33 | 25-10 9/16 | 8 1/4 | 6 1/2 |
| 34 | 26- 7 15/16 | 8 1/2 | 6 11/16 |
| 35 | 27- 5 3/8 | 8 3/4 | 6 7/8 |
| 36 | 28- 2 13/16 | 9 | 7 1/16 |
| 37 | 29- 0 3/16 | 9 1/4 | 7 1/4 |
| 38 | 29- 9 5/8 | 9 1/2 | 7 7/16 |
| 39 | 30- 7 | 9 3/4 | 7 5/8 |
| 40 | 31- 4 7/16 | 10 | 7 13/16 |
| 41 | 32- 1 13/16 | 10 1/4 | 8 1/16 |
| 42 | 32-11 1/4 | 10 1/2 | 8 1/4 |
| 43 | 33- 8 11/16 | 10 3/4 | 8 7/16 |
| 44 | 34- 6 1/16 | 11 | 8 5/8 |
| 45 | 35- 3 1/2 | 11 1/4 | 8 13/16 |
| 46 | 36- 0 7/8 | 11 1/2 | 9 |
| 47 | 36-10 5/16 | 11 3/4 | 9 3/16 |
| 48 | 37- 7 11/16 | 12 | 9 7/16 |
| 49 | 38- 5 1/8 | | |
| 50 | 39- 2 9/16 | | |

## Hip or Valley Rafters
### 14 1/2 in 17 – 40 1/2°

| SPAN Feet | LENGTH Feet Inches | SPAN Inches | LENGTH Inches |
|---|---|---|---|
| 1 | 0-11 3/16 | 1/4 | 1/4 |
| 2 | 1-10 5/16 | 1/2 | 7/16 |
| 3 | 2- 9 1/2 | 3/4 | 11/16 |
| 4 | 3- 8 5/8 | 1 | 15/16 |
| 5 | 4- 7 13/16 | 1 1/4 | 1 3/16 |
| 6 | 5- 6 15/16 | 1 1/2 | 1 3/8 |
| 7 | 6- 6 1/8 | 1 3/4 | 1 5/8 |
| 8 | 7- 5 5/16 | 2 | 1 7/8 |
| 9 | 8- 4 7/16 | 2 1/4 | 2 1/16 |
| 10 | 9- 3 5/8 | 2 1/2 | 2 5/16 |
| 11 | 10- 2 3/4 | 2 3/4 | 2 9/16 |
| 12 | 11- 1 15/16 | 3 | 2 13/16 |
| 13 | 12- 1 1/16 | 3 1/4 | 3 |
| 14 | 13- 0 1/4 | 3 1/2 | 3 1/4 |
| 15 | 13-11 7/16 | 3 3/4 | 3 1/2 |
| 16 | 14-10 9/16 | 4 | 3 3/4 |
| 17 | 15- 9 3/4 | 4 1/4 | 3 15/16 |
| 18 | 16- 8 7/8 | 4 1/2 | 4 3/16 |
| 19 | 17- 8 1/16 | 4 3/4 | 4 7/16 |
| 20 | 18- 7 3/16 | 5 | 4 5/8 |
| 21 | 19- 6 3/8 | 5 1/4 | 4 7/8 |
| 22 | 20- 5 9/16 | 5 1/2 | 5 1/8 |
| 23 | 21- 4 11/16 | 5 3/4 | 5 3/8 |
| 24 | 22- 3 7/8 | 6 | 5 9/16 |
| 25 | 23- 3 | 6 1/4 | 5 13/16 |
| 26 | 24- 2 3/16 | 6 1/2 | 6 1/16 |
| 27 | 25- 1 5/16 | 6 3/4 | 6 1/4 |
| 28 | 26- 0 1/2 | 7 | 6 1/2 |
| 29 | 26-11 11/16 | 7 1/4 | 6 3/4 |
| 30 | 27-10 13/16 | 7 1/2 | 7 |
| 31 | 28-10 | 7 3/4 | 7 3/16 |
| 32 | 29- 9 1/8 | 8 | 7 7/16 |
| 33 | 30- 8 5/16 | 8 1/4 | 7 11/16 |
| 34 | 31- 7 7/16 | 8 1/2 | 7 7/8 |
| 35 | 32- 6 5/8 | 8 3/4 | 8 1/8 |
| 36 | 33- 5 13/16 | 9 | 8 3/8 |
| 37 | 34- 4 15/16 | 9 1/4 | 8 5/8 |
| 38 | 35- 4 1/8 | 9 1/2 | 8 13/16 |
| 39 | 36- 3 1/4 | 9 3/4 | 9 1/16 |
| 40 | 37- 2 7/16 | 10 | 9 5/16 |
| 41 | 38- 1 9/16 | 10 1/4 | 9 9/16 |
| 42 | 39- 0 3/4 | 10 1/2 | 9 3/4 |
| 43 | 39-11 15/16 | 10 3/4 | 10 |
| 44 | 40-11 1/16 | 11 | 10 1/4 |
| 45 | 41-10 1/4 | 11 1/4 | 10 7/16 |
| 46 | 42- 9 3/8 | 11 1/2 | 10 11/16 |
| 47 | 43- 8 9/16 | 11 3/4 | 10 15/16 |
| 48 | 44- 7 11/16 | 12 | 11 3/16 |
| 49 | 45- 6 7/8 | | |
| 50 | 46- 6 1/16 | | |

# 15 in 12 – 51¼°

## *Cuts and Saw Settings*

**Tail Cut, Hip or Valley Rafters,** with Square Cut Common Rafters:
Framing Square: **8¼ & 24**; set saw at **26¼°**
(Compound Miter or Radial Arm Saws: set bevel at 26¼°, miter at 19°)
Explanation: Section 3(d), pages 12–13.

**Chamfer Cut, Hip or Valley Rafters:**
Set saw at **33½°**. Explanation: Section 3(d), page 10.

**Fascia Board Miter Cut at Hip or Valley:**
Framing Square: **9⅜ & 12**; set saw at **26¼°**
(Compound Miter or Radial Arm Saws: set bevel at 26¼°, miter at 38°)
Explanation: Section 5, pages 24–25.

**Frieze Block against Hip or Valley Rafter:**
Framing Square: **9⅜ & 12**; set saw at **26¼°**
(Compound Miter or Radial Arm Saws: set bevel at 26¼°, miter at 38°)
Explanation: Section 6, page 26.

**Roof Sheathing at Hip or Valley:**
Framing Square: **12 & 19 3/16**; set saw at **33½°**
(Compound Miter or Radial Arm Saws: set bevel at 33½°, miter at 32°)
Distance "d" = **30"**. Explanation: Section 7, pages 26–27.

## *Calculating the Length of a Rafter*

**General Rule:** Span of building
less Thickness of ridge board (if any)
gives Length of rafter to plumb line of seat cut

*and* Span of building
less Thickness of ridge board (if any)
plus 2 × projection of overhang
gives Total length of rafter to tail cut

Explanation: Section 4, pages 13–19.

| *Jack Rafters* | | | | | |
|---|---|---|---|---|---|
| SPACED *Inches* | DIFFERENCE *Inches* | SPACED *Inches* | DIFFERENCE *Inches* | SPACED *Inches* | DIFFERENCE *Inches* |
| 1/4 | 3/8 | 15 | 24 | **32** | **51 1/4** |
| 1/2 | 13/16 | **16** | **25 5/8** | 33 | 52 13/16 |
| 3/4 | 1 3/16 | 17 | 27 3/16 | 34 | 54 7/16 |
| 1 | 1 5/8 | 18 | 28 13/16 | 35 | 56 |
| 2 | 3 3/16 | 19 | 30 7/16 | **36** | **57 5/8** |
| 3 | 4 13/16 | 20 | 32 | 37 | 59 1/4 |
| 4 | 6 3/8 | 21 | 33 5/8 | 38 | 60 13/16 |
| 5 | 8 | 22 | 35 3/16 | 39 | 62 7/16 |
| 6 | 9 5/8 | 23 | 36 13/16 | 40 | 64 |
| 7 | 11 3/16 | **24** | **38 7/16** | 41 | 65 5/8 |
| 8 | 12 13/16 | 25 | 40 | 42 | 67 1/4 |
| 9 | 14 7/16 | 26 | 41 5/8 | 43 | 68 13/16 |
| 10 | 16 | 27 | 43 1/4 | 44 | 70 7/16 |
| 11 | 17 5/8 | 28 | 44 13/16 | 45 | 72 1/16 |
| **12** | **19 3/16** | 29 | 46 7/16 | 46 | 73 5/8 |
| 13 | 20 13/16 | 30 | 48 | 47 | 75 1/4 |
| 14 | 22 7/16 | 31 | 49 5/8 | **48** | **76 13/16** |

| Common Rafters<br>15 in 12 – 51 1/4° | | | |
|---|---|---|---|
| SPAN<br>*Feet* | LENGTH<br>*Feet Inches* | SPAN<br>*Inches* | LENGTH<br>*Inches* |
| 1 | 0 - 9 5/8 | 1/4 | 3/16 |
| 2 | 1 - 7 3/16 | 1/2 | 3/8 |
| 3 | 2 - 4 13/16 | 3/4 | 5/8 |
| 4 | 3 - 2 7/16 | 1 | 13/16 |
| 5 | 4 - 0 | 1 1/4 | 1 |
| 6 | 4 - 9 5/8 | 1 1/2 | 1 3/16 |
| 7 | 5 - 7 1/4 | 1 3/4 | 1 3/8 |
| 8 | 6 - 4 13/16 | 2 | 1 5/8 |
| 9 | 7 - 2 7/16 | 2 1/4 | 1 13/16 |
| 10 | 8 - 0 1/16 | 2 1/2 | 2 |
| 11 | 8 - 9 5/8 | 2 3/4 | 2 3/16 |
| 12 | 9 - 7 1/4 | 3 | 2 3/8 |
| 13 | 10 - 4 7/8 | 3 1/4 | 2 5/8 |
| 14 | 11 - 2 7/16 | 3 1/2 | 2 13/16 |
| 15 | 12 - 0 1/16 | 3 3/4 | 3 |
| 16 | 12 - 9 11/16 | 4 | 3 3/16 |
| 17 | 13 - 7 1/4 | 4 1/4 | 3 3/8 |
| 18 | 14 - 4 7/8 | 4 1/2 | 3 5/8 |
| 19 | 15 - 2 1/2 | 4 3/4 | 3 13/16 |
| 20 | 16 - 0 1/16 | 5 | 4 |
| 21 | 16 - 9 11/16 | 5 1/4 | 4 3/16 |
| 22 | 17 - 7 5/16 | 5 1/2 | 4 3/8 |
| 23 | 18 - 4 15/16 | 5 3/4 | 4 5/8 |
| 24 | 19 - 2 1/2 | 6 | 4 13/16 |
| 25 | 20 - 0 1/8 | 6 1/4 | 5 |
| 26 | 20 - 9 3/4 | 6 1/2 | 5 3/16 |
| 27 | 21 - 7 5/16 | 6 3/4 | 5 3/8 |
| 28 | 22 - 4 15/16 | 7 | 5 5/8 |
| 29 | 23 - 2 9/16 | 7 1/4 | 5 13/16 |
| 30 | 24 - 0 1/8 | 7 1/2 | 6 |
| 31 | 24 - 9 3/4 | 7 3/4 | 6 3/16 |
| 32 | 25 - 7 3/8 | 8 | 6 3/8 |
| 33 | 26 - 4 15/16 | 8 1/4 | 6 5/8 |
| 34 | 27 - 2 9/16 | 8 1/2 | 6 13/16 |
| 35 | 28 - 0 3/16 | 8 3/4 | 7 |
| 36 | 28 - 9 3/4 | 9 | 7 3/16 |
| 37 | 29 - 7 3/8 | 9 1/4 | 7 3/8 |
| 38 | 30 - 5 | 9 1/2 | 7 5/8 |
| 39 | 31 - 2 9/16 | 9 3/4 | 7 13/16 |
| 40 | 32 - 0 3/16 | 10 | 8 |
| 41 | 32 - 9 13/16 | 10 1/4 | 8 3/16 |
| 42 | 33 - 7 3/8 | 10 1/2 | 8 3/8 |
| 43 | 34 - 5 | 10 3/4 | 8 5/8 |
| 44 | 35 - 2 5/8 | 11 | 8 13/16 |
| 45 | 36 - 0 3/16 | 11 1/4 | 9 |
| 46 | 36 - 9 13/16 | 11 1/2 | 9 3/16 |
| 47 | 37 - 7 7/16 | 11 3/4 | 9 3/8 |
| 48 | 38 - 5 | 12 | 9 5/8 |
| 49 | 39 - 2 5/8 | | |
| 50 | 40 - 0 1/4 | | |

| Hip or Valley Rafters<br>15 in 17 – 41 1/2° | | | |
|---|---|---|---|
| SPAN<br>*Feet* | LENGTH<br>*Feet Inches* | SPAN<br>*Inches* | LENGTH<br>*Inches* |
| 1 | 0-11 5/16 | 1/4 | 1/4 |
| 2 | 1-10 5/8 | 1/2 | 1/2 |
| 3 | 2-10 | 3/4 | 11/16 |
| 4 | 3- 9 5/16 | 1 | 15/16 |
| 5 | 4- 8 5/8 | 1 1/4 | 1 3/16 |
| 6 | 5- 7 15/16 | 1 1/2 | 1 7/16 |
| 7 | 6- 7 1/4 | 1 3/4 | 1 5/8 |
| 8 | 7- 6 5/8 | 2 | 1 7/8 |
| 9 | 8- 5 15/16 | 2 1/4 | 2 1/8 |
| 10 | 9- 5 1/4 | 2 1/2 | 2 3/8 |
| 11 | 10- 4 9/16 | 2 3/4 | 2 5/8 |
| 12 | 11- 3 7/8 | 3 | 2 13/16 |
| 13 | 12- 3 1/4 | 3 1/4 | 3 1/16 |
| 14 | 13- 2 9/16 | 3 1/2 | 3 5/16 |
| 15 | 14- 1 7/8 | 3 3/4 | 3 9/16 |
| 16 | 15- 1 3/16 | 4 | 3 3/4 |
| 17 | 16- 0 1/2 | 4 1/4 | 4 |
| 18 | 16-11 7/8 | 4 1/2 | 4 1/4 |
| 19 | 17-11 3/16 | 4 3/4 | 4 1/2 |
| 20 | 18-10 1/2 | 5 | 4 11/16 |
| 21 | 19- 9 13/16 | 5 1/4 | 4 15/16 |
| 22 | 20- 9 1/8 | 5 1/2 | 5 3/16 |
| 23 | 21- 8 1/2 | 5 3/4 | 5 7/16 |
| 24 | 22- 7 13/16 | 6 | 5 11/16 |
| 25 | 23- 7 1/8 | 6 1/4 | 5 7/8 |
| 26 | 24- 6 7/16 | 6 1/2 | 6 1/8 |
| 27 | 25- 5 3/4 | 6 3/4 | 6 3/8 |
| 28 | 26- 5 1/16 | 7 | 6 5/8 |
| 29 | 27- 4 7/16 | 7 1/4 | 6 13/16 |
| 30 | 28- 3 3/4 | 7 1/2 | 7 1/16 |
| 31 | 29- 3 1/16 | 7 3/4 | 7 5/16 |
| 32 | 30- 2 3/8 | 8 | 7 9/16 |
| 33 | 31- 1 11/16 | 8 1/4 | 7 13/16 |
| 34 | 32- 1 1/16 | 8 1/2 | 8 |
| 35 | 33- 0 3/8 | 8 3/4 | 8 1/4 |
| 36 | 33-11 11/16 | 9 | 8 1/2 |
| 37 | 34-11 | 9 1/4 | 8 3/4 |
| 38 | 35-10 5/16 | 9 1/2 | 8 15/16 |
| 39 | 36- 9 11/16 | 9 3/4 | 9 3/16 |
| 40 | 37- 9 | 10 | 9 7/16 |
| 41 | 38- 8 5/16 | 10 1/4 | 9 11/16 |
| 42 | 39- 7 5/8 | 10 1/2 | 9 15/16 |
| 43 | 40- 6 15/16 | 10 3/4 | 10 1/8 |
| 44 | 41- 6 5/16 | 11 | 10 3/8 |
| 45 | 42- 5 5/8 | 11 1/4 | 10 5/8 |
| 46 | 43- 4 15/16 | 11 1/2 | 10 7/8 |
| 47 | 44- 4 1/4 | 11 3/4 | 11 1/16 |
| 48 | 45- 3 9/16 | 12 | 11 5/16 |
| 49 | 46- 2 15/16 | | |
| 50 | 47- 2 1/4 | | |

# $15^1/_2$ in 12 – $52^1/_4$°

## *Cuts and Saw Settings*

**Tail Cut, Hip or Valley Rafters,** with Square Cut Common Rafters:
Framing Square: **8 3/16 & 24**; set saw at **25 3/4**°
(Compound Miter or Radial Arm Saws: set bevel at 25 3/4°, miter at 19°)
Explanation: Section 3(d), pages 12–13.

**Chamfer Cut, Hip or Valley Rafters:**
Set saw at **34°**. Explanation: Section 3(d), page 10.

**Fascia Board Miter Cut at Hip or Valley:**
Framing Square: **9 1/2 & 12**; set saw at **25 3/4**°
(Compound Miter or Radial Arm Saws: set bevel at 25 3/4°, miter at 38 1/4°)
Explanation: Section 5, pages 24–25.

**Frieze Block against Hip or Valley Rafter:**
Framing Square: **9 1/2 & 12**; set saw at **25 3/4**°
(Compound Miter or Radial Arm Saws: set bevel at 25 3/4°, miter at 38 1/4°)
Explanation: Section 6, page 26.

**Roof Sheathing at Hip or Valley:**
Framing Square: **12 & 19 5/8**; set saw at **34°**
(Compound Miter or Radial Arm Saws: set bevel at 34°, miter at 31 1/2°)
Distance "d" = **29 3/8"**. Explanation: Section 7, pages 26–27.

## *Calculating the Length of a Rafter*

**General Rule:** Span of building
less Thickness of ridge board (if any)
gives Length of rafter to plumb line of seat cut

*and* Span of building
less Thickness of ridge board (if any)
plus 2 × projection of overhang
gives Total length of rafter to tail cut

Explanation: Section 4, pages 13–19.

| *Jack Rafters* | | | | | |
|---|---|---|---|---|---|
| SPACED *Inches* | DIFFERENCE *Inches* | SPACED *Inches* | DIFFERENCE *Inches* | SPACED *Inches* | DIFFERENCE *Inches* |
| 1/4 | 7/16 | 15 | 24 1/2 | **32** | **52 1/4** |
| 1/2 | 13/16 | **16** | **26 1/8** | 33 | 53 15/16 |
| 3/4 | 1 1/4 | 17 | 27 3/4 | 34 | 55 9/16 |
| 1 | 1 5/8 | 18 | 29 3/8 | 35 | 57 3/16 |
| 2 | 3 1/4 | 19 | 31 1/16 | **36** | **58 13/16** |
| 3 | 4 7/8 | 20 | 32 11/16 | 37 | 60 7/16 |
| 4 | 6 9/16 | 21 | 34 5/16 | 38 | 62 1/16 |
| 5 | 8 3/16 | 22 | 35 15/16 | 39 | 63 11/16 |
| 6 | 9 13/16 | 23 | 37 9/16 | 40 | 65 5/16 |
| 7 | 11 7/16 | **24** | **39 3/16** | 41 | 67 |
| 8 | 13 1/16 | 25 | 40 13/16 | 42 | 68 5/8 |
| 9 | 14 11/16 | 26 | 42 1/2 | 43 | 70 1/4 |
| 10 | 16 5/16 | 27 | 44 1/8 | 44 | 71 7/8 |
| 11 | 18 | 28 | 45 3/4 | 45 | 73 1/2 |
| **12** | **19 5/8** | 29 | 47 3/8 | 46 | 75 1/8 |
| 13 | 21 1/4 | 30 | 49 | 47 | 76 3/4 |
| 14 | 22 7/8 | 31 | 50 5/8 | **48** | **78 7/16** |

| ***Common Rafters*** **15 1/2 in 12 – 52 1/4°** | | | |
|---|---|---|---|
| **SPAN** ***Feet*** | **LENGTH** ***Feet Inches*** | **SPAN** ***Inches*** | **LENGTH** ***Inches*** |
| 1 | 0- 9 13/16 | 1/4 | 3/16 |
| 2 | 1- 7 5/8 | 1/2 | 7/16 |
| 3 | 2- 5 3/8 | 3/4 | 5/8 |
| 4 | 3- 3 3/16 | 1 | 13/16 |
| 5 | 4- 1 | 1 1/4 | 1 |
| 6 | 4-10 13/16 | 1 1/2 | 1 1/4 |
| 7 | 5- 8 5/8 | 1 3/4 | 1 7/16 |
| 8 | 6- 6 7/16 | 2 | 1 5/8 |
| 9 | 7- 4 3/16 | 2 1/4 | 1 13/16 |
| 10 | 8- 2 | 2 1/2 | 2 1/16 |
| 11 | 8-11 13/16 | 2 3/4 | 2 1/4 |
| 12 | 9- 9 5/8 | 3 | 2 7/16 |
| 13 | 10- 7 7/16 | 3 1/4 | 2 5/8 |
| 14 | 11- 5 3/16 | 3 1/2 | 2 7/8 |
| 15 | 12- 3 | 3 3/4 | 3 1/16 |
| 16 | 13- 0 13/16 | 4 | 3 1/4 |
| 17 | 13-10 5/8 | 4 1/4 | 3 1/2 |
| 18 | 14- 8 7/16 | 4 1/2 | 3 11/16 |
| 19 | 15- 6 1/4 | 4 3/4 | 3 7/8 |
| 20 | 16- 4 | 5 | 4 1/16 |
| 21 | 17- 1 13/16 | 5 1/4 | 4 5/16 |
| 22 | 17-11 5/8 | 5 1/2 | 4 1/2 |
| 23 | 18- 9 7/16 | 5 3/4 | 4 11/16 |
| 24 | 19- 7 1/4 | 6 | 4 7/8 |
| 25 | 20- 5 | 6 1/4 | 5 1/8 |
| 26 | 21- 2 13/16 | 6 1/2 | 5 5/16 |
| 27 | 22- 0 5/8 | 6 3/4 | 5 1/2 |
| 28 | 22-10 7/16 | 7 | 5 11/16 |
| 29 | 23- 8 1/4 | 7 1/4 | 5 15/16 |
| 30 | 24- 6 1/16 | 7 1/2 | 6 1/8 |
| 31 | 25- 3 13/16 | 7 3/4 | 6 5/16 |
| 32 | 26- 1 5/8 | 8 | 6 9/16 |
| 33 | 26-11 7/16 | 8 1/4 | 6 3/4 |
| 34 | 27- 9 1/4 | 8 1/2 | 6 15/16 |
| 35 | 28- 7 1/16 | 8 3/4 | 7 1/8 |
| 36 | 29- 4 13/16 | 9 | 7 3/8 |
| 37 | 30- 2 5/8 | 9 1/4 | 7 9/16 |
| 38 | 31- 0 7/16 | 9 1/2 | 7 3/4 |
| 39 | 31-10 1/4 | 9 3/4 | 7 15/16 |
| 40 | 32- 8 1/16 | 10 | 8 3/16 |
| 41 | 33- 5 7/8 | 10 1/4 | 8 3/8 |
| 42 | 34- 3 5/8 | 10 1/2 | 8 9/16 |
| 43 | 35- 1 7/16 | 10 3/4 | 8 3/4 |
| 44 | 35-11 1/4 | 11 | 9 |
| 45 | 36- 9 1/16 | 11 1/4 | 9 3/16 |
| 46 | 37- 6 7/8 | 11 1/2 | 9 3/8 |
| 47 | 38- 4 5/8 | 11 3/4 | 9 5/8 |
| 48 | 39- 2 7/16 | 12 | 9 13/16 |
| 49 | 40- 0 1/4 | | |
| 50 | 40-10 1/16 | | |

| ***Hip or Valley Rafters*** **15 1/2 in 17 – 42 1/2°** | | | |
|---|---|---|---|
| **SPAN** ***Feet*** | **LENGTH** ***Feet Inches*** | **SPAN** ***Inches*** | **LENGTH** ***Inches*** |
| 1 | 0-11 1/2 | 1/4 | 1/4 |
| 2 | 1-11 | 1/2 | 1/2 |
| 3 | 2-10 1/2 | 3/4 | 11/16 |
| 4 | 3- 9 15/16 | 1 | 15/16 |
| 5 | 4- 9 7/16 | 1 1/4 | 1 3/16 |
| 6 | 5- 8 15/16 | 1 1/2 | 1 7/16 |
| 7 | 6- 8 7/16 | 1 3/4 | 1 11/16 |
| 8 | 7- 7 15/16 | 2 | 1 15/16 |
| 9 | 8- 7 7/16 | 2 1/4 | 2 1/8 |
| 10 | 9- 6 15/16 | 2 1/2 | 2 3/8 |
| 11 | 10- 6 7/16 | 2 3/4 | 2 5/8 |
| 12 | 11- 5 7/8 | 3 | 2 7/8 |
| 13 | 12- 5 3/8 | 3 1/4 | 3 1/8 |
| 14 | 13- 4 7/8 | 3 1/2 | 3 3/8 |
| 15 | 14- 4 3/8 | 3 3/4 | 3 9/16 |
| 16 | 15- 3 7/8 | 4 | 3 13/16 |
| 17 | 16- 3 3/8 | 4 1/4 | 4 1/16 |
| 18 | 17- 2 7/8 | 4 1/2 | 4 5/16 |
| 19 | 18- 2 3/8 | 4 3/4 | 4 9/16 |
| 20 | 19- 1 13/16 | 5 | 4 13/16 |
| 21 | 20- 1 5/16 | 5 1/4 | 5 |
| 22 | 21- 0 13/16 | 5 1/2 | 5 1/4 |
| 23 | 22- 0 5/16 | 5 3/4 | 5 1/2 |
| 24 | 22-11 13/16 | 6 | 5 3/4 |
| 25 | 23-11 5/16 | 6 1/4 | 6 |
| 26 | 24-10 13/16 | 6 1/2 | 6 1/4 |
| 27 | 25-10 1/4 | 6 3/4 | 6 7/16 |
| 28 | 26- 9 3/4 | 7 | 6 11/16 |
| 29 | 27- 9 1/4 | 7 1/4 | 6 15/16 |
| 30 | 28- 8 3/4 | 7 1/2 | 7 3/16 |
| 31 | 29- 8 1/4 | 7 3/4 | 7 7/16 |
| 32 | 30- 7 3/4 | 8 | 7 11/16 |
| 33 | 31- 7 1/4 | 8 1/4 | 7 7/8 |
| 34 | 32- 6 3/4 | 8 1/2 | 8 1/8 |
| 35 | 33- 6 3/16 | 8 3/4 | 8 3/8 |
| 36 | 34- 5 11/16 | 9 | 8 5/8 |
| 37 | 35- 5 3/16 | 9 1/4 | 8 7/8 |
| 38 | 36- 4 11/16 | 9 1/2 | 9 1/8 |
| 39 | 37- 4 3/16 | 9 3/4 | 9 5/16 |
| 40 | 38- 3 11/16 | 10 | 9 9/16 |
| 41 | 39- 3 3/16 | 10 1/4 | 9 13/16 |
| 42 | 40- 2 11/16 | 10 1/2 | 10 1/16 |
| 43 | 41- 2 1/8 | 10 3/4 | 10 5/16 |
| 44 | 42- 1 5/8 | 11 | 10 9/16 |
| 45 | 43- 1 1/8 | 11 1/4 | 10 3/4 |
| 46 | 44- 0 5/8 | 11 1/2 | 11 |
| 47 | 45- 0 1/8 | 11 3/4 | 11 1/4 |
| 48 | 45-11 5/8 | 12 | 11 1/2 |
| 49 | 46-11 1/8 | | |
| 50 | 47-10 9/16 | | |

# 16 in 12 – 53 1/4°

## *Cuts and Saw Settings*

**Tail Cut, Hip or Valley Rafters,** with Square Cut Common Rafters:
Framing Square: **8 1/8 & 24**; set saw at **25°**
(Compound Miter or Radial Arm Saws: set bevel at 25°, miter at 18 3/4°)
Explanation: Section 3(d), pages 12–13.

**Chamfer Cut, Hip or Valley Rafters:**
Set saw at **34 1/2°**. Explanation: Section 3(d), page 10.

**Fascia Board Miter Cut at Hip or Valley:**
Framing Square: **9 5/8 & 12**; set saw at **25°**
(Compound Miter or Radial Arm Saws: set bevel at 25°, miter at 38 3/4°)
Explanation: Section 5, pages 24–25.

**Frieze Block against Hip or Valley Rafter:**
Framing Square: **9 5/8 & 12**; set saw at **25°**
(Compound Miter or Radial Arm Saws: set bevel at 25°, miter at 38 3/4°)
Explanation: Section 6, page 26.

**Roof Sheathing at Hip or Valley:**
Framing Square: **12 & 20**; set saw at **34 1/2°**
(Compound Miter or Radial Arm Saws: set bevel at 34 1/2°, miter at 31°)
Distance "d" = **28 13/16"**. Explanation: Section 7, pages 26–27.

## *Calculating the Length of a Rafter*

**General Rule:** Span of building
less Thickness of ridge board (if any)
gives Length of rafter to plumb line of seat cut

*and* Span of building
less Thickness of ridge board (if any)
plus 2 × projection of overhang
gives Total length of rafter to tail cut

Explanation: Section 4, pages 13–19.

| *Jack Rafters* | | | | | |
|---|---|---|---|---|---|
| SPACED *Inches* | DIFFERENCE *Inches* | SPACED *Inches* | DIFFERENCE *Inches* | SPACED *Inches* | DIFFERENCE *Inches* |
| 1/4 | 7/16 | 15 | 25 | **32** | **53 5/16** |
| 1/2 | 13/16 | **16** | **26 11/16** | 33 | 55 |
| 3/4 | 1 1/4 | 17 | 28 5/16 | 34 | 56 11/16 |
| 1 | 1 11/16 | 18 | 30 | 35 | 58 5/16 |
| 2 | 3 5/16 | 19 | 31 11/16 | **36** | **60** |
| 3 | 5 | 20 | 33 5/16 | 37 | 61 11/16 |
| 4 | 6 11/16 | 21 | 35 | 38 | 63 5/16 |
| 5 | 8 5/16 | 22 | 36 11/16 | 39 | 65 |
| 6 | 10 | 23 | 38 5/16 | 40 | 66 11/16 |
| 7 | 11 11/16 | **24** | **40** | 41 | 68 5/16 |
| 8 | 13 5/16 | 25 | 41 11/16 | 42 | 70 |
| 9 | 15 | 26 | 43 5/16 | 43 | 71 11/16 |
| 10 | 16 11/16 | 27 | 45 | 44 | 73 5/16 |
| 11 | 18 5/16 | 28 | 46 11/16 | 45 | 75 |
| **12** | **20** | 29 | 48 5/16 | 46 | 76 11/16 |
| 13 | 21 11/16 | 30 | 50 | 47 | 78 5/16 |
| 14 | 23 5/16 | 31 | 51 11/16 | **48** | **80** |

| ***Common Rafters* 16 in 12 – 53 1/4°** | | | |
|---|---|---|---|
| **SPAN** *Feet* | **LENGTH** *Feet Inches* | **SPAN** *Inches* | **LENGTH** *Inches* |
| 1 | 0 - 10 | 1/4 | 3/16 |
| 2 | 1 - 8 | 1/2 | 7/16 |
| 3 | 2 - 6 | 3/4 | 5/8 |
| 4 | 3 - 4 | 1 | 13/16 |
| 5 | 4 - 2 | 1 1/4 | 1 1/16 |
| 6 | 5 - 0 | 1 1/2 | 1 1/4 |
| 7 | 5 - 10 | 1 3/4 | 1 7/16 |
| 8 | 6 - 8 | 2 | 1 11/16 |
| 9 | 7 - 6 | 2 1/4 | 1 7/8 |
| 10 | 8 - 4 | 2 1/2 | 2 1/16 |
| 11 | 9 - 2 | 2 3/4 | 2 5/16 |
| 12 | 10 - 0 | 3 | 2 1/2 |
| 13 | 10 - 10 | 3 1/4 | 2 11/16 |
| 14 | 11 - 8 | 3 1/2 | 2 15/16 |
| 15 | 12 - 6 | 3 3/4 | 3 1/8 |
| 16 | 13 - 4 | 4 | 3 5/16 |
| 17 | 14 - 2 | 4 1/4 | 3 9/16 |
| 18 | 15 - 0 | 4 1/2 | 3 3/4 |
| 19 | 15 - 10 | 4 3/4 | 3 15/16 |
| 20 | 16 - 8 | 5 | 4 3/16 |
| 21 | 17 - 6 | 5 1/4 | 4 3/8 |
| 22 | 18 - 4 | 5 1/2 | 4 9/16 |
| 23 | 19 - 2 | 5 3/4 | 4 13/16 |
| 24 | 20 - 0 | 6 | 5 |
| 25 | 20 - 10 | 6 1/4 | 5 3/16 |
| 26 | 21 - 8 | 6 1/2 | 5 7/16 |
| 27 | 22 - 6 | 6 3/4 | 5 5/8 |
| 28 | 23 - 4 | 7 | 5 13/16 |
| 29 | 24 - 2 | 7 1/4 | 6 1/16 |
| 30 | 25 - 0 | 7 1/2 | 6 1/4 |
| 31 | 25 - 10 | 7 3/4 | 6 7/16 |
| 32 | 26 - 8 | 8 | 6 11/16 |
| 33 | 27 - 6 | 8 1/4 | 6 7/8 |
| 34 | 28 - 4 | 8 1/2 | 7 1/16 |
| 35 | 29 - 2 | 8 3/4 | 7 5/16 |
| 36 | 30 - 0 | 9 | 7 1/2 |
| 37 | 30 - 10 | 9 1/4 | 7 11/16 |
| 38 | 31 - 8 | 9 1/2 | 7 15/16 |
| 39 | 32 - 6 | 9 3/4 | 8 1/8 |
| 40 | 33 - 4 | 10 | 8 5/16 |
| 41 | 34 - 2 | 10 1/4 | 8 9/16 |
| 42 | 35 - 0 | 10 1/2 | 8 3/4 |
| 43 | 35 - 10 | 10 3/4 | 8 15/16 |
| 44 | 36 - 8 | 11 | 9 3/16 |
| 45 | 37 - 6 | 11 1/4 | 9 3/8 |
| 46 | 38 - 4 | 11 1/2 | 9 9/16 |
| 47 | 39 - 2 | 11 3/4 | 9 13/16 |
| 48 | 40 - 0 | 12 | 10 |
| 49 | 40 - 10 | | |
| 50 | 41 - 8 | | |

| ***Hip or Valley Rafters* 16 in 17 – 43 1/4°** | | | |
|---|---|---|---|
| **SPAN** *Feet* | **LENGTH** *Feet Inches* | **SPAN** *Inches* | **LENGTH** *Inches* |
| 1 | 0-11 11/16 | 1/4 | 1/4 |
| 2 | 1-11 5/16 | 1/2 | 1/2 |
| 3 | 2-11 | 3/4 | 3/4 |
| 4 | 3-10 5/8 | 1 | 1 |
| 5 | 4-10 5/16 | 1 1/4 | 1 3/16 |
| 6 | 5-10 | 1 1/2 | 1 7/16 |
| 7 | 6- 9 5/8 | 1 3/4 | 1 11/16 |
| 8 | 7- 9 5/16 | 2 | 1 15/16 |
| 9 | 8- 8 15/16 | 2 1/4 | 2 3/16 |
| 10 | 9- 8 5/8 | 2 1/2 | 2 7/16 |
| 11 | 10- 8 1/4 | 2 3/4 | 2 11/16 |
| 12 | 11- 7 15/16 | 3 | 2 15/16 |
| 13 | 12- 7 5/8 | 3 1/4 | 3 3/16 |
| 14 | 13- 7 1/4 | 3 1/2 | 3 3/8 |
| 15 | 14- 6 15/16 | 3 3/4 | 3 5/8 |
| 16 | 15- 6 9/16 | 4 | 3 7/8 |
| 17 | 16- 6 1/4 | 4 1/4 | 4 1/8 |
| 18 | 17- 5 15/16 | 4 1/2 | 4 3/8 |
| 19 | 18- 5 9/16 | 4 3/4 | 4 5/8 |
| 20 | 19- 5 1/4 | 5 | 4 7/8 |
| 21 | 20- 4 7/8 | 5 1/4 | 5 1/8 |
| 22 | 21- 4 9/16 | 5 1/2 | 5 3/8 |
| 23 | 22- 4 1/4 | 5 3/4 | 5 9/16 |
| 24 | 23- 3 7/8 | 6 | 5 13/16 |
| 25 | 24- 3 9/16 | 6 1/4 | 6 1/16 |
| 26 | 25- 3 3/16 | 6 1/2 | 6 5/16 |
| 27 | 26- 2 7/8 | 6 3/4 | 6 9/16 |
| 28 | 27- 2 9/16 | 7 | 6 13/16 |
| 29 | 28- 2 3/16 | 7 1/4 | 7 1/16 |
| 30 | 29- 1 7/8 | 7 1/2 | 7 5/16 |
| 31 | 30- 1 1/2 | 7 3/4 | 7 9/16 |
| 32 | 31- 1 3/16 | 8 | 7 3/4 |
| 33 | 32- 0 13/16 | 8 1/4 | 8 |
| 34 | 33- 0 1/2 | 8 1/2 | 8 1/4 |
| 35 | 34- 0 3/16 | 8 3/4 | 8 1/2 |
| 36 | 34-11 13/16 | 9 | 8 3/4 |
| 37 | 35-11 1/2 | 9 1/4 | 9 |
| 38 | 36-11 1/8 | 9 1/2 | 9 1/4 |
| 39 | 37-10 13/16 | 9 3/4 | 9 1/2 |
| 40 | 38-10 1/2 | 10 | 9 11/16 |
| 41 | 39-10 1/8 | 10 1/4 | 9 15/16 |
| 42 | 40- 9 13/16 | 10 1/2 | 10 3/16 |
| 43 | 41- 9 7/16 | 10 3/4 | 10 7/16 |
| 44 | 42- 9 1/8 | 11 | 10 11/16 |
| 45 | 43- 8 13/16 | 11 1/4 | 10 15/16 |
| 46 | 44- 8 7/16 | 11 1/2 | 11 3/16 |
| 47 | 45- 8 1/8 | 11 3/4 | 11 7/16 |
| 48 | 46- 7 3/4 | 12 | 11 11/16 |
| 49 | 47- 7 7/16 | | |
| 50 | 48- 7 1/8 | | |

# $16^1/_2$ in 12 – 54°

## *Cuts and Saw Settings*

**Tail Cut, Hip or Valley Rafters,** with Square Cut Common Rafters:
Framing Square: **8 1/16 & 24**; set saw at **24 1/2°**
(Compound Miter or Radial Arm Saws: set bevel at 24 1/2°, miter at 18 1/2°)
Explanation: Section 3(d), pages 12–13.

**Chamfer Cut, Hip or Valley Rafters:**
Set saw at **35°**. Explanation: Section 3(d), page 10.

**Fascia Board Miter Cut at Hip or Valley:**
Framing Square: **9 11/16 & 12**; set saw at **24 1/2°**
(Compound Miter or Radial Arm Saws: set bevel at 24 1/2°, miter at 39°)
Explanation: Section 5, pages 24–25.

**Frieze Block against Hip or Valley Rafter:**
Framing Square: **9 11/16 & 12**; set saw at **24 1/2°**
(Compound Miter or Radial Arm Saws: set bevel at 24 1/2°, miter at 39°)
Explanation: Section 6, page 26.

**Roof Sheathing at Hip or Valley:**
Framing Square: **12 & 20 3/8**; set saw at **35°**
(Compound Miter or Radial Arm Saws: set bevel at 35°, miter at 30 1/2°)
Distance "d" = **28 1/4"**. Explanation: Section 7, pages 26–27.

## *Calculating the Length of a Rafter*

| **General Rule:** | | Span of building |
|---|---|---|
| | less | Thickness of ridge board (if any) |
| | gives | Length of rafter to plumb line of seat cut |
| *and* | | Span of building |
| | less | Thickness of ridge board (if any) |
| | plus | 2 × projection of overhang |
| | gives | Total length of rafter to tail cut |

Explanation: Section 4, pages 13–19.

| *Jack Rafters* | | | | | |
|---|---|---|---|---|---|
| SPACED *Inches* | DIFFERENCE *Inches* | SPACED *Inches* | DIFFERENCE *Inches* | SPACED *Inches* | DIFFERENCE *Inches* |
| 1/4 | 7/16 | 15 | 25 1/2 | **32** | **54 3/8** |
| 1/2 | 7/8 | **16** | **27 3/16** | 33 | 56 1/8 |
| 3/4 | 1 1/4 | 17 | 28 7/8 | 34 | 57 13/16 |
| 1 | 1 11/16 | 18 | 30 5/8 | 35 | 59 1/2 |
| 2 | 3 3/8 | 19 | 32 5/16 | **36** | **61 3/16** |
| 3 | 5 1/8 | 20 | 34 | 37 | 62 15/16 |
| 4 | 6 13/16 | 21 | 35 11/16 | 38 | 64 5/8 |
| 5 | 8 1/2 | 22 | 37 3/8 | 39 | 66 5/16 |
| 6 | 10 3/16 | 23 | 39 1/8 | 40 | 68 |
| 7 | 11 7/8 | **24** | **40 13/16** | 41 | 69 11/16 |
| 8 | 13 5/8 | 25 | 42 1/2 | 42 | 71 7/16 |
| 9 | 15 5/16 | 26 | 44 3/16 | 43 | 73 1/8 |
| 10 | 17 | 27 | 45 7/8 | 44 | 74 13/16 |
| 11 | 18 11/16 | 28 | 47 5/8 | 45 | 76 1/2 |
| **12** | **20 3/8** | 29 | 49 5/16 | 46 | 78 3/16 |
| 13 | 22 1/8 | 30 | 51 | 47 | 79 15/16 |
| 14 | 23 13/16 | 31 | 52 11/16 | **48** | **81 5/8** |

| Common Rafters 16 1/2 in 12 – 54° | | | |
|---|---|---|---|
| SPAN Feet | LENGTH Feet Inches | SPAN Inches | LENGTH Inches |
| 1 | 0-10 3/16 | 1/4 | 3/16 |
| 2 | 1- 8 3/8 | 1/2 | 7/16 |
| 3 | 2- 6 5/8 | 3/4 | 5/8 |
| 4 | 3- 4 13/16 | 1 | 7/8 |
| 5 | 4- 3 | 1 1/4 | 1 1/16 |
| 6 | 5- 1 3/16 | 1 1/2 | 1 1/4 |
| 7 | 5-11 7/16 | 1 3/4 | 1 1/2 |
| 8 | 6- 9 5/8 | 2 | 1 11/16 |
| 9 | 7- 7 13/16 | 2 1/4 | 1 15/16 |
| 10 | 8- 6 | 2 1/2 | 2 1/8 |
| 11 | 9- 4 3/16 | 2 3/4 | 2 5/16 |
| 12 | 10- 2 7/16 | 3 | 2 9/16 |
| 13 | 11- 0 5/8 | 3 1/4 | 2 3/4 |
| 14 | 11-10 13/16 | 3 1/2 | 3 |
| 15 | 12- 9 | 3 3/4 | 3 3/16 |
| 16 | 13- 7 3/16 | 4 | 3 3/8 |
| 17 | 14- 5 7/16 | 4 1/4 | 3 5/8 |
| 18 | 15- 3 5/8 | 4 1/2 | 3 13/16 |
| 19 | 16- 1 13/16 | 4 3/4 | 4 1/16 |
| 20 | 17- 0 | 5 | 4 1/4 |
| 21 | 17-10 1/4 | 5 1/4 | 4 7/16 |
| 22 | 18- 8 7/16 | 5 1/2 | 4 11/16 |
| 23 | 19- 6 5/8 | 5 3/4 | 4 7/8 |
| 24 | 20- 4 13/16 | 6 | 5 1/8 |
| 25 | 21- 3 | 6 1/4 | 5 5/16 |
| 26 | 22- 1 1/4 | 6 1/2 | 5 1/2 |
| 27 | 22-11 7/16 | 6 3/4 | 5 3/4 |
| 28 | 23- 9 5/8 | 7 | 5 15/16 |
| 29 | 24- 7 13/16 | 7 1/4 | 6 3/16 |
| 30 | 25- 6 1/16 | 7 1/2 | 6 3/8 |
| 31 | 26- 4 1/4 | 7 3/4 | 6 9/16 |
| 32 | 27- 2 7/16 | 8 | 6 13/16 |
| 33 | 28- 0 5/8 | 8 1/4 | 7 |
| 34 | 28-10 13/16 | 8 1/2 | 7 1/4 |
| 35 | 29- 9 1/16 | 8 3/4 | 7 7/16 |
| 36 | 30- 7 1/4 | 9 | 7 5/8 |
| 37 | 31- 5 7/16 | 9 1/4 | 7 7/8 |
| 38 | 32- 3 5/8 | 9 1/2 | 8 1/16 |
| 39 | 33- 1 13/16 | 9 3/4 | 8 5/16 |
| 40 | 34- 0 1/16 | 10 | 8 1/2 |
| 41 | 34-10 1/4 | 10 1/4 | 8 11/16 |
| 42 | 35- 8 7/16 | 10 1/2 | 8 15/16 |
| 43 | 36- 6 5/8 | 10 3/4 | 9 1/8 |
| 44 | 37- 4 7/8 | 11 | 9 3/8 |
| 45 | 38- 3 1/16 | 11 1/4 | 9 9/16 |
| 46 | 39- 1 1/4 | 11 1/2 | 9 3/4 |
| 47 | 39-11 7/16 | 11 3/4 | 10 |
| 48 | 40- 9 5/8 | 12 | 10 3/16 |
| 49 | 41- 7 7/8 | | |
| 50 | 42- 6 1/16 | | |

| Hip or Valley Rafters 16 1/2 in 17 – 44 1/4° | | | |
|---|---|---|---|
| SPAN Feet | LENGTH Feet Inches | SPAN Inches | LENGTH Inches |
| 1 | 0-11 13/16 | 1/4 | 1/4 |
| 2 | 1-11 11/16 | 1/2 | 1/2 |
| 3 | 2-11 1/2 | 3/4 | 3/4 |
| 4 | 3-11 5/16 | 1 | 1 |
| 5 | 4-11 3/16 | 1 1/4 | 1 1/4 |
| 6 | 5-11 | 1 1/2 | 1 1/2 |
| 7 | 6-10 13/16 | 1 3/4 | 1 3/4 |
| 8 | 7-10 11/16 | 2 | 2 |
| 9 | 8-10 1/2 | 2 1/4 | 2 1/4 |
| 10 | 9-10 3/8 | 2 1/2 | 2 7/16 |
| 11 | 10-10 3/16 | 2 3/4 | 2 11/16 |
| 12 | 11-10 | 3 | 2 15/16 |
| 13 | 12- 9 7/8 | 3 1/4 | 3 3/16 |
| 14 | 13- 9 11/16 | 3 1/2 | 3 7/16 |
| 15 | 14- 9 1/2 | 3 3/4 | 3 11/16 |
| 16 | 15- 9 3/8 | 4 | 3 15/16 |
| 17 | 16- 9 3/16 | 4 1/4 | 4 3/16 |
| 18 | 17- 9 | 4 1/2 | 4 7/16 |
| 19 | 18- 8 7/8 | 4 3/4 | 4 11/16 |
| 20 | 19- 8 11/16 | 5 | 4 15/16 |
| 21 | 20- 8 1/2 | 5 1/4 | 5 3/16 |
| 22 | 21- 8 3/8 | 5 1/2 | 5 7/16 |
| 23 | 22- 8 3/16 | 5 3/4 | 5 11/16 |
| 24 | 23- 8 1/16 | 6 | 5 15/16 |
| 25 | 24- 7 7/8 | 6 1/4 | 6 3/16 |
| 26 | 25- 7 11/16 | 6 1/2 | 6 7/16 |
| 27 | 26- 7 9/16 | 6 3/4 | 6 11/16 |
| 28 | 27- 7 3/8 | 7 | 6 7/8 |
| 29 | 28- 7 3/16 | 7 1/4 | 7 1/8 |
| 30 | 29- 7 1/16 | 7 1/2 | 7 3/8 |
| 31 | 30- 6 7/8 | 7 3/4 | 7 5/8 |
| 32 | 31- 6 11/16 | 8 | 7 7/8 |
| 33 | 32- 6 9/16 | 8 1/4 | 8 1/8 |
| 34 | 33- 6 3/8 | 8 1/2 | 8 3/8 |
| 35 | 34- 6 3/16 | 8 3/4 | 8 5/8 |
| 36 | 35- 6 1/16 | 9 | 8 7/8 |
| 37 | 36- 5 7/8 | 9 1/4 | 9 1/8 |
| 38 | 37- 5 3/4 | 9 1/2 | 9 3/8 |
| 39 | 38- 5 9/16 | 9 3/4 | 9 5/8 |
| 40 | 39- 5 3/8 | 10 | 9 7/8 |
| 41 | 40- 5 1/4 | 10 1/4 | 10 1/8 |
| 42 | 41- 5 1/16 | 10 1/2 | 10 3/8 |
| 43 | 42- 4 7/8 | 10 3/4 | 10 5/8 |
| 44 | 43- 4 3/4 | 11 | 10 7/8 |
| 45 | 44- 4 9/16 | 11 1/4 | 11 1/8 |
| 46 | 45- 4 3/8 | 11 1/2 | 11 5/16 |
| 47 | 46- 4 1/4 | 11 3/4 | 11 9/16 |
| 48 | 47- 4 1/16 | 12 | 11 13/16 |
| 49 | 48- 3 7/8 | | |
| 50 | 49- 3 3/4 | | |

# 17 in 12 – 54 3/4°

## *Cuts and Saw Settings*

**Tail Cut, Hip or Valley Rafters,** with Square Cut Common Rafters:
Framing Square: **8 & 24**; set saw at **24**°
(Compound Miter or Radial Arm Saws: set bevel at 24°, miter at 18 1/2°)
Explanation: Section 3(d), pages 12–13.

**Chamfer Cut, Hip or Valley Rafters:**
Set saw at **35 1/4°**. Explanation: Section 3(d), page 10.

**Fascia Board Miter Cut at Hip or Valley:**
Framing Square: **9 13/16 & 12**; set saw at **24**°
(Compound Miter or Radial Arm Saws: set bevel at 24°, miter at 39 1/4°)
Explanation: Section 5, pages 24–25.

**Frieze Block against Hip or Valley Rafter:**
Framing Square: **9 13/16 & 12**; set saw at **24**°
(Compound Miter or Radial Arm Saws: set bevel at 24°, miter at 39 1/4°)
Explanation: Section 6, page 26.

**Roof Sheathing at Hip or Valley:**
Framing Square: **12 & 20 13/16**; set saw at **35 1/4°**
(Compound Miter or Radial Arm Saws: set bevel at 35 1/4°, miter at 30°)
Distance "d" = **27 11/16"**. Explanation: Section 7, pages 26–27.

## *Calculating the Length of a Rafter*

**General Rule:** Span of building
less Thickness of ridge board (if any)
gives Length of rafter to plumb line of seat cut

*and* Span of building
less Thickness of ridge board (if any)
plus 2 × projection of overhang
gives Total length of rafter to tail cut

Explanation: Section 4, pages 13–19.

| *Jack Rafters* | | | | | |
|---|---|---|---|---|---|
| SPACED *Inches* | DIFFERENCE *Inches* | SPACED *Inches* | DIFFERENCE *Inches* | SPACED *Inches* | DIFFERENCE *Inches* |
| 1/4 | 7/16 | 15 | 26 | **32** | **55 1/2** |
| 1/2 | 7/8 | **16** | **27 3/4** | 33 | 57 1/4 |
| 3/4 | 1 5/16 | 17 | 29 1/2 | 34 | 58 15/16 |
| 1 | 1 3/4 | 18 | 31 3/16 | 35 | 60 11/16 |
| 2 | 3 7/16 | 19 | 32 15/16 | **36** | **62 7/16** |
| 3 | 5 3/16 | 20 | 34 11/16 | 37 | 64 3/16 |
| 4 | 6 15/16 | 21 | 36 7/16 | 38 | 65 7/8 |
| 5 | 8 11/16 | 22 | 38 1/8 | 39 | 67 5/8 |
| 6 | 10 3/8 | 23 | 39 7/8 | 40 | 69 3/8 |
| 7 | 12 1/8 | **24** | **41 5/8** | 41 | 71 1/8 |
| 8 | 13 7/8 | 25 | 43 3/8 | 42 | 72 13/16 |
| 9 | 15 5/8 | 26 | 45 1/16 | 43 | 74 9/16 |
| 10 | 17 5/16 | 27 | 46 13/16 | 44 | 76 5/16 |
| 11 | 19 1/16 | 28 | 48 9/16 | 45 | 78 1/16 |
| **12** | **20 13/16** | 29 | 50 5/16 | 46 | 79 3/4 |
| 13 | 22 9/16 | 30 | 52 | 47 | 81 1/2 |
| 14 | 24 1/4 | 31 | 53 3/4 | **48** | **83 1/4** |

| Common Rafters<br>17 in 12 – 54 3/4° | | | |
|---|---|---|---|
| **SPAN** *Feet* | **LENGTH** *Feet Inches* | **SPAN** *Inches* | **LENGTH** *Inches* |
| 1 | 0-10 3/8 | 1/4 | 3/16 |
| 2 | 1- 8 13/16 | 1/2 | 7/16 |
| 3 | 2- 7 3/16 | 3/4 | 5/8 |
| 4 | 3- 5 5/8 | 1 | 7/8 |
| 5 | 4- 4 | 1 1/4 | 1 1/16 |
| 6 | 5- 2 7/16 | 1 1/2 | 1 5/16 |
| 7 | 6- 0 13/16 | 1 3/4 | 1 1/2 |
| 8 | 6-11 1/4 | 2 | 1 3/4 |
| 9 | 7- 9 5/8 | 2 1/4 | 1 15/16 |
| 10 | 8- 8 1/16 | 2 1/2 | 2 3/16 |
| 11 | 9- 6 7/16 | 2 3/4 | 2 3/8 |
| 12 | 10- 4 7/8 | 3 | 2 5/8 |
| 13 | 11- 3 1/4 | 3 1/4 | 2 13/16 |
| 14 | 12- 1 11/16 | 3 1/2 | 3 1/16 |
| 15 | 13- 0 1/16 | 3 3/4 | 3 1/4 |
| 16 | 13-10 1/2 | 4 | 3 7/16 |
| 17 | 14- 8 7/8 | 4 1/4 | 3 11/16 |
| 18 | 15- 7 1/4 | 4 1/2 | 3 7/8 |
| 19 | 16- 5 11/16 | 4 3/4 | 4 1/8 |
| 20 | 17- 4 1/16 | 5 | 4 5/16 |
| 21 | 18- 2 1/2 | 5 1/4 | 4 9/16 |
| 22 | 19- 0 7/8 | 5 1/2 | 4 3/4 |
| 23 | 19-11 5/16 | 5 3/4 | 5 |
| 24 | 20- 9 11/16 | 6 | 5 3/16 |
| 25 | 21- 8 1/8 | 6 1/4 | 5 7/16 |
| 26 | 22- 6 1/2 | 6 1/2 | 5 5/8 |
| 27 | 23- 4 15/16 | 6 3/4 | 5 7/8 |
| 28 | 24- 3 5/16 | 7 | 6 1/16 |
| 29 | 25- 1 3/4 | 7 1/4 | 6 5/16 |
| 30 | 26- 0 1/8 | 7 1/2 | 6 1/2 |
| 31 | 26-10 9/16 | 7 3/4 | 6 3/4 |
| 32 | 27- 8 15/16 | 8 | 6 15/16 |
| 33 | 28- 7 5/16 | 8 1/4 | 7 1/8 |
| 34 | 29- 5 3/4 | 8 1/2 | 7 3/8 |
| 35 | 30- 4 1/8 | 8 3/4 | 7 9/16 |
| 36 | 31- 2 9/16 | 9 | 7 13/16 |
| 37 | 32- 0 15/16 | 9 1/4 | 8 |
| 38 | 32-11 3/8 | 9 1/2 | 8 1/4 |
| 39 | 33- 9 3/4 | 9 3/4 | 8 7/16 |
| 40 | 34- 8 3/16 | 10 | 8 11/16 |
| 41 | 35- 6 9/16 | 10 1/4 | 8 7/8 |
| 42 | 36- 5 | 10 1/2 | 9 1/8 |
| 43 | 37- 3 3/8 | 10 3/4 | 9 5/16 |
| 44 | 38- 1 13/16 | 11 | 9 9/16 |
| 45 | 39- 0 3/16 | 11 1/4 | 9 3/4 |
| 46 | 39-10 5/8 | 11 1/2 | 10 |
| 47 | 40- 9 | 11 3/4 | 10 3/16 |
| 48 | 41- 7 7/16 | 12 | 10 3/8 |
| 49 | 42- 5 13/16 | | |
| 50 | 43- 4 3/16 | | |

| Hip or Valley Rafters<br>17 in 17 – 45° | | | |
|---|---|---|---|
| **SPAN** *Feet* | **LENGTH** *Feet Inches* | **SPAN** *Inches* | **LENGTH** *Inches* |
| 1 | 1 - 0 | 1/4 | 1/4 |
| 2 | 2 - 0 | 1/2 | 1/2 |
| 3 | 3 - 0 | 3/4 | 3/4 |
| 4 | 4 - 0 1/16 | 1 | 1 |
| 5 | 5 - 0 1/16 | 1 1/4 | 1 1/4 |
| 6 | 6 - 0 1/16 | 1 1/2 | 1 1/2 |
| 7 | 7 - 0 1/16 | 1 3/4 | 1 3/4 |
| 8 | 8 - 0 1/16 | 2 | 2 |
| 9 | 9 - 0 1/16 | 2 1/4 | 2 1/4 |
| 10 | 10 - 0 1/8 | 2 1/2 | 2 1/2 |
| 11 | 11 - 0 1/8 | 2 3/4 | 2 3/4 |
| 12 | 12 - 0 1/8 | 3 | 3 |
| 13 | 13 - 0 1/8 | 3 1/4 | 3 1/4 |
| 14 | 14 - 0 1/8 | 3 1/2 | 3 1/2 |
| 15 | 15 - 0 1/8 | 3 3/4 | 3 3/4 |
| 16 | 16 - 0 3/16 | 4 | 4 |
| 17 | 17 - 0 3/16 | 4 1/4 | 4 1/4 |
| 18 | 18 - 0 3/16 | 4 1/2 | 4 1/2 |
| 19 | 19 - 0 3/16 | 4 3/4 | 4 3/4 |
| 20 | 20 - 0 3/16 | 5 | 5 |
| 21 | 21 - 0 3/16 | 5 1/4 | 5 1/4 |
| 22 | 22 - 0 1/4 | 5 1/2 | 5 1/2 |
| 23 | 23 - 0 1/4 | 5 3/4 | 5 3/4 |
| 24 | 24 - 0 1/4 | 6 | 6 |
| 25 | 25 - 0 1/4 | 6 1/4 | 6 1/4 |
| 26 | 26 - 0 1/4 | 6 1/2 | 6 1/2 |
| 27 | 27 - 0 1/4 | 6 3/4 | 6 3/4 |
| 28 | 28 - 0 5/16 | 7 | 7 |
| 29 | 29 - 0 5/16 | 7 1/4 | 7 1/4 |
| 30 | 30 - 0 5/16 | 7 1/2 | 7 1/2 |
| 31 | 31 - 0 5/16 | 7 3/4 | 7 3/4 |
| 32 | 32 - 0 5/16 | 8 | 8 |
| 33 | 33 - 0 5/16 | 8 1/4 | 8 1/4 |
| 34 | 34 - 0 3/8 | 8 1/2 | 8 1/2 |
| 35 | 35 - 0 3/8 | 8 3/4 | 8 3/4 |
| 36 | 36 - 0 3/8 | 9 | 9 |
| 37 | 37 - 0 3/8 | 9 1/4 | 9 1/4 |
| 38 | 38 - 0 3/8 | 9 1/2 | 9 1/2 |
| 39 | 39 - 0 3/8 | 9 3/4 | 9 3/4 |
| 40 | 40 - 0 7/16 | 10 | 10 |
| 41 | 41 - 0 7/16 | 10 1/4 | 10 1/4 |
| 42 | 42 - 0 7/16 | 10 1/2 | 10 1/2 |
| 43 | 43 - 0 7/16 | 10 3/4 | 10 3/4 |
| 44 | 44 - 0 7/16 | 11 | 11 |
| 45 | 45 - 0 7/16 | 11 1/4 | 11 1/4 |
| 46 | 46 - 0 1/2 | 11 1/2 | 11 1/2 |
| 47 | 47 - 0 1/2 | 11 3/4 | 11 3/4 |
| 48 | 48 - 0 1/2 | 12 | 12 |
| 49 | 49 - 0 1/2 | | |
| 50 | 50 - 0 1/2 | | |

# 17 1/2 in 12 – 55 1/2°

## *Cuts and Saw Settings*

**Tail Cut, Hip or Valley Rafters,** with Square Cut Common Rafters:
Framing Square: **7 15/16 & 24**; set saw at **23 1/2°**
(Compound Miter or Radial Arm Saws: set bevel at 23 1/2°, miter at 18 1/4°)
Explanation: Section 3(d), pages 12–13.

**Chamfer Cut, Hip or Valley Rafters:**
Set saw at **35 3/4°**. Explanation: Section 3(d), page 10.

**Fascia Board Miter Cut at Hip or Valley:**
Framing Square: **9 7/8 & 12**; set saw at **23 1/2°**
(Compound Miter or Radial Arm Saws: set bevel at 23 1/2°, miter at 39 1/2°)
Explanation: Section 5, pages 24–25.

**Frieze Block against Hip or Valley Rafter:**
Framing Square: **9 7/8 & 12**; set saw at **23 1/2°**
(Compound Miter or Radial Arm Saws: set bevel at 23 1/2°, miter at 39 1/2°)
Explanation: Section 6, page 26.

**Roof Sheathing at Hip or Valley:**
Framing Square: **12 & 21 1/4**; set saw at **35 3/4°**
(Compound Miter or Radial Arm Saws: set bevel at 35 3/4°, miter at 29 1/2°)
Distance "d" = **27 1/8"**. Explanation: Section 7, pages 26–27.

## *Calculating the Length of a Rafter*

**General Rule:** Span of building
less Thickness of ridge board (if any)
gives Length of rafter to plumb line of seat cut

*and* Span of building
less Thickness of ridge board (if any)
plus 2 × projection of overhang
gives Total length of rafter to tail cut

Explanation: Section 4, pages 13–19.

### *Jack Rafters*

| SPACED *Inches* | DIFFERENCE *Inches* | SPACED *Inches* | DIFFERENCE *Inches* | SPACED *Inches* | DIFFERENCE *Inches* |
|---|---|---|---|---|---|
| 1/4 | 7/16 | 15 | 26 1/2 | **32** | **56 9/16** |
| 1/2 | 7/8 | **16** | **28 5/16** | 33 | 58 3/8 |
| 3/4 | 1 5/16 | 17 | 30 1/16 | 34 | 60 1/8 |
| 1 | 1 3/4 | 18 | 31 13/16 | 35 | 61 7/8 |
| 2 | 3 9/16 | 19 | 33 5/8 | **36** | **63 11/16** |
| 3 | 5 5/16 | 20 | 35 3/8 | 37 | 65 7/16 |
| 4 | 7 1/16 | 21 | 37 1/8 | 38 | 67 3/16 |
| 5 | 8 13/16 | 22 | 38 7/8 | 39 | 68 15/16 |
| 6 | 10 5/8 | 23 | 40 11/16 | 40 | 70 3/4 |
| 7 | 12 3/8 | **24** | **42 7/16** | 41 | 72 1/2 |
| 8 | 14 1/8 | 25 | 44 3/16 | 42 | 74 1/4 |
| 9 | 15 15/16 | 26 | 46 | 43 | 76 1/16 |
| 10 | 17 11/16 | 27 | 47 3/4 | 44 | 77 13/16 |
| 11 | 19 7/16 | 28 | 49 1/2 | 45 | 79 9/16 |
| **12** | **21 1/4** | 29 | 51 1/4 | 46 | 81 5/16 |
| 13 | 23 | 30 | 53 1/16 | 47 | 83 1/8 |
| 14 | 24 3/4 | 31 | 54 13/16 | **48** | **84 7/8** |

## Common Rafters
### 17 1/2 in 12 – 55 1/2°

| SPAN Feet | LENGTH Feet Inches | SPAN Inches | LENGTH Inches |
|---|---|---|---|
| 1 | 0-10 5/8 | 1/4 | 1/4 |
| 2 | 1- 9 1/4 | 1/2 | 7/16 |
| 3 | 2- 7 13/16 | 3/4 | 11/16 |
| 4 | 3- 6 7/16 | 1 | 7/8 |
| 5 | 4- 5 1/16 | 1 1/4 | 1 1/8 |
| 6 | 5- 3 11/16 | 1 1/2 | 1 5/16 |
| 7 | 6- 2 1/4 | 1 3/4 | 1 9/16 |
| 8 | 7- 0 7/8 | 2 | 1 3/4 |
| 9 | 7-11 1/2 | 2 1/4 | 2 |
| 10 | 8-10 1/8 | 2 1/2 | 2 3/16 |
| 11 | 9- 8 11/16 | 2 3/4 | 2 7/16 |
| 12 | 10- 7 5/16 | 3 | 2 5/8 |
| 13 | 11- 5 15/16 | 3 1/4 | 2 7/8 |
| 14 | 12- 4 9/16 | 3 1/2 | 3 1/8 |
| 15 | 13- 3 1/8 | 3 3/4 | 3 5/16 |
| 16 | 14- 1 3/4 | 4 | 3 9/16 |
| 17 | 15- 0 3/8 | 4 1/4 | 3 3/4 |
| 18 | 15-11 | 4 1/2 | 4 |
| 19 | 16- 9 9/16 | 4 3/4 | 4 3/16 |
| 20 | 17- 8 3/16 | 5 | 4 7/16 |
| 21 | 18- 6 13/16 | 5 1/4 | 4 5/8 |
| 22 | 19- 5 7/16 | 5 1/2 | 4 7/8 |
| 23 | 20- 4 | 5 3/4 | 5 1/16 |
| 24 | 21- 2 5/8 | 6 | 5 5/16 |
| 25 | 22- 1 1/4 | 6 1/4 | 5 1/2 |
| 26 | 22-11 7/8 | 6 1/2 | 5 3/4 |
| 27 | 23-10 7/16 | 6 3/4 | 5 15/16 |
| 28 | 24- 9 1/16 | 7 | 6 3/16 |
| 29 | 25- 7 11/16 | 7 1/4 | 6 7/16 |
| 30 | 26- 6 5/16 | 7 1/2 | 6 5/8 |
| 31 | 27- 4 7/8 | 7 3/4 | 6 7/8 |
| 32 | 28- 3 1/2 | 8 | 7 1/16 |
| 33 | 29- 2 1/8 | 8 1/4 | 7 5/16 |
| 34 | 30- 0 3/4 | 8 1/2 | 7 1/2 |
| 35 | 30-11 5/16 | 8 3/4 | 7 3/4 |
| 36 | 31- 9 15/16 | 9 | 7 15/16 |
| 37 | 32- 8 9/16 | 9 1/4 | 8 3/16 |
| 38 | 33- 7 3/16 | 9 1/2 | 8 3/8 |
| 39 | 34- 5 3/4 | 9 3/4 | 8 5/8 |
| 40 | 35- 4 3/8 | 10 | 8 13/16 |
| 41 | 36- 3 | 10 1/4 | 9 1/16 |
| 42 | 37- 1 5/8 | 10 1/2 | 9 5/16 |
| 43 | 38- 0 3/16 | 10 3/4 | 9 1/2 |
| 44 | 38-10 13/16 | 11 | 9 3/4 |
| 45 | 39- 9 7/16 | 11 1/4 | 9 15/16 |
| 46 | 40- 8 1/16 | 11 1/2 | 10 3/16 |
| 47 | 41- 6 5/8 | 11 3/4 | 10 3/8 |
| 48 | 42- 5 1/4 | 12 | 10 5/8 |
| 49 | 43- 3 7/8 | | |
| 50 | 44- 2 1/2 | | |

## Hip or Valley Rafters
### 17 1/2 in 17 – 46°

| SPAN Feet | LENGTH Feet Inches | SPAN Inches | LENGTH Inches |
|---|---|---|---|
| 1 | 1 - 0 3/16 | 1/4 | 1/4 |
| 2 | 2 - 0 3/8 | 1/2 | 1/2 |
| 3 | 3 - 0 9/16 | 3/4 | 3/4 |
| 4 | 4 - 0 3/4 | 1 | 1 |
| 5 | 5 - 0 15/16 | 1 1/4 | 1 1/4 |
| 6 | 6 - 1 1/8 | 1 1/2 | 1 1/2 |
| 7 | 7 - 1 5/16 | 1 3/4 | 1 3/4 |
| 8 | 8 - 1 1/2 | 2 | 2 1/16 |
| 9 | 9 - 1 11/16 | 2 1/4 | 2 5/16 |
| 10 | 10 - 1 7/8 | 2 1/2 | 2 9/16 |
| 11 | 11 - 2 1/16 | 2 3/4 | 2 13/16 |
| 12 | 12 - 2 1/4 | 3 | 3 1/16 |
| 13 | 13 - 2 7/16 | 3 1/4 | 3 5/16 |
| 14 | 14 - 2 5/8 | 3 1/2 | 3 9/16 |
| 15 | 15 - 2 13/16 | 3 3/4 | 3 13/16 |
| 16 | 16 - 3 | 4 | 4 1/16 |
| 17 | 17 - 3 3/16 | 4 1/4 | 4 5/16 |
| 18 | 18 - 3 3/8 | 4 1/2 | 4 9/16 |
| 19 | 19 - 3 9/16 | 4 3/4 | 4 13/16 |
| 20 | 20 - 3 3/4 | 5 | 5 1/16 |
| 21 | 21 - 3 15/16 | 5 1/4 | 5 5/16 |
| 22 | 22 - 4 1/8 | 5 1/2 | 5 9/16 |
| 23 | 23 - 4 5/16 | 5 3/4 | 5 13/16 |
| 24 | 24 - 4 1/2 | 6 | 6 1/8 |
| 25 | 25 - 4 11/16 | 6 1/4 | 6 3/8 |
| 26 | 26 - 4 7/8 | 6 1/2 | 6 5/8 |
| 27 | 27 - 5 1/16 | 6 3/4 | 6 7/8 |
| 28 | 28 - 5 5/16 | 7 | 7 1/8 |
| 29 | 29 - 5 1/2 | 7 1/4 | 7 3/8 |
| 30 | 30 - 5 11/16 | 7 1/2 | 7 5/8 |
| 31 | 31 - 5 7/8 | 7 3/4 | 7 7/8 |
| 32 | 32 - 6 1/16 | 8 | 8 1/8 |
| 33 | 33 - 6 1/4 | 8 1/4 | 8 3/8 |
| 34 | 34 - 6 7/16 | 8 1/2 | 8 5/8 |
| 35 | 35 - 6 5/8 | 8 3/4 | 8 7/8 |
| 36 | 36 - 6 13/16 | 9 | 9 1/8 |
| 37 | 37 - 7 | 9 1/4 | 9 3/8 |
| 38 | 38 - 7 3/16 | 9 1/2 | 9 5/8 |
| 39 | 39 - 7 3/8 | 9 3/4 | 9 7/8 |
| 40 | 40 - 7 9/16 | 10 | 10 3/16 |
| 41 | 41 - 7 3/4 | 10 1/4 | 10 7/16 |
| 42 | 42 - 7 15/16 | 10 1/2 | 10 11/16 |
| 43 | 43 - 8 1/8 | 10 3/4 | 10 15/16 |
| 44 | 44 - 8 5/16 | 11 | 11 3/16 |
| 45 | 45 - 8 1/2 | 11 1/4 | 11 7/16 |
| 46 | 46 - 8 11/16 | 11 1/2 | 11 11/16 |
| 47 | 47 - 8 7/8 | 11 3/4 | 11 15/16 |
| 48 | 48 - 9 1/16 | 12 | 12 3/16 |
| 49 | 49 - 9 1/4 | | |
| 50 | 50 - 9 7/16 | | |

# 18 in 12 – 56 1/4°

## *Cuts and Saw Settings*

**Tail Cut, Hip or Valley Rafters,** with Square Cut Common Rafters:
Framing Square: **7 13/16 & 24**; set saw at **23°**
(Compound Miter or Radial Arm Saws: set bevel at 23°, miter at 18°)
Explanation: Section 3(d), pages 12–13.

**Chamfer Cut, Hip or Valley Rafters:**
Set saw at **36°**. Explanation: Section 3(d), page 10.

**Fascia Board Miter Cut at Hip or Valley:**
Framing Square: **10 & 12**; set saw at **23°**
(Compound Miter or Radial Arm Saws: set bevel at 23°, miter at 39 3/4°)
Explanation: Section 5, pages 24–25.

**Frieze Block against Hip or Valley Rafter:**
Framing Square: **10 & 12**; set saw at **23°**
(Compound Miter or Radial Arm Saws: set bevel at 23°, miter at 39 3/4°)
Explanation: Section 6, page 26.

**Roof Sheathing at Hip or Valley:**
Framing Square: **12 & 21 5/8**; set saw at **36°**
(Compound Miter or Radial Arm Saws: set bevel at 36°, miter at 29°)
Distance "d" = **26 5/8"**. Explanation: Section 7, pages 26–27.

## *Calculating the Length of a Rafter*

**General Rule:** Span of building
less Thickness of ridge board (if any)
gives Length of rafter to plumb line of seat cut

*and* Span of building
less Thickness of ridge board (if any)
plus 2 × projection of overhang
gives Total length of rafter to tail cut

Explanation: Section 4, pages 13–19.

| *Jack Rafters* | | | | | |
|---|---|---|---|---|---|
| SPACED *Inches* | DIFFERENCE *Inches* | SPACED *Inches* | DIFFERENCE *Inches* | SPACED *Inches* | DIFFERENCE *Inches* |
| 1/4 | 7/16 | 15 | 27 1/16 | **32** | **57 11/16** |
| 1/2 | 7/8 | **16** | **28 7/8** | 33 | 59 1/2 |
| 3/4 | 1 3/8 | 17 | 30 5/8 | 34 | 61 5/16 |
| 1 | 1 13/16 | 18 | 32 7/16 | 35 | 63 1/8 |
| 2 | 3 5/8 | 19 | 34 1/4 | **36** | **64 7/8** |
| 3 | 5 7/16 | 20 | 36 1/16 | 37 | 66 11/16 |
| 4 | 7 3/16 | 21 | 37 7/8 | 38 | 68 1/2 |
| 5 | 9 | 22 | 39 11/16 | 39 | 70 5/16 |
| 6 | 10 13/16 | 23 | 41 7/16 | 40 | 72 1/8 |
| 7 | 12 5/8 | **24** | **43 1/4** | 41 | 73 15/16 |
| 8 | 14 7/16 | 25 | 45 1/16 | 42 | 75 11/16 |
| 9 | 16 1/4 | 26 | 46 7/8 | 43 | 77 1/2 |
| 10 | 18 | 27 | 48 11/16 | 44 | 79 5/16 |
| 11 | 19 13/16 | 28 | 50 1/2 | 45 | 81 1/8 |
| **12** | **21 5/8** | 29 | 52 1/4 | 46 | 82 15/16 |
| 13 | 23 7/16 | 30 | 54 1/16 | 47 | 84 3/4 |
| 14 | 25 1/4 | 31 | 55 7/8 | **48** | **86 9/16** |

## Common Rafters
## 18 in 12 – 56 1/4°

| SPAN Feet | LENGTH Feet Inches | SPAN Inches | LENGTH Inches |
|---|---|---|---|
| 1 | 0-10 13/16 | 1/4 | 1/4 |
| 2 | 1- 9 5/8 | 1/2 | 7/16 |
| 3 | 2- 8 7/16 | 3/4 | 11/16 |
| 4 | 3- 7 1/4 | 1 | 7/8 |
| 5 | 4- 6 1/16 | 1 1/4 | 1 1/8 |
| 6 | 5- 4 7/8 | 1 1/2 | 1 3/8 |
| 7 | 6- 3 11/16 | 1 3/4 | 1 9/16 |
| 8 | 7- 2 9/16 | 2 | 1 13/16 |
| 9 | 8- 1 3/8 | 2 1/4 | 2 |
| 10 | 9- 0 3/16 | 2 1/2 | 2 1/4 |
| 11 | 9-11 | 2 3/4 | 2 1/2 |
| 12 | 10- 9 13/16 | 3 | 2 11/16 |
| 13 | 11- 8 5/8 | 3 1/4 | 2 15/16 |
| 14 | 12- 7 7/16 | 3 1/2 | 3 1/8 |
| 15 | 13- 6 1/4 | 3 3/4 | 3 3/8 |
| 16 | 14- 5 1/16 | 4 | 3 5/8 |
| 17 | 15- 3 7/8 | 4 1/4 | 3 13/16 |
| 18 | 16- 2 11/16 | 4 1/2 | 4 1/16 |
| 19 | 17- 1 1/2 | 4 3/4 | 4 5/16 |
| 20 | 18- 0 5/16 | 5 | 4 1/2 |
| 21 | 18-11 1/8 | 5 1/4 | 4 3/4 |
| 22 | 19- 9 15/16 | 5 1/2 | 4 15/16 |
| 23 | 20- 8 13/16 | 5 3/4 | 5 3/16 |
| 24 | 21- 7 5/8 | 6 | 5 7/16 |
| 25 | 22- 6 7/16 | 6 1/4 | 5 5/8 |
| 26 | 23- 5 1/4 | 6 1/2 | 5 7/8 |
| 27 | 24- 4 1/16 | 6 3/4 | 6 1/16 |
| 28 | 25- 2 7/8 | 7 | 6 5/16 |
| 29 | 26- 1 11/16 | 7 1/4 | 6 9/16 |
| 30 | 27- 0 1/2 | 7 1/2 | 6 3/4 |
| 31 | 27-11 5/16 | 7 3/4 | 7 |
| 32 | 28-10 1/8 | 8 | 7 3/16 |
| 33 | 29- 8 15/16 | 8 1/4 | 7 7/16 |
| 34 | 30- 7 3/4 | 8 1/2 | 7 11/16 |
| 35 | 31- 6 9/16 | 8 3/4 | 7 7/8 |
| 36 | 32- 5 3/8 | 9 | 8 1/8 |
| 37 | 33- 4 3/16 | 9 1/4 | 8 5/16 |
| 38 | 34- 3 1/16 | 9 1/2 | 8 9/16 |
| 39 | 35- 1 7/8 | 9 3/4 | 8 13/16 |
| 40 | 36- 0 11/16 | 10 | 9 |
| 41 | 36-11 1/2 | 10 1/4 | 9 1/4 |
| 42 | 37-10 5/16 | 10 1/2 | 9 7/16 |
| 43 | 38- 9 1/8 | 10 3/4 | 9 11/16 |
| 44 | 39- 7 15/16 | 11 | 9 15/16 |
| 45 | 40- 6 3/4 | 11 1/4 | 10 1/8 |
| 46 | 41- 5 9/16 | 11 1/2 | 10 3/8 |
| 47 | 42- 4 3/8 | 11 3/4 | 10 9/16 |
| 48 | 43- 3 3/16 | 12 | 10 13/16 |
| 49 | 44- 2 | | |
| 50 | 45- 0 13/16 | | |

## Hip or Valley Rafters
## 18 in 17 – 46 3/4°

| SPAN Feet | LENGTH Feet Inches | SPAN Inches | LENGTH Inches |
|---|---|---|---|
| 1 | 1- 0 3/8 | 1/4 | 1/4 |
| 2 | 2- 0 3/4 | 1/2 | 1/2 |
| 3 | 3- 1 1/8 | 3/4 | 3/4 |
| 4 | 4- 1 1/2 | 1 | 1 |
| 5 | 5- 1 7/8 | 1 1/4 | 1 5/16 |
| 6 | 6- 2 3/16 | 1 1/2 | 1 9/16 |
| 7 | 7- 2 9/16 | 1 3/4 | 1 13/16 |
| 8 | 8- 2 15/16 | 2 | 2 1/16 |
| 9 | 9- 3 5/16 | 2 1/4 | 2 5/16 |
| 10 | 10- 3 11/16 | 2 1/2 | 2 9/16 |
| 11 | 11- 4 1/16 | 2 3/4 | 2 13/16 |
| 12 | 12- 4 7/16 | 3 | 3 1/16 |
| 13 | 13- 4 13/16 | 3 1/4 | 3 3/8 |
| 14 | 14- 5 3/16 | 3 1/2 | 3 5/8 |
| 15 | 15- 5 9/16 | 3 3/4 | 3 7/8 |
| 16 | 16- 5 15/16 | 4 | 4 1/8 |
| 17 | 17- 6 1/4 | 4 1/4 | 4 3/8 |
| 18 | 18- 6 5/8 | 4 1/2 | 4 5/8 |
| 19 | 19- 7 | 4 3/4 | 4 7/8 |
| 20 | 20- 7 3/8 | 5 | 5 1/8 |
| 21 | 21- 7 3/4 | 5 1/4 | 5 7/16 |
| 22 | 22- 8 1/8 | 5 1/2 | 5 11/16 |
| 23 | 23- 8 1/2 | 5 3/4 | 5 15/16 |
| 24 | 24- 8 7/8 | 6 | 6 3/16 |
| 25 | 25- 9 1/4 | 6 1/4 | 6 7/16 |
| 26 | 26- 9 5/8 | 6 1/2 | 6 11/16 |
| 27 | 27-10 | 6 3/4 | 6 15/16 |
| 28 | 28-10 5/16 | 7 | 7 3/16 |
| 29 | 29-10 11/16 | 7 1/4 | 7 1/2 |
| 30 | 30-11 1/16 | 7 1/2 | 7 3/4 |
| 31 | 31-11 7/16 | 7 3/4 | 8 |
| 32 | 32-11 13/16 | 8 | 8 1/4 |
| 33 | 34- 0 3/16 | 8 1/4 | 8 1/2 |
| 34 | 35- 0 9/16 | 8 1/2 | 8 3/4 |
| 35 | 36- 0 15/16 | 8 3/4 | 9 |
| 36 | 37- 1 5/16 | 9 | 9 1/4 |
| 37 | 38- 1 11/16 | 9 1/4 | 9 9/16 |
| 38 | 39- 2 1/16 | 9 1/2 | 9 13/16 |
| 39 | 40- 2 3/8 | 9 3/4 | 10 1/16 |
| 40 | 41- 2 3/4 | 10 | 10 5/16 |
| 41 | 42- 3 1/8 | 10 1/4 | 10 9/16 |
| 42 | 43- 3 1/2 | 10 1/2 | 10 13/16 |
| 43 | 44- 3 7/8 | 10 3/4 | 11 1/16 |
| 44 | 45- 4 1/4 | 11 | 11 5/16 |
| 45 | 46- 4 5/8 | 11 1/4 | 11 5/8 |
| 46 | 47- 5 | 11 1/2 | 11 7/8 |
| 47 | 48- 5 3/8 | 11 3/4 | 12 1/8 |
| 48 | 49- 5 3/4 | 12 | 12 3/8 |
| 49 | 50- 6 1/8 | | |
| 50 | 51- 6 7/16 | | |

# 18 1/2 in 12 – 57°

## *Cuts and Saw Settings*

**Tail Cut, Hip or Valley Rafters,** with Square Cut Common Rafters:
Framing Square: **7 3/4 & 24**; set saw at **22 3/4°**
(Compound Miter or Radial Arm Saws: set bevel at 22 3/4°, miter at 18°)
Explanation: Section 3(d), pages 12–13.

**Chamfer Cut, Hip or Valley Rafters:**
Set saw at **36 1/2°**. Explanation: Section 3(d), page 10.

**Fascia Board Miter Cut at Hip or Valley:**
Framing Square: **10 1/16 & 12**; set saw at **22 3/4°**
(Compound Miter or Radial Arm Saws: set bevel at 22 3/4°, miter at 40°)
Explanation: Section 5, pages 24–25.

**Frieze Block against Hip or Valley Rafter:**
Framing Square: **10 1/16 & 12**; set saw at **22 3/4°**
(Compound Miter or Radial Arm Saws: set bevel at 22 3/4°, miter at 40°)
Explanation: Section 6, page 26.

**Roof Sheathing at Hip or Valley:**
Framing Square: **12 & 22 1/16**; set saw at **36 1/2°**
(Compound Miter or Radial Arm Saws: set bevel at 36 1/2°, miter at 28 1/2°)
Distance "d" = **26 1/8"**. Explanation: Section 7, pages 26–27.

## *Calculating the Length of a Rafter*

**General Rule:** Span of building
less Thickness of ridge board (if any)
gives Length of rafter to plumb line of seat cut

*and* Span of building
less Thickness of ridge board (if any)
plus 2 × projection of overhang
gives Total length of rafter to tail cut

Explanation: Section 4, pages 13–19.

| *Jack Rafters* | | | | | |
|---|---|---|---|---|---|
| SPACED *Inches* | DIFFERENCE *Inches* | SPACED *Inches* | DIFFERENCE *Inches* | SPACED *Inches* | DIFFERENCE *Inches* |
| 1/4 | 7/16 | 15 | 27 9/16 | **32** | **58 13/16** |
| 1/2 | 15/16 | **16** | **29 3/8** | 33 | 60 5/8 |
| 3/4 | 1 3/8 | 17 | 31 1/4 | 34 | 62 1/2 |
| 1 | 1 13/16 | 18 | 33 1/16 | 35 | 64 5/16 |
| 2 | 3 11/16 | 19 | 34 15/16 | **36** | **66 1/8** |
| 3 | 5 1/2 | 20 | 36 3/4 | 37 | 68 |
| 4 | 7 3/8 | 21 | 38 9/16 | 38 | 69 13/16 |
| 5 | 9 3/16 | 22 | 40 7/16 | 39 | 71 11/16 |
| 6 | 11 | 23 | 42 1/4 | 40 | 73 1/2 |
| 7 | 12 7/8 | **24** | **44 1/8** | 41 | 75 5/16 |
| 8 | 14 11/16 | 25 | 45 15/16 | 42 | 77 3/16 |
| 9 | 16 9/16 | 26 | 47 3/4 | 43 | 79 |
| 10 | 18 3/8 | 27 | 49 5/8 | 44 | 80 7/8 |
| 11 | 20 3/16 | 28 | 51 7/16 | 45 | 82 11/16 |
| **12** | **22 1/16** | 29 | 53 5/16 | 46 | 84 1/2 |
| 13 | 23 7/8 | 30 | 55 1/8 | 47 | 86 3/8 |
| 14 | 25 3/4 | 31 | 56 15/16 | **48** | **88 3/16** |

| Common Rafters 18 1/2 in 12 – 57° | | | | |
|---|---|---|---|---|
| SPAN Feet | LENGTH Feet Inches | | SPAN Inches | LENGTH Inches |
| 1 | 0-11 | | 1/4 | 1/4 |
| 2 | 1-10 1/16 | | 1/2 | 7/16 |
| 3 | 2- 9 1/16 | | 3/4 | 11/16 |
| 4 | 3- 8 1/8 | | 1 | 15/16 |
| 5 | 4- 7 1/8 | | 1 1/4 | 1 1/8 |
| 6 | 5- 6 1/8 | | 1 1/2 | 1 3/8 |
| 7 | 6- 5 3/16 | | 1 3/4 | 1 5/8 |
| 8 | 7- 4 3/16 | | 2 | 1 13/16 |
| 9 | 8- 3 1/4 | | 2 1/4 | 2 1/16 |
| 10 | 9- 2 1/4 | | 2 1/2 | 2 5/16 |
| 11 | 10- 1 1/4 | | 2 3/4 | 2 1/2 |
| 12 | 11- 0 5/16 | | 3 | 2 3/4 |
| 13 | 11-11 5/16 | | 3 1/4 | 3 |
| 14 | 12-10 3/8 | | 3 1/2 | 3 3/16 |
| 15 | 13- 9 3/8 | | 3 3/4 | 3 7/16 |
| 16 | 14- 8 7/16 | | 4 | 3 11/16 |
| 17 | 15- 7 7/16 | | 4 1/4 | 3 7/8 |
| 18 | 16- 6 7/16 | | 4 1/2 | 4 1/8 |
| 19 | 17- 5 1/2 | | 4 3/4 | 4 3/8 |
| 20 | 18- 4 1/2 | | 5 | 4 5/8 |
| 21 | 19- 3 9/16 | | 5 1/4 | 4 13/16 |
| 22 | 20- 2 9/16 | | 5 1/2 | 5 1/16 |
| 23 | 21- 1 9/16 | | 5 3/4 | 5 5/16 |
| 24 | 22- 0 5/8 | | 6 | 5 1/2 |
| 25 | 22-11 5/8 | | 6 1/4 | 5 3/4 |
| 26 | 23-10 11/16 | | 6 1/2 | 6 |
| 27 | 24- 9 11/16 | | 6 3/4 | 6 3/16 |
| 28 | 25- 8 11/16 | | 7 | 6 7/16 |
| 29 | 26- 7 3/4 | | 7 1/4 | 6 11/16 |
| 30 | 27- 6 3/4 | | 7 1/2 | 6 7/8 |
| 31 | 28- 5 13/16 | | 7 3/4 | 7 1/8 |
| 32 | 29- 4 13/16 | | 8 | 7 3/8 |
| 33 | 30- 3 13/16 | | 8 1/4 | 7 9/16 |
| 34 | 31- 2 7/8 | | 8 1/2 | 7 13/16 |
| 35 | 32- 1 7/8 | | 8 3/4 | 8 1/16 |
| 36 | 33- 0 15/16 | | 9 | 8 1/4 |
| 37 | 33-11 15/16 | | 9 1/4 | 8 1/2 |
| 38 | 34-11 | | 9 1/2 | 8 3/4 |
| 39 | 35-10 | | 9 3/4 | 8 15/16 |
| 40 | 36- 9 | | 10 | 9 3/16 |
| 41 | 37- 8 1/16 | | 10 1/4 | 9 7/16 |
| 42 | 38- 7 1/16 | | 10 1/2 | 9 5/8 |
| 43 | 39- 6 1/8 | | 10 3/4 | 9 7/8 |
| 44 | 40- 5 1/8 | | 11 | 10 1/8 |
| 45 | 41- 4 1/8 | | 11 1/4 | 10 5/16 |
| 46 | 42- 3 3/16 | | 11 1/2 | 10 9/16 |
| 47 | 43- 2 3/16 | | 11 3/4 | 10 13/16 |
| 48 | 44- 1 1/4 | | 12 | 11 |
| 49 | 45- 0 1/4 | | | |
| 50 | 45-11 1/4 | | | |

| Hip or Valley Rafters 18 1/2 in 17 – 47 1/2° | | | | |
|---|---|---|---|---|
| SPAN Feet | LENGTH Feet Inches | | SPAN Inches | LENGTH Inches |
| 1 | 1- 0 9/16 | | 1/4 | 1/4 |
| 2 | 2- 1 1/8 | | 1/2 | 1/2 |
| 3 | 3- 1 11/16 | | 3/4 | 13/16 |
| 4 | 4- 2 3/16 | | 1 | 1 1/16 |
| 5 | 5- 2 3/4 | | 1 1/4 | 1 5/16 |
| 6 | 6- 3 5/16 | | 1 1/2 | 1 9/16 |
| 7 | 7- 3 7/8 | | 1 3/4 | 1 13/16 |
| 8 | 8- 4 7/16 | | 2 | 2 1/16 |
| 9 | 9- 5 | | 2 1/4 | 2 3/8 |
| 10 | 10- 5 1/2 | | 2 1/2 | 2 5/8 |
| 11 | 11- 6 1/16 | | 2 3/4 | 2 7/8 |
| 12 | 12- 6 5/8 | | 3 | 3 1/8 |
| 13 | 13- 7 3/16 | | 3 1/4 | 3 3/8 |
| 14 | 14- 7 3/4 | | 3 1/2 | 3 11/16 |
| 15 | 15- 8 5/16 | | 3 3/4 | 3 15/16 |
| 16 | 16- 8 13/16 | | 4 | 4 3/16 |
| 17 | 17- 9 3/8 | | 4 1/4 | 4 7/16 |
| 18 | 18- 9 15/16 | | 4 1/2 | 4 11/16 |
| 19 | 19-10 1/2 | | 4 3/4 | 4 15/16 |
| 20 | 20-11 1/16 | | 5 | 5 1/4 |
| 21 | 21-11 5/8 | | 5 1/4 | 5 1/2 |
| 22 | 23- 0 1/8 | | 5 1/2 | 5 3/4 |
| 23 | 24- 0 11/16 | | 5 3/4 | 6 |
| 24 | 25- 1 1/4 | | 6 | 6 1/4 |
| 25 | 26- 1 13/16 | | 6 1/4 | 6 9/16 |
| 26 | 27- 2 3/8 | | 6 1/2 | 6 13/16 |
| 27 | 28- 2 15/16 | | 6 3/4 | 7 1/16 |
| 28 | 29- 3 7/16 | | 7 | 7 5/16 |
| 29 | 30- 4 | | 7 1/4 | 7 9/16 |
| 30 | 31- 4 9/16 | | 7 1/2 | 7 7/8 |
| 31 | 32- 5 1/8 | | 7 3/4 | 8 1/8 |
| 32 | 33- 5 11/16 | | 8 | 8 3/8 |
| 33 | 34- 6 1/4 | | 8 1/4 | 8 5/8 |
| 34 | 35- 6 13/16 | | 8 1/2 | 8 7/8 |
| 35 | 36- 7 5/16 | | 8 3/4 | 9 1/8 |
| 36 | 37- 7 7/8 | | 9 | 9 7/16 |
| 37 | 38- 8 7/16 | | 9 1/4 | 9 11/16 |
| 38 | 39- 9 | | 9 1/2 | 9 15/16 |
| 39 | 40- 9 9/16 | | 9 3/4 | 10 3/16 |
| 40 | 41-10 1/8 | | 10 | 10 7/16 |
| 41 | 42-10 5/8 | | 10 1/4 | 10 3/4 |
| 42 | 43-11 3/16 | | 10 1/2 | 11 |
| 43 | 44-11 3/4 | | 10 3/4 | 11 1/4 |
| 44 | 46- 0 5/16 | | 11 | 11 1/2 |
| 45 | 47- 0 7/8 | | 11 1/4 | 11 3/4 |
| 46 | 48- 1 7/16 | | 11 1/2 | 12 |
| 47 | 49- 1 15/16 | | 11 3/4 | 12 5/16 |
| 48 | 50- 2 1/2 | | 12 | 12 9/16 |
| 49 | 51- 3 1/16 | | | |
| 50 | 52- 3 5/8 | | | |

# 19 in 12 – 57 3/4°

## *Cuts and Saw Settings*

**Tail Cut, Hip or Valley Rafters,** with Square Cut Common Rafters:
Framing Square: **7 11/16 & 24**; set saw at **22 1/4°**
(Compound Miter or Radial Arm Saws: set bevel at 22 1/4°, miter at 17 3/4°)
Explanation: Section 3(d), pages 12–13.

**Chamfer Cut, Hip or Valley Rafters:**
Set saw at **36 3/4°**. Explanation: Section 3(d), page 10.

**Fascia Board Miter Cut at Hip or Valley:**
Framing Square: **10 1/8 & 12**; set saw at **22 1/4°**
(Compound Miter or Radial Arm Saws: set bevel at 22 1/4°, miter at 40 1/4°)
Explanation: Section 5, pages 24–25.

**Frieze Block against Hip or Valley Rafter:**
Framing Square: **10 1/8 & 12**; set saw at **22 1/4°**
(Compound Miter or Radial Arm Saws: set bevel at 22 1/4°, miter at 40 1/4°)
Explanation: Section 6, page 26.

**Roof Sheathing at Hip or Valley:**
Framing Square: **12 & 22 1/2**; set saw at **36 3/4°**
(Compound Miter or Radial Arm Saws: set bevel at 36 3/4°, miter at 28°)
Distance "d" = **25 5/8"**. Explanation: Section 7, pages 26–27.

## *Calculating the Length of a Rafter*

**General Rule:** Span of building
less Thickness of ridge board (if any)
gives Length of rafter to plumb line of seat cut

*and* Span of building
less Thickness of ridge board (if any)
plus 2 × projection of overhang
gives Total length of rafter to tail cut

Explanation: Section 4, pages 13–19.

| *Jack Rafters* | | | | | |
|---|---|---|---|---|---|
| SPACED *Inches* | DIFFERENCE *Inches* | SPACED *Inches* | DIFFERENCE *Inches* | SPACED *Inches* | DIFFERENCE *Inches* |
| 1/4 | 7/16 | 15 | 28 1/16 | **32** | **59 15/16** |
| 1/2 | 15/16 | **16** | **29 15/16** | 33 | 61 13/16 |
| 3/4 | 1 3/8 | 17 | 31 13/16 | 34 | 63 11/16 |
| 1 | 1 7/8 | 18 | 33 11/16 | 35 | 65 9/16 |
| 2 | 3 3/4 | 19 | 35 9/16 | **36** | **67 7/16** |
| 3 | 5 5/8 | 20 | 37 7/16 | 37 | 69 5/16 |
| 4 | 7 1/2 | 21 | 39 5/16 | 38 | 71 3/16 |
| 5 | 9 3/8 | 22 | 41 3/16 | 39 | 73 1/16 |
| 6 | 11 1/4 | 23 | 43 1/16 | 40 | 74 15/16 |
| 7 | 13 1/8 | **24** | **44 15/16** | 41 | 76 3/4 |
| 8 | 15 | 25 | 46 13/16 | 42 | 78 5/8 |
| 9 | 16 7/8 | 26 | 48 11/16 | 43 | 80 1/2 |
| 10 | 18 3/4 | 27 | 50 9/16 | 44 | 82 3/8 |
| 11 | 20 5/8 | 28 | 52 7/16 | 45 | 84 1/4 |
| **12** | **22 1/2** | 29 | 54 5/16 | 46 | 86 1/8 |
| 13 | 24 3/8 | 30 | 56 3/16 | 47 | 88 |
| 14 | 26 3/16 | 31 | 58 1/16 | **48** | **89 7/8** |

## Common Rafters
## 19 in 12 – 57 3/4°

| SPAN Feet | LENGTH Feet Inches | SPAN Inches | LENGTH Inches |
|---|---|---|---|
| 1 | 0-11 1/4 | 1/4 | 1/4 |
| 2 | 1-10 1/2 | 1/2 | 7/16 |
| 3 | 2- 9 11/16 | 3/4 | 11/16 |
| 4 | 3- 8 15/16 | 1 | 15/16 |
| 5 | 4- 8 3/16 | 1 1/4 | 1 3/16 |
| 6 | 5- 7 7/16 | 1 1/2 | 1 3/8 |
| 7 | 6- 6 5/8 | 1 3/4 | 1 5/8 |
| 8 | 7- 5 7/8 | 2 | 1 7/8 |
| 9 | 8- 5 1/8 | 2 1/4 | 2 1/8 |
| 10 | 9- 4 3/8 | 2 1/2 | 2 5/16 |
| 11 | 10- 3 5/8 | 2 3/4 | 2 9/16 |
| 12 | 11- 2 13/16 | 3 | 2 13/16 |
| 13 | 12- 2 1/16 | 3 1/4 | 3 1/16 |
| 14 | 13- 1 5/16 | 3 1/2 | 3 1/4 |
| 15 | 14- 0 9/16 | 3 3/4 | 3 1/2 |
| 16 | 14-11 3/4 | 4 | 3 3/4 |
| 17 | 15-11 | 4 1/4 | 4 |
| 18 | 16-10 1/4 | 4 1/2 | 4 3/16 |
| 19 | 17- 9 1/2 | 4 3/4 | 4 7/16 |
| 20 | 18- 8 3/4 | 5 | 4 11/16 |
| 21 | 19- 7 15/16 | 5 1/4 | 4 15/16 |
| 22 | 20- 7 3/16 | 5 1/2 | 5 1/8 |
| 23 | 21- 6 7/16 | 5 3/4 | 5 3/8 |
| 24 | 22- 5 11/16 | 6 | 5 5/8 |
| 25 | 23- 4 7/8 | 6 1/4 | 5 7/8 |
| 26 | 24- 4 1/8 | 6 1/2 | 6 1/16 |
| 27 | 25- 3 3/8 | 6 3/4 | 6 5/16 |
| 28 | 26- 2 5/8 | 7 | 6 9/16 |
| 29 | 27- 1 7/8 | 7 1/4 | 6 13/16 |
| 30 | 28- 1 1/16 | 7 1/2 | 7 |
| 31 | 29- 0 5/16 | 7 3/4 | 7 1/4 |
| 32 | 29-11 9/16 | 8 | 7 1/2 |
| 33 | 30-10 13/16 | 8 1/4 | 7 3/4 |
| 34 | 31-10 | 8 1/2 | 7 15/16 |
| 35 | 32- 9 1/4 | 8 3/4 | 8 3/16 |
| 36 | 33- 8 1/2 | 9 | 8 7/16 |
| 37 | 34- 7 3/4 | 9 1/4 | 8 11/16 |
| 38 | 35- 7 | 9 1/2 | 8 7/8 |
| 39 | 36- 6 3/16 | 9 3/4 | 9 1/8 |
| 40 | 37- 5 7/16 | 10 | 9 3/8 |
| 41 | 38- 4 11/16 | 10 1/4 | 9 5/8 |
| 42 | 39- 3 15/16 | 10 1/2 | 9 13/16 |
| 43 | 40- 3 1/8 | 10 3/4 | 10 1/16 |
| 44 | 41- 2 3/8 | 11 | 10 5/16 |
| 45 | 42- 1 5/8 | 11 1/4 | 10 9/16 |
| 46 | 43- 0 7/8 | 11 1/2 | 10 3/4 |
| 47 | 44- 0 1/8 | 11 3/4 | 11 |
| 48 | 44-11 5/16 | 12 | 11 1/4 |
| 49 | 45-10 9/16 | | |
| 50 | 46- 9 13/16 | | |

## Hip or Valley Rafters
## 19 in 17 – 48 1/4°

| SPAN Feet | LENGTH Feet Inches | SPAN Inches | LENGTH Inches |
|---|---|---|---|
| 1 | 1- 0 3/4 | 1/4 | 1/4 |
| 2 | 2- 1 1/2 | 1/2 | 1/2 |
| 3 | 3- 2 3/16 | 3/4 | 13/16 |
| 4 | 4- 2 15/16 | 1 | 1 1/16 |
| 5 | 5- 3 11/16 | 1 1/4 | 1 5/16 |
| 6 | 6- 4 7/16 | 1 1/2 | 1 9/16 |
| 7 | 7- 5 3/16 | 1 3/4 | 1 7/8 |
| 8 | 8- 5 7/8 | 2 | 2 1/8 |
| 9 | 9- 6 5/8 | 2 1/4 | 2 3/8 |
| 10 | 10- 7 3/8 | 2 1/2 | 2 5/8 |
| 11 | 11- 8 1/8 | 2 3/4 | 2 15/16 |
| 12 | 12- 8 7/8 | 3 | 3 3/16 |
| 13 | 13- 9 9/16 | 3 1/4 | 3 7/16 |
| 14 | 14-10 5/16 | 3 1/2 | 3 11/16 |
| 15 | 15-11 1/16 | 3 3/4 | 4 |
| 16 | 16-11 13/16 | 4 | 4 1/4 |
| 17 | 18- 0 9/16 | 4 1/4 | 4 1/2 |
| 18 | 19- 1 1/4 | 4 1/2 | 4 3/4 |
| 19 | 20- 2 | 4 3/4 | 5 1/16 |
| 20 | 21- 2 3/4 | 5 | 5 5/16 |
| 21 | 22- 3 1/2 | 5 1/4 | 5 9/16 |
| 22 | 23- 4 1/4 | 5 1/2 | 5 13/16 |
| 23 | 24- 4 15/16 | 5 3/4 | 6 1/8 |
| 24 | 25- 5 11/16 | 6 | 6 3/8 |
| 25 | 26- 6 7/16 | 6 1/4 | 6 5/8 |
| 26 | 27- 7 3/16 | 6 1/2 | 6 7/8 |
| 27 | 28- 7 15/16 | 6 3/4 | 7 3/16 |
| 28 | 29- 8 11/16 | 7 | 7 7/16 |
| 29 | 30- 9 3/8 | 7 1/4 | 7 11/16 |
| 30 | 31-10 1/8 | 7 1/2 | 7 15/16 |
| 31 | 32-10 7/8 | 7 3/4 | 8 1/4 |
| 32 | 33-11 5/8 | 8 | 8 1/2 |
| 33 | 35- 0 3/8 | 8 1/4 | 8 3/4 |
| 34 | 36- 1 1/16 | 8 1/2 | 9 |
| 35 | 37- 1 13/16 | 8 3/4 | 9 5/16 |
| 36 | 38- 2 9/16 | 9 | 9 9/16 |
| 37 | 39- 3 5/16 | 9 1/4 | 9 13/16 |
| 38 | 40- 4 1/16 | 9 1/2 | 10 1/16 |
| 39 | 41- 4 3/4 | 9 3/4 | 10 3/8 |
| 40 | 42- 5 1/2 | 10 | 10 5/8 |
| 41 | 43- 6 1/4 | 10 1/4 | 10 7/8 |
| 42 | 44- 7 | 10 1/2 | 11 1/8 |
| 43 | 45- 7 3/4 | 10 3/4 | 11 7/16 |
| 44 | 46- 8 7/16 | 11 | 11 11/16 |
| 45 | 47- 9 3/16 | 11 1/4 | 11 15/16 |
| 46 | 48- 9 15/16 | 11 1/2 | 12 3/16 |
| 47 | 49-10 11/16 | 11 3/4 | 12 1/2 |
| 48 | 50-11 7/16 | 12 | 12 3/4 |
| 49 | 52- 0 1/8 | | |
| 50 | 53- 0 7/8 | | |

# 19½ in 12 – 58½°

## *Cuts and Saw Settings*

**Tail Cut, Hip or Valley Rafters,** with Square Cut Common Rafters:
Framing Square: **7 9/16 & 24**; set saw at **21 3/4°**
(Compound Miter or Radial Arm Saws: set bevel at 21 3/4°, miter at 17 1/2°)
Explanation: Section 3(d), pages 12–13.

**Chamfer Cut, Hip or Valley Rafters:**
Set saw at **37°**. Explanation: Section 3(d), page 10.

**Fascia Board Miter Cut at Hip or Valley:**
Framing Square: **10 1/4 & 12**; set saw at **21 3/4°**
(Compound Miter or Radial Arm Saws: set bevel at 21 3/4°, miter at 40 1/2°)
Explanation: Section 5, pages 24–25.

**Frieze Block against Hip or Valley Rafter:**
Framing Square: **10 1/4 & 12**; set saw at **21 3/4°**
(Compound Miter or Radial Arm Saws: set bevel at 21 3/4°, miter at 40 1/2°)
Explanation: Section 6, page 26.

**Roof Sheathing at Hip or Valley:**
Framing Square: **12 & 22 7/8**; set saw at **37°**
(Compound Miter or Radial Arm Saws: set bevel at 37°, miter at 27 3/4°)
Distance "d" = **25 3/16"**. Explanation: Section 7, pages 26–27.

## *Calculating the Length of a Rafter*

**General Rule:** Span of building
less Thickness of ridge board (if any)
gives Length of rafter to plumb line of seat cut

*and* Span of building
less Thickness of ridge board (if any)
plus 2 × projection of overhang
gives Total length of rafter to tail cut

Explanation: Section 4, pages 13–19.

| *Jack Rafters* | | | | | |
|---|---|---|---|---|---|
| SPACED *Inches* | DIFFERENCE *Inches* | SPACED *Inches* | DIFFERENCE *Inches* | SPACED *Inches* | DIFFERENCE *Inches* |
| 1/4 | 1/2 | 15 | 28 5/8 | **32** | **61 1/16** |
| 1/2 | 15/16 | **16** | **30 1/2** | 33 | 62 15/16 |
| 3/4 | 1 7/16 | 17 | 32 7/16 | 34 | 64 7/8 |
| 1 | 1 15/16 | 18 | 34 3/8 | 35 | 66 13/16 |
| 2 | 3 13/16 | 19 | 36 1/4 | **36** | **68 11/16** |
| 3 | 5 3/4 | 20 | 38 3/16 | 37 | 70 5/8 |
| 4 | 7 5/8 | 21 | 40 1/16 | 38 | 72 1/2 |
| 5 | 9 9/16 | 22 | 42 | 39 | 74 7/16 |
| 6 | 11 7/16 | 23 | 43 7/8 | 40 | 76 5/16 |
| 7 | 13 3/8 | **24** | **45 13/16** | 41 | 78 1/4 |
| 8 | 15 1/4 | 25 | 47 11/16 | 42 | 80 1/8 |
| 9 | 17 3/16 | 26 | 49 5/8 | 43 | 82 1/16 |
| 10 | 19 1/16 | 27 | 51 1/2 | 44 | 83 15/16 |
| 11 | 21 | 28 | 53 7/16 | 45 | 85 7/8 |
| **12** | **22 7/8** | 29 | 55 5/16 | 46 | 87 3/4 |
| 13 | 24 13/16 | 30 | 57 1/4 | 47 | 89 11/16 |
| 14 | 26 11/16 | 31 | 59 1/8 | **48** | **91 9/16** |

| **Common Rafters 19 1/2 in 12 – 58 1/2°** | | | |
|---|---|---|---|
| **SPAN** *Feet* | **LENGTH** *Feet Inches* | **SPAN** *Inches* | **LENGTH** *Inches* |
| 1 | 0-11 7/16 | 1/4 | 1/4 |
| 2 | 1-10 7/8 | 1/2 | 1/2 |
| 3 | 2-10 3/8 | 3/4 | 11/16 |
| 4 | 3- 9 13/16 | 1 | 15/16 |
| 5 | 4- 9 1/4 | 1 1/4 | 1 3/16 |
| 6 | 5- 8 11/16 | 1 1/2 | 1 7/16 |
| 7 | 6- 8 1/8 | 1 3/4 | 1 11/16 |
| 8 | 7- 7 9/16 | 2 | 1 15/16 |
| 9 | 8- 7 1/16 | 2 1/4 | 2 1/8 |
| 10 | 9- 6 1/2 | 2 1/2 | 2 3/8 |
| 11 | 10- 5 15/16 | 2 3/4 | 2 5/8 |
| 12 | 11- 5 3/8 | 3 | 2 7/8 |
| 13 | 12- 4 13/16 | 3 1/4 | 3 1/8 |
| 14 | 13- 4 1/4 | 3 1/2 | 3 5/16 |
| 15 | 14- 3 3/4 | 3 3/4 | 3 9/16 |
| 16 | 15- 3 3/16 | 4 | 3 13/16 |
| 17 | 16- 2 5/8 | 4 1/4 | 4 1/16 |
| 18 | 17- 2 1/16 | 4 1/2 | 4 5/16 |
| 19 | 18- 1 1/2 | 4 3/4 | 4 9/16 |
| 20 | 19- 0 15/16 | 5 | 4 3/4 |
| 21 | 20- 0 7/16 | 5 1/4 | 5 |
| 22 | 20-11 7/8 | 5 1/2 | 5 1/4 |
| 23 | 21-11 5/16 | 5 3/4 | 5 1/2 |
| 24 | 22-10 3/4 | 6 | 5 3/4 |
| 25 | 23-10 3/16 | 6 1/4 | 5 15/16 |
| 26 | 24- 9 5/8 | 6 1/2 | 6 3/16 |
| 27 | 25- 9 1/8 | 6 3/4 | 6 7/16 |
| 28 | 26- 8 9/16 | 7 | 6 11/16 |
| 29 | 27- 8 | 7 1/4 | 6 15/16 |
| 30 | 28- 7 7/16 | 7 1/2 | 7 1/8 |
| 31 | 29- 6 7/8 | 7 3/4 | 7 3/8 |
| 32 | 30- 6 3/8 | 8 | 7 5/8 |
| 33 | 31- 5 13/16 | 8 1/4 | 7 7/8 |
| 34 | 32- 5 1/4 | 8 1/2 | 8 1/8 |
| 35 | 33- 4 11/16 | 8 3/4 | 8 3/8 |
| 36 | 34- 4 1/8 | 9 | 8 9/16 |
| 37 | 35- 3 9/16 | 9 1/4 | 8 13/16 |
| 38 | 36- 3 1/16 | 9 1/2 | 9 1/16 |
| 39 | 37- 2 1/2 | 9 3/4 | 9 5/16 |
| 40 | 38- 1 15/16 | 10 | 9 9/16 |
| 41 | 39- 1 3/8 | 10 1/4 | 9 3/4 |
| 42 | 40- 0 13/16 | 10 1/2 | 10 |
| 43 | 41- 0 1/4 | 10 3/4 | 10 1/4 |
| 44 | 41-11 3/4 | 11 | 10 1/2 |
| 45 | 42-11 3/16 | 11 1/4 | 10 3/4 |
| 46 | 43-10 5/8 | 11 1/2 | 11 |
| 47 | 44-10 1/16 | 11 3/4 | 11 3/16 |
| 48 | 45- 9 1/2 | 12 | 11 7/16 |
| 49 | 46- 8 15/16 | | |
| 50 | 47- 8 7/16 | | |

| **Hip or Valley Rafters 19 1/2 in 17 – 49°** | | | |
|---|---|---|---|
| **SPAN** *Feet* | **LENGTH** *Feet Inches* | **SPAN** *Inches* | **LENGTH** *Inches* |
| 1 | 1- 0 15/16 | 1/4 | 1/4 |
| 2 | 2- 1 7/8 | 1/2 | 9/16 |
| 3 | 3- 2 3/4 | 3/4 | 13/16 |
| 4 | 4- 3 11/16 | 1 | 1 1/16 |
| 5 | 5- 4 5/8 | 1 1/4 | 1 3/8 |
| 6 | 6- 5 9/16 | 1 1/2 | 1 5/8 |
| 7 | 7- 6 1/2 | 1 3/4 | 1 7/8 |
| 8 | 8- 7 3/8 | 2 | 2 1/8 |
| 9 | 9- 8 5/16 | 2 1/4 | 2 7/16 |
| 10 | 10- 9 1/4 | 2 1/2 | 2 11/16 |
| 11 | 11-10 3/16 | 2 3/4 | 2 15/16 |
| 12 | 12-11 1/8 | 3 | 3 1/4 |
| 13 | 14- 0 | 3 1/4 | 3 1/2 |
| 14 | 15- 0 15/16 | 3 1/2 | 3 3/4 |
| 15 | 16- 1 7/8 | 3 3/4 | 4 1/16 |
| 16 | 17- 2 13/16 | 4 | 4 5/16 |
| 17 | 18- 3 3/4 | 4 1/4 | 4 9/16 |
| 18 | 19- 4 5/8 | 4 1/2 | 4 7/8 |
| 19 | 20- 5 9/16 | 4 3/4 | 5 1/8 |
| 20 | 21- 6 1/2 | 5 | 5 3/8 |
| 21 | 22- 7 7/16 | 5 1/4 | 5 5/8 |
| 22 | 23- 8 3/8 | 5 1/2 | 5 15/16 |
| 23 | 24- 9 1/4 | 5 3/4 | 6 3/16 |
| 24 | 25-10 3/16 | 6 | 6 7/16 |
| 25 | 26-11 1/8 | 6 1/4 | 6 3/4 |
| 26 | 28- 0 1/16 | 6 1/2 | 7 |
| 27 | 29- 1 | 6 3/4 | 7 1/4 |
| 28 | 30- 1 15/16 | 7 | 7 9/16 |
| 29 | 31- 2 13/16 | 7 1/4 | 7 13/16 |
| 30 | 32- 3 3/4 | 7 1/2 | 8 1/16 |
| 31 | 33- 4 11/16 | 7 3/4 | 8 3/8 |
| 32 | 34- 5 5/8 | 8 | 8 5/8 |
| 33 | 35- 6 9/16 | 8 1/4 | 8 7/8 |
| 34 | 36- 7 7/16 | 8 1/2 | 9 1/8 |
| 35 | 37- 8 3/8 | 8 3/4 | 9 7/16 |
| 36 | 38- 9 5/16 | 9 | 9 11/16 |
| 37 | 39-10 1/4 | 9 1/4 | 9 15/16 |
| 38 | 40-11 3/16 | 9 1/2 | 10 1/4 |
| 39 | 42- 0 1/16 | 9 3/4 | 10 1/2 |
| 40 | 43- 1 | 10 | 10 3/4 |
| 41 | 44- 1 15/16 | 10 1/4 | 11 1/16 |
| 42 | 45- 2 7/8 | 10 1/2 | 11 5/16 |
| 43 | 46- 3 13/16 | 10 3/4 | 11 9/16 |
| 44 | 47- 4 11/16 | 11 | 11 7/8 |
| 45 | 48- 5 5/8 | 11 1/4 | 12 1/8 |
| 46 | 49- 6 9/16 | 11 1/2 | 12 3/8 |
| 47 | 50- 7 1/2 | 11 3/4 | 12 5/8 |
| 48 | 51- 8 7/16 | 12 | 12 15/16 |
| 49 | 52- 9 5/16 | | |
| 50 | 53-10 1/4 | | |

# 20 in 12 – 59°

## *Cuts and Saw Settings*

**Tail Cut, Hip or Valley Rafters,** with Square Cut Common Rafters:
Framing Square: **7 1/2 & 24**; set saw at **21 1/4°**
(Compound Miter or Radial Arm Saws: set bevel at 21 1/4°, miter at 17 1/4°)
Explanation: Section 3(d), pages 12–13.

**Chamfer Cut, Hip or Valley Rafters:**
Set saw at **37 1/4°**. Explanation: Section 3(d), page 10.

**Fascia Board Miter Cut at Hip or Valley:**
Framing Square: **10 5/16 & 12**; set saw at **21 1/4°**
(Compound Miter or Radial Arm Saws: set bevel at 21 1/4°, miter at 40 1/2°)
Explanation: Section 5, pages 24–25.

**Frieze Block against Hip or Valley Rafter:**
Framing Square: **10 5/16 & 12**; set saw at **21 1/4°**
(Compound Miter or Radial Arm Saws: set bevel at 21 1/4°, miter at 40 1/2°)
Explanation: Section 6, page 26.

**Roof Sheathing at Hip or Valley:**
Framing Square: **12 & 23 5/16**; set saw at **37 1/4°**
(Compound Miter or Radial Arm Saws: set bevel at 37 1/4°, miter at 27 1/4°)
Distance "d" = **24 11/16"**. Explanation: Section 7, pages 26–27.

## *Calculating the Length of a Rafter*

**General Rule:** Span of building
less Thickness of ridge board (if any)
gives Length of rafter to plumb line of seat cut

*and* Span of building
less Thickness of ridge board (if any)
plus 2 × projection of overhang
gives Total length of rafter to tail cut

Explanation: Section 4, pages 13–19.

| *Jack Rafters* | | | | | |
|---|---|---|---|---|---|
| SPACED *Inches* | DIFFERENCE *Inches* | SPACED *Inches* | DIFFERENCE *Inches* | SPACED *Inches* | DIFFERENCE *Inches* |
| 1/4 | 1/2 | 15 | 29 1/8 | **32** | **62 3/16** |
| 1/2 | 1 | **16** | **31 1/8** | 33 | 64 1/8 |
| 3/4 | 1 7/16 | 17 | 33 1/16 | 34 | 66 1/16 |
| 1 | 1 15/16 | 18 | 35 | 35 | 68 |
| 2 | 3 7/8 | 19 | 36 15/16 | **36** | **70** |
| 3 | 5 13/16 | 20 | 38 7/8 | 37 | 71 15/16 |
| 4 | 7 3/4 | 21 | 40 13/16 | 38 | 73 7/8 |
| 5 | 9 11/16 | 22 | 42 3/4 | 39 | 75 13/16 |
| 6 | 11 11/16 | 23 | 44 11/16 | 40 | 77 3/4 |
| 7 | 13 5/8 | **24** | **46 5/8** | 41 | 79 11/16 |
| 8 | 15 9/16 | 25 | 48 9/16 | 42 | 81 5/8 |
| 9 | 17 1/2 | 26 | 50 9/16 | 43 | 83 9/16 |
| 10 | 19 7/16 | 27 | 52 1/2 | 44 | 85 1/2 |
| 11 | 21 3/8 | 28 | 54 7/16 | 45 | 87 7/16 |
| **12** | **23 5/16** | 29 | 56 3/8 | 46 | 89 7/16 |
| 13 | 25 1/4 | 30 | 58 5/16 | 47 | 91 3/8 |
| 14 | 27 3/16 | 31 | 60 1/4 | **48** | **93 5/16** |

## *Common Rafters* 20 in 12 – 59°

| SPAN *Feet* | LENGTH *Feet Inches* | SPAN *Inches* | LENGTH *Inches* |
|---|---|---|---|
| 1 | 0-11 11/16 | 1/4 | 1/4 |
| 2 | 1-11 5/16 | 1/2 | 1/2 |
| 3 | 2-11 | 3/4 | 3/4 |
| 4 | 3-10 5/8 | 1 | 1 |
| 5 | 4-10 5/16 | 1 1/4 | 1 3/16 |
| 6 | 5-10 | 1 1/2 | 1 7/16 |
| 7 | 6- 9 5/8 | 1 3/4 | 1 11/16 |
| 8 | 7- 9 5/16 | 2 | 1 15/16 |
| 9 | 8- 8 15/16 | 2 1/4 | 2 3/16 |
| 10 | 9- 8 5/8 | 2 1/2 | 2 7/16 |
| 11 | 10- 8 1/4 | 2 3/4 | 2 11/16 |
| 12 | 11- 7 15/16 | 3 | 2 15/16 |
| 13 | 12- 7 5/8 | 3 1/4 | 3 3/16 |
| 14 | 13- 7 1/4 | 3 1/2 | 3 3/8 |
| 15 | 14- 6 15/16 | 3 3/4 | 3 5/8 |
| 16 | 15- 6 9/16 | 4 | 3 7/8 |
| 17 | 16- 6 1/4 | 4 1/4 | 4 1/8 |
| 18 | 17- 5 15/16 | 4 1/2 | 4 3/8 |
| 19 | 18- 5 9/16 | 4 3/4 | 4 5/8 |
| 20 | 19- 5 1/4 | 5 | 4 7/8 |
| 21 | 20- 4 7/8 | 5 1/4 | 5 1/8 |
| 22 | 21- 4 9/16 | 5 1/2 | 5 3/8 |
| 23 | 22- 4 1/4 | 5 3/4 | 5 9/16 |
| 24 | 23- 3 7/8 | 6 | 5 13/16 |
| 25 | 24- 3 9/16 | 6 1/4 | 6 1/16 |
| 26 | 25- 3 3/16 | 6 1/2 | 6 5/16 |
| 27 | 26- 2 7/8 | 6 3/4 | 6 9/16 |
| 28 | 27- 2 9/16 | 7 | 6 13/16 |
| 29 | 28- 2 3/16 | 7 1/4 | 7 1/16 |
| 30 | 29- 1 7/8 | 7 1/2 | 7 5/16 |
| 31 | 30- 1 1/2 | 7 3/4 | 7 9/16 |
| 32 | 31- 1 3/16 | 8 | 7 3/4 |
| 33 | 32- 0 13/16 | 8 1/4 | 8 |
| 34 | 33- 0 1/2 | 8 1/2 | 8 1/4 |
| 35 | 34- 0 3/16 | 8 3/4 | 8 1/2 |
| 36 | 34-11 13/16 | 9 | 8 3/4 |
| 37 | 35-11 1/2 | 9 1/4 | 9 |
| 38 | 36-11 1/8 | 9 1/2 | 9 1/4 |
| 39 | 37-10 13/16 | 9 3/4 | 9 1/2 |
| 40 | 38-10 1/2 | 10 | 9 11/16 |
| 41 | 39-10 1/8 | 10 1/4 | 9 15/16 |
| 42 | 40- 9 13/16 | 10 1/2 | 10 3/16 |
| 43 | 41- 9 7/16 | 10 3/4 | 10 7/16 |
| 44 | 42- 9 1/8 | 11 | 10 11/16 |
| 45 | 43- 8 13/16 | 11 1/4 | 10 15/16 |
| 46 | 44- 8 7/16 | 11 1/2 | 11 3/16 |
| 47 | 45- 8 1/8 | 11 3/4 | 11 7/16 |
| 48 | 46- 7 3/4 | 12 | 11 11/16 |
| 49 | 47- 7 7/16 | | |
| 50 | 48- 7 1/8 | | |

## *Hip or Valley Rafters* 20 in 17 – 49 3/4°

| SPAN *Feet* | LENGTH *Feet Inches* | SPAN *Inches* | LENGTH *Inches* |
|---|---|---|---|
| 1 | 1- 1 1/8 | 1/4 | 1/4 |
| 2 | 2- 2 1/4 | 1/2 | 9/16 |
| 3 | 3- 3 3/8 | 3/4 | 13/16 |
| 4 | 4- 4 7/16 | 1 | 1 1/16 |
| 5 | 5- 5 9/16 | 1 1/4 | 1 3/8 |
| 6 | 6- 6 11/16 | 1 1/2 | 1 5/8 |
| 7 | 7- 7 13/16 | 1 3/4 | 1 15/16 |
| 8 | 8- 8 15/16 | 2 | 2 3/16 |
| 9 | 9-10 1/16 | 2 1/4 | 2 7/16 |
| 10 | 10-11 1/8 | 2 1/2 | 2 3/4 |
| 11 | 12- 0 1/4 | 2 3/4 | 3 |
| 12 | 13- 1 3/8 | 3 | 3 1/4 |
| 13 | 14- 2 1/2 | 3 1/4 | 3 9/16 |
| 14 | 15- 3 5/8 | 3 1/2 | 3 13/16 |
| 15 | 16- 4 3/4 | 3 3/4 | 4 1/8 |
| 16 | 17- 5 13/16 | 4 | 4 3/8 |
| 17 | 18- 6 15/16 | 4 1/4 | 4 5/8 |
| 18 | 19- 8 1/16 | 4 1/2 | 4 15/16 |
| 19 | 20- 9 3/16 | 4 3/4 | 5 3/16 |
| 20 | 21-10 5/16 | 5 | 5 7/16 |
| 21 | 22-11 7/16 | 5 1/4 | 5 3/4 |
| 22 | 24- 0 1/2 | 5 1/2 | 6 |
| 23 | 25- 1 5/8 | 5 3/4 | 6 5/16 |
| 24 | 26- 2 3/4 | 6 | 6 9/16 |
| 25 | 27- 3 7/8 | 6 1/4 | 6 13/16 |
| 26 | 28- 5 | 6 1/2 | 7 1/8 |
| 27 | 29- 6 1/8 | 6 3/4 | 7 3/8 |
| 28 | 30- 7 3/16 | 7 | 7 5/8 |
| 29 | 31- 8 5/16 | 7 1/4 | 7 15/16 |
| 30 | 32- 9 7/16 | 7 1/2 | 8 3/16 |
| 31 | 33-10 9/16 | 7 3/4 | 8 1/2 |
| 32 | 34-11 11/16 | 8 | 8 3/4 |
| 33 | 36- 0 13/16 | 8 1/4 | 9 |
| 34 | 37- 1 7/8 | 8 1/2 | 9 5/16 |
| 35 | 38- 3 | 8 3/4 | 9 9/16 |
| 36 | 39- 4 1/8 | 9 | 9 13/16 |
| 37 | 40- 5 1/4 | 9 1/4 | 10 1/8 |
| 38 | 41- 6 3/8 | 9 1/2 | 10 3/8 |
| 39 | 42- 7 1/2 | 9 3/4 | 10 5/8 |
| 40 | 43- 8 5/8 | 10 | 10 15/16 |
| 41 | 44- 9 11/16 | 10 1/4 | 11 3/16 |
| 42 | 45-10 13/16 | 10 1/2 | 11 1/2 |
| 43 | 46-11 15/16 | 10 3/4 | 11 3/4 |
| 44 | 48- 1 1/16 | 11 | 12 |
| 45 | 49- 2 3/16 | 11 1/4 | 12 5/16 |
| 46 | 50- 3 5/16 | 11 1/2 | 12 9/16 |
| 47 | 51- 4 3/8 | 11 3/4 | 12 13/16 |
| 48 | 52- 5 1/2 | 12 | 13 1/8 |
| 49 | 53- 6 5/8 | | |
| 50 | 54- 7 3/4 | | |

# 20 1/2 in 12 – 59 3/4°

## *Cuts and Saw Settings*

**Tail Cut, Hip or Valley Rafters,** with Square Cut Common Rafters:
Framing Square: **7 3/8 & 24**; set saw at **21°**
(Compound Miter or Radial Arm Saws: set bevel at 21°, miter at 17 1/4°)
Explanation: Section 3(d), pages 12–13.

**Chamfer Cut, Hip or Valley Rafters:**
Set saw at **37 1/2°**. Explanation: Section 3(d), page 10.

**Fascia Board Miter Cut at Hip or Valley:**
Framing Square: **10 3/8 & 12**; set saw at **21°**
(Compound Miter or Radial Arm Saws: set bevel at 21°, miter at 40 3/4°)
Explanation: Section 5, pages 24–25.

**Frieze Block against Hip or Valley Rafter:**
Framing Square: **10 3/8 & 12**; set saw at **21°**
(Compound Miter or Radial Arm Saws: set bevel at 21°, miter at 40 3/4°)
Explanation: Section 6, page 26.

**Roof Sheathing at Hip or Valley:**
Framing Square: **12 & 23 3/4**; set saw at **37 1/2°**
(Compound Miter or Radial Arm Saws: set bevel at 37 1/2°, miter at 26 3/4°)
Distance "d" = **24 1/4"**. Explanation: Section 7, pages 26–27.

## *Calculating the Length of a Rafter*

| **General Rule:** | | Span of building |
|---|---|---|
| | less | Thickness of ridge board (if any) |
| | gives | Length of rafter to plumb line of seat cut |
| *and* | | Span of building |
| | less | Thickness of ridge board (if any) |
| | plus | 2 × projection of overhang |
| | gives | Total length of rafter to tail cut |

Explanation: Section 4, pages 13–19.

| *Jack Rafters* | | | | | |
|---|---|---|---|---|---|
| SPACED *Inches* | DIFFERENCE *Inches* | SPACED *Inches* | DIFFERENCE *Inches* | SPACED *Inches* | DIFFERENCE *Inches* |
| 1/4 | 1/2 | 15 | 29 11/16 | **32** | **63 3/8** |
| 1/2 | 1 | **16** | **31 11/16** | 33 | 65 5/16 |
| 3/4 | 1 1/2 | 17 | 33 5/8 | 34 | 67 5/16 |
| 1 | 2 | 18 | 35 5/8 | 35 | 69 5/16 |
| 2 | 3 15/16 | 19 | 37 5/8 | **36** | **71 1/4** |
| 3 | 5 15/16 | 20 | 39 9/16 | 37 | 73 1/4 |
| 4 | 7 15/16 | 21 | 41 9/16 | 38 | 75 1/4 |
| 5 | 9 7/8 | 22 | 43 9/16 | 39 | 77 3/16 |
| 6 | 11 7/8 | 23 | 45 1/2 | 40 | 79 3/16 |
| 7 | 13 7/8 | **24** | **47 1/2** | 41 | 81 3/16 |
| 8 | 15 13/16 | 25 | 49 1/2 | 42 | 83 1/8 |
| 9 | 17 13/16 | 26 | 51 7/16 | 43 | 85 1/8 |
| 10 | 19 13/16 | 27 | 53 7/16 | 44 | 87 1/8 |
| 11 | 21 3/4 | 28 | 55 7/16 | 45 | 89 1/16 |
| **12** | **23 3/4** | 29 | 57 3/8 | 46 | 91 1/16 |
| 13 | 25 3/4 | 30 | 59 3/8 | 47 | 93 1/16 |
| 14 | 27 11/16 | 31 | 61 3/8 | **48** | **95** |

### Common Rafters
### 20 1/2 in 12 – 59 3/4°

| SPAN Feet | LENGTH Feet Inches | SPAN Inches | LENGTH Inches |
|---|---|---|---|
| 1 | 0-11 7/8 | 1/4 | 1/4 |
| 2 | 1-11 3/4 | 1/2 | 1/2 |
| 3 | 2-11 5/8 | 3/4 | 3/4 |
| 4 | 3-11 1/2 | 1 | 1 |
| 5 | 4-11 3/8 | 1 1/4 | 1 1/4 |
| 6 | 5-11 1/4 | 1 1/2 | 1 1/2 |
| 7 | 6-11 1/8 | 1 3/4 | 1 3/4 |
| 8 | 7-11 | 2 | 2 |
| 9 | 8-10 7/8 | 2 1/4 | 2 1/4 |
| 10 | 9-10 3/4 | 2 1/2 | 2 1/2 |
| 11 | 10-10 5/8 | 2 3/4 | 2 3/4 |
| 12 | 11-10 1/2 | 3 | 3 |
| 13 | 12-10 3/8 | 3 1/4 | 3 3/16 |
| 14 | 13-10 1/4 | 3 1/2 | 3 7/16 |
| 15 | 14-10 1/8 | 3 3/4 | 3 11/16 |
| 16 | 15-10 1/16 | 4 | 3 15/16 |
| 17 | 16- 9 15/16 | 4 1/4 | 4 3/16 |
| 18 | 17- 9 13/16 | 4 1/2 | 4 7/16 |
| 19 | 18- 9 11/16 | 4 3/4 | 4 11/16 |
| 20 | 19- 9 9/16 | 5 | 4 15/16 |
| 21 | 20- 9 7/16 | 5 1/4 | 5 3/16 |
| 22 | 21- 9 5/16 | 5 1/2 | 5 7/16 |
| 23 | 22- 9 3/16 | 5 3/4 | 5 11/16 |
| 24 | 23- 9 1/16 | 6 | 5 15/16 |
| 25 | 24- 8 15/16 | 6 1/4 | 6 3/16 |
| 26 | 25- 8 13/16 | 6 1/2 | 6 7/16 |
| 27 | 26- 8 11/16 | 6 3/4 | 6 11/16 |
| 28 | 27- 8 9/16 | 7 | 6 15/16 |
| 29 | 28- 8 7/16 | 7 1/4 | 7 3/16 |
| 30 | 29- 8 5/16 | 7 1/2 | 7 7/16 |
| 31 | 30- 8 3/16 | 7 3/4 | 7 11/16 |
| 32 | 31- 8 1/16 | 8 | 7 15/16 |
| 33 | 32- 7 15/16 | 8 1/4 | 8 3/16 |
| 34 | 33- 7 13/16 | 8 1/2 | 8 7/16 |
| 35 | 34- 7 11/16 | 8 3/4 | 8 11/16 |
| 36 | 35- 7 9/16 | 9 | 8 15/16 |
| 37 | 36- 7 7/16 | 9 1/4 | 9 1/8 |
| 38 | 37- 7 5/16 | 9 1/2 | 9 3/8 |
| 39 | 38- 7 3/16 | 9 3/4 | 9 5/8 |
| 40 | 39- 7 1/16 | 10 | 9 7/8 |
| 41 | 40- 6 15/16 | 10 1/4 | 10 1/8 |
| 42 | 41- 6 13/16 | 10 1/2 | 10 3/8 |
| 43 | 42- 6 11/16 | 10 3/4 | 10 5/8 |
| 44 | 43- 6 9/16 | 11 | 10 7/8 |
| 45 | 44- 6 7/16 | 11 1/4 | 11 1/8 |
| 46 | 45- 6 5/16 | 11 1/2 | 11 3/8 |
| 47 | 46- 6 3/16 | 11 3/4 | 11 5/8 |
| 48 | 47- 6 1/8 | 12 | 11 7/8 |
| 49 | 48- 6 | | |
| 50 | 49- 5 7/8 | | |

### Hip or Valley Rafters
### 20 1/2 in 17 – 50 1/2°

| SPAN Feet | LENGTH Feet Inches | SPAN Inches | LENGTH Inches |
|---|---|---|---|
| 1 | 1- 1 5/16 | 1/4 | 1/4 |
| 2 | 2- 2 5/8 | 1/2 | 9/16 |
| 3 | 3- 3 15/16 | 3/4 | 13/16 |
| 4 | 4- 5 1/4 | 1 | 1 1/8 |
| 5 | 5- 6 9/16 | 1 1/4 | 1 3/8 |
| 6 | 6- 7 13/16 | 1 1/2 | 1 11/16 |
| 7 | 7- 9 1/8 | 1 3/4 | 1 15/16 |
| 8 | 8-10 7/16 | 2 | 2 3/16 |
| 9 | 9-11 3/4 | 2 1/4 | 2 1/2 |
| 10 | 11- 1 1/16 | 2 1/2 | 2 3/4 |
| 11 | 12- 2 3/8 | 2 3/4 | 3 1/16 |
| 12 | 13- 3 11/16 | 3 | 3 5/16 |
| 13 | 14- 5 | 3 1/4 | 3 5/8 |
| 14 | 15- 6 5/16 | 3 1/2 | 3 7/8 |
| 15 | 16- 7 5/8 | 3 3/4 | 4 3/16 |
| 16 | 17- 8 7/8 | 4 | 4 7/16 |
| 17 | 18-10 3/16 | 4 1/4 | 4 11/16 |
| 18 | 19-11 1/2 | 4 1/2 | 5 |
| 19 | 21- 0 13/16 | 4 3/4 | 5 1/4 |
| 20 | 22- 2 1/8 | 5 | 5 9/16 |
| 21 | 23- 3 7/16 | 5 1/4 | 5 13/16 |
| 22 | 24- 4 3/4 | 5 1/2 | 6 1/8 |
| 23 | 25- 6 1/16 | 5 3/4 | 6 3/8 |
| 24 | 26- 7 3/8 | 6 | 6 5/8 |
| 25 | 27- 8 11/16 | 6 1/4 | 6 15/16 |
| 26 | 28- 9 15/16 | 6 1/2 | 7 3/16 |
| 27 | 29-11 1/4 | 6 3/4 | 7 1/2 |
| 28 | 31- 0 9/16 | 7 | 7 3/4 |
| 29 | 32- 1 7/8 | 7 1/4 | 8 1/16 |
| 30 | 33- 3 3/16 | 7 1/2 | 8 5/16 |
| 31 | 34- 4 1/2 | 7 3/4 | 8 5/8 |
| 32 | 35- 5 13/16 | 8 | 8 7/8 |
| 33 | 36- 7 1/8 | 8 1/4 | 9 1/8 |
| 34 | 37- 8 7/16 | 8 1/2 | 9 7/16 |
| 35 | 38- 9 3/4 | 8 3/4 | 9 11/16 |
| 36 | 39-11 1/16 | 9 | 10 |
| 37 | 41- 0 5/16 | 9 1/4 | 10 1/4 |
| 38 | 42- 1 5/8 | 9 1/2 | 10 9/16 |
| 39 | 43- 2 15/16 | 9 3/4 | 10 13/16 |
| 40 | 44- 4 1/4 | 10 | 11 1/16 |
| 41 | 45- 5 9/16 | 10 1/4 | 11 3/8 |
| 42 | 46- 6 7/8 | 10 1/2 | 11 5/8 |
| 43 | 47- 8 3/16 | 10 3/4 | 11 15/16 |
| 44 | 48- 9 1/2 | 11 | 12 3/16 |
| 45 | 49-10 13/16 | 11 1/4 | 12 1/2 |
| 46 | 51- 0 1/8 | 11 1/2 | 12 3/4 |
| 47 | 52- 1 3/8 | 11 3/4 | 13 |
| 48 | 53- 2 11/16 | 12 | 13 5/16 |
| 49 | 54- 4 | | |
| 50 | 55- 5 5/16 | | |

## 21 in 12 – 60 1/4°

### *Cuts and Saw Settings*

**Tail Cut, Hip or Valley Rafters,** with Square Cut Common Rafters:
Framing Square: **7 5/16 & 24**; set saw at **20 1/2°**
(Compound Miter or Radial Arm Saws: set bevel at 20 1/2°, miter at 17°)
Explanation: Section 3(d), pages 12–13.

**Chamfer Cut, Hip or Valley Rafters:**
Set saw at **37 3/4°**. Explanation: Section 3(d), page 10.

**Fascia Board Miter Cut at Hip or Valley:**
Framing Square: **10 7/16 & 12**; set saw at **20 1/2°**
(Compound Miter or Radial Arm Saws: set bevel at 20 1/2°, miter at 41°)
Explanation: Section 5, pages 24–25.

**Frieze Block against Hip or Valley Rafter:**
Framing Square: **10 7/16 & 12**; set saw at **20 1/2°**
(Compound Miter or Radial Arm Saws: set bevel at 20 1/2°, miter at 41°)
Explanation: Section 6, page 26.

**Roof Sheathing at Hip or Valley:**
Framing Square: **6 & 12 1/16**; set saw at **37 3/4°**
(Compound Miter or Radial Arm Saws: set bevel at 37 3/4°, miter at 26 1/2°)
Distance "d" = **23 13/16"**. Explanation: Section 7, pages 26–27.

### *Calculating the Length of a Rafter*

**General Rule:** Span of building
less Thickness of ridge board (if any)
gives Length of rafter to plumb line of seat cut

*and* Span of building
less Thickness of ridge board (if any)
plus 2 × projection of overhang
gives Total length of rafter to tail cut

Explanation: Section 4, pages 13–19.

| *Jack Rafters* | | | | | |
|---|---|---|---|---|---|
| SPACED *Inches* | DIFFERENCE *Inches* | SPACED *Inches* | DIFFERENCE *Inches* | SPACED *Inches* | DIFFERENCE *Inches* |
| 1/4 | 1/2 | 15 | 30 1/4 | **32** | **64 1/2** |
| 1/2 | 1 | **16** | **32 1/4** | 33 | 66 1/2 |
| 3/4 | 1 1/2 | 17 | 34 1/4 | 34 | 68 1/2 |
| 1 | 2 | 18 | 36 1/4 | 35 | 70 9/16 |
| 2 | 4 | 19 | 38 5/16 | **36** | **72 9/16** |
| 3 | 6 1/16 | 20 | 40 5/16 | 37 | 74 9/16 |
| 4 | 8 1/16 | 21 | 42 5/16 | 38 | 76 9/16 |
| 5 | 10 1/16 | 22 | 44 5/16 | 39 | 78 5/8 |
| 6 | 12 1/16 | 23 | 46 3/8 | 40 | 80 5/8 |
| 7 | 14 1/8 | **24** | **48 3/8** | 41 | 82 5/8 |
| 8 | 16 1/8 | 25 | 50 3/8 | 42 | 84 5/8 |
| 9 | 18 1/8 | 26 | 52 3/8 | 43 | 86 11/16 |
| 10 | 20 1/8 | 27 | 54 7/16 | 44 | 88 11/16 |
| 11 | 22 3/16 | 28 | 56 7/16 | 45 | 90 11/16 |
| **12** | **24 3/16** | 29 | 58 7/16 | 46 | 92 11/16 |
| 13 | 26 3/16 | 30 | 60 7/16 | 47 | 94 3/4 |
| 14 | 28 3/16 | 31 | 62 1/2 | **48** | **96 3/4** |

## *Common Rafters*
## 21 in 12 – 60 1/4°

| SPAN *Feet* | LENGTH *Feet Inches* | SPAN *Inches* | LENGTH *Inches* |
|---|---|---|---|
| 1 | 1 - 0 1/16 | 1/4 | 1/4 |
| 2 | 2 - 0 3/16 | 1/2 | 1/2 |
| 3 | 3 - 0 1/4 | 3/4 | 3/4 |
| 4 | 4 - 0 3/8 | 1 | 1 |
| 5 | 5 - 0 7/16 | 1 1/4 | 1 1/4 |
| 6 | 6 - 0 9/16 | 1 1/2 | 1 1/2 |
| 7 | 7 - 0 5/8 | 1 3/4 | 1 3/4 |
| 8 | 8 - 0 3/4 | 2 | 2 |
| 9 | 9 - 0 13/16 | 2 1/4 | 2 1/4 |
| 10 | 10 - 0 15/16 | 2 1/2 | 2 1/2 |
| 11 | 11 - 1 | 2 3/4 | 2 3/4 |
| 12 | 12 - 1 1/8 | 3 | 3 |
| 13 | 13 - 1 3/16 | 3 1/4 | 3 1/4 |
| 14 | 14 - 1 5/16 | 3 1/2 | 3 1/2 |
| 15 | 15 - 1 3/8 | 3 3/4 | 3 3/4 |
| 16 | 16 - 1 1/2 | 4 | 4 |
| 17 | 17 - 1 9/16 | 4 1/4 | 4 5/16 |
| 18 | 18 - 1 11/16 | 4 1/2 | 4 9/16 |
| 19 | 19 - 1 3/4 | 4 3/4 | 4 13/16 |
| 20 | 20 - 1 7/8 | 5 | 5 1/16 |
| 21 | 21 - 1 15/16 | 5 1/4 | 5 5/16 |
| 22 | 22 - 2 1/16 | 5 1/2 | 5 9/16 |
| 23 | 23 - 2 1/8 | 5 3/4 | 5 13/16 |
| 24 | 24 - 2 1/4 | 6 | 6 1/16 |
| 25 | 25 - 2 5/16 | 6 1/4 | 6 5/16 |
| 26 | 26 - 2 7/16 | 6 1/2 | 6 9/16 |
| 27 | 27 - 2 1/2 | 6 3/4 | 6 13/16 |
| 28 | 28 - 2 5/8 | 7 | 7 1/16 |
| 29 | 29 - 2 11/16 | 7 1/4 | 7 5/16 |
| 30 | 30 - 2 13/16 | 7 1/2 | 7 9/16 |
| 31 | 31 - 2 7/8 | 7 3/4 | 7 13/16 |
| 32 | 32 - 3 | 8 | 8 1/16 |
| 33 | 33 - 3 1/16 | 8 1/4 | 8 5/16 |
| 34 | 34 - 3 3/16 | 8 1/2 | 8 9/16 |
| 35 | 35 - 3 1/4 | 8 3/4 | 8 13/16 |
| 36 | 36 - 3 3/8 | 9 | 9 1/16 |
| 37 | 37 - 3 7/16 | 9 1/4 | 9 5/16 |
| 38 | 38 - 3 9/16 | 9 1/2 | 9 9/16 |
| 39 | 39 - 3 5/8 | 9 3/4 | 9 13/16 |
| 40 | 40 - 3 3/4 | 10 | 10 1/16 |
| 41 | 41 - 3 13/16 | 10 1/4 | 10 5/16 |
| 42 | 42 - 3 15/16 | 10 1/2 | 10 9/16 |
| 43 | 43 - 4 | 10 3/4 | 10 13/16 |
| 44 | 44 - 4 1/8 | 11 | 11 1/16 |
| 45 | 45 - 4 3/16 | 11 1/4 | 11 5/16 |
| 46 | 46 - 4 5/16 | 11 1/2 | 11 9/16 |
| 47 | 47 - 4 3/8 | 11 3/4 | 11 13/16 |
| 48 | 48 - 4 1/2 | 12 | 12 1/16 |
| 49 | 49 - 4 9/16 | | |
| 50 | 50 - 4 11/16 | | |

## *Hip or Valley Rafters*
## 21 in 17 – 51°

| SPAN *Feet* | LENGTH *Feet Inches* | SPAN *Inches* | LENGTH *Inches* |
|---|---|---|---|
| 1 | 1 - 1 1/2 | 1/4 | 5/16 |
| 2 | 2 - 3 | 1/2 | 9/16 |
| 3 | 3 - 4 1/2 | 3/4 | 7/8 |
| 4 | 4 - 6 | 1 | 1 1/8 |
| 5 | 5 - 7 1/2 | 1 1/4 | 1 7/16 |
| 6 | 6 - 9 | 1 1/2 | 1 11/16 |
| 7 | 7 - 10 1/2 | 1 3/4 | 2 |
| 8 | 9 - 0 | 2 | 2 1/4 |
| 9 | 10 - 1 1/2 | 2 1/4 | 2 9/16 |
| 10 | 11 - 3 | 2 1/2 | 2 13/16 |
| 11 | 12 - 4 1/2 | 2 3/4 | 3 1/8 |
| 12 | 13 - 6 | 3 | 3 3/8 |
| 13 | 14 - 7 1/2 | 3 1/4 | 3 11/16 |
| 14 | 15 - 9 | 3 1/2 | 3 15/16 |
| 15 | 16 - 10 1/2 | 3 3/4 | 4 1/4 |
| 16 | 18 - 0 | 4 | 4 1/2 |
| 17 | 19 - 1 1/2 | 4 1/4 | 4 13/16 |
| 18 | 20 - 3 | 4 1/2 | 5 1/16 |
| 19 | 21 - 4 1/2 | 4 3/4 | 5 3/8 |
| 20 | 22 - 6 | 5 | 5 5/8 |
| 21 | 23 - 7 1/2 | 5 1/4 | 5 15/16 |
| 22 | 24 - 9 | 5 1/2 | 6 3/16 |
| 23 | 25 - 10 1/2 | 5 3/4 | 6 1/2 |
| 24 | 27 - 0 | 6 | 6 3/4 |
| 25 | 28 - 1 1/2 | 6 1/4 | 7 1/16 |
| 26 | 29 - 3 | 6 1/2 | 7 5/16 |
| 27 | 30 - 4 1/2 | 6 3/4 | 7 5/8 |
| 28 | 31 - 6 | 7 | 7 7/8 |
| 29 | 32 - 7 1/2 | 7 1/4 | 8 3/16 |
| 30 | 33 - 9 | 7 1/2 | 8 7/16 |
| 31 | 34 - 10 1/2 | 7 3/4 | 8 3/4 |
| 32 | 36 - 0 | 8 | 9 |
| 33 | 37 - 1 1/2 | 8 1/4 | 9 5/16 |
| 34 | 38 - 3 | 8 1/2 | 9 9/16 |
| 35 | 39 - 4 1/2 | 8 3/4 | 9 7/8 |
| 36 | 40 - 6 | 9 | 10 1/8 |
| 37 | 41 - 7 1/2 | 9 1/4 | 10 7/16 |
| 38 | 42 - 9 | 9 1/2 | 10 11/16 |
| 39 | 43 - 10 1/2 | 9 3/4 | 11 |
| 40 | 45 - 0 | 10 | 11 1/4 |
| 41 | 46 - 1 1/2 | 10 1/4 | 11 9/16 |
| 42 | 47 - 3 | 10 1/2 | 11 13/16 |
| 43 | 48 - 4 1/2 | 10 3/4 | 12 1/8 |
| 44 | 49 - 6 | 11 | 12 3/8 |
| 45 | 50 - 7 1/2 | 11 1/4 | 12 11/16 |
| 46 | 51 - 9 | 11 1/2 | 12 15/16 |
| 47 | 52 - 10 1/2 | 11 3/4 | 13 1/4 |
| 48 | 54 - 0 | 12 | 13 1/2 |
| 49 | 55 - 1 1/2 | | |
| 50 | 56 - 3 | | |

# 21 1/2 in 12 – 60 3/4°

## *Cuts and Saw Settings*

**Tail Cut, Hip or Valley Rafters,** with Square Cut Common Rafters:
Framing Square: **7 1/4 & 24**; set saw at **20 1/4°**
(Compound Miter or Radial Arm Saws: set bevel at 20 1/4°, miter at 16 3/4°)
Explanation: Section 3(d), pages 12–13.

**Chamfer Cut, Hip or Valley Rafters:**
Set saw at **38 1/4°**. Explanation: Section 3(d), page 10.

**Fascia Board Miter Cut at Hip or Valley:**
Framing Square: **10 1/2 & 12**; set saw at **20 1/4°**
(Compound Miter or Radial Arm Saws: set bevel at 20 1/4°, miter at 41 1/4°)
Explanation: Section 5, pages 24–25.

**Frieze Block against Hip or Valley Rafter:**
Framing Square: **10 1/2 & 12**; set saw at **20 1/4°**
(Compound Miter or Radial Arm Saws: set bevel at 20 1/4°, miter at 41 1/4°)
Explanation: Section 6, page 26.

**Roof Sheathing at Hip or Valley:**
Framing Square: **6 & 12 5/16**; set saw at **38 1/4°**
(Compound Miter or Radial Arm Saws: set bevel at 38 1/4°, miter at 26°)
Distance "d" = **23 3/8"**. Explanation: Section 7, pages 26–27.

## *Calculating the Length of a Rafter*

**General Rule:** Span of building
less Thickness of ridge board (if any)
gives Length of rafter to plumb line of seat cut

*and* Span of building
less Thickness of ridge board (if any)
plus 2 × projection of overhang
gives Total length of rafter to tail cut

Explanation: Section 4, pages 13–19.

### *Jack Rafters*

| SPACED *Inches* | DIFFERENCE *Inches* | SPACED *Inches* | DIFFERENCE *Inches* | SPACED *Inches* | DIFFERENCE *Inches* |
|---|---|---|---|---|---|
| 1/4 | 1/2 | 15 | 30 3/4 | **32** | **65 11/16** |
| 1/2 | 1 | **16** | **32 13/16** | 33 | 67 11/16 |
| 3/4 | 1 9/16 | 17 | 34 7/8 | 34 | 69 3/4 |
| 1 | 2 1/16 | 18 | 36 15/16 | 35 | 71 13/16 |
| 2 | 4 1/8 | 19 | 39 | **36** | **73 7/8** |
| 3 | 6 1/8 | 20 | 41 1/16 | 37 | 75 15/16 |
| 4 | 8 3/16 | 21 | 43 1/16 | 38 | 78 |
| 5 | 10 1/4 | 22 | 45 1/8 | 39 | 80 |
| 6 | 12 5/16 | 23 | 47 3/16 | 40 | 82 1/16 |
| 7 | 14 3/8 | **24** | **49 1/4** | 41 | 84 1/8 |
| 8 | 16 7/16 | 25 | 51 5/16 | 42 | 86 3/16 |
| 9 | 18 7/16 | 26 | 53 3/8 | 43 | 88 1/4 |
| 10 | 20 1/2 | 27 | 55 3/8 | 44 | 90 1/4 |
| 11 | 22 9/16 | 28 | 57 7/16 | 45 | 92 5/16 |
| **12** | **24 5/8** | 29 | 59 1/2 | 46 | 94 3/8 |
| 13 | 26 11/16 | 30 | 61 9/16 | 47 | 96 7/16 |
| 14 | 28 3/4 | 31 | 63 5/8 | **48** | **98 1/2** |

### *Common Rafters*
### 21 1/2 in 12 – 60 3/4°

| SPAN *Feet* | LENGTH *Feet Inches* | SPAN *Inches* | LENGTH *Inches* |
|---|---|---|---|
| 1 | 1- 0 5/16 | 1/4 | 1/4 |
| 2 | 2- 0 5/8 | 1/2 | 1/2 |
| 3 | 3- 0 15/16 | 3/4 | 3/4 |
| 4 | 4- 1 1/4 | 1 | 1 |
| 5 | 5- 1 9/16 | 1 1/4 | 1 5/16 |
| 6 | 6- 1 7/8 | 1 1/2 | 1 9/16 |
| 7 | 7- 2 3/16 | 1 3/4 | 1 13/16 |
| 8 | 8- 2 1/2 | 2 | 2 1/16 |
| 9 | 9- 2 13/16 | 2 1/4 | 2 5/16 |
| 10 | 10- 3 1/8 | 2 1/2 | 2 9/16 |
| 11 | 11- 3 7/16 | 2 3/4 | 2 13/16 |
| 12 | 12- 3 3/4 | 3 | 3 1/16 |
| 13 | 13- 4 1/16 | 3 1/4 | 3 5/16 |
| 14 | 14- 4 3/8 | 3 1/2 | 3 9/16 |
| 15 | 15- 4 11/16 | 3 3/4 | 3 7/8 |
| 16 | 16- 5 | 4 | 4 1/8 |
| 17 | 17- 5 5/16 | 4 1/4 | 4 3/8 |
| 18 | 18- 5 5/8 | 4 1/2 | 4 5/8 |
| 19 | 19- 5 15/16 | 4 3/4 | 4 7/8 |
| 20 | 20- 6 1/4 | 5 | 5 1/8 |
| 21 | 21- 6 9/16 | 5 1/4 | 5 3/8 |
| 22 | 22- 6 13/16 | 5 1/2 | 5 5/8 |
| 23 | 23- 7 1/8 | 5 3/4 | 5 7/8 |
| 24 | 24- 7 7/16 | 6 | 6 1/8 |
| 25 | 25- 7 3/4 | 6 1/4 | 6 7/16 |
| 26 | 26- 8 1/16 | 6 1/2 | 6 11/16 |
| 27 | 27- 8 3/8 | 6 3/4 | 6 15/16 |
| 28 | 28- 8 11/16 | 7 | 7 3/16 |
| 29 | 29- 9 | 7 1/4 | 7 7/16 |
| 30 | 30- 9 5/16 | 7 1/2 | 7 11/16 |
| 31 | 31- 9 5/8 | 7 3/4 | 7 15/16 |
| 32 | 32- 9 15/16 | 8 | 8 3/16 |
| 33 | 33-10 1/4 | 8 1/4 | 8 7/16 |
| 34 | 34-10 9/16 | 8 1/2 | 8 3/4 |
| 35 | 35-10 7/8 | 8 3/4 | 9 |
| 36 | 36-11 3/16 | 9 | 9 1/4 |
| 37 | 37-11 1/2 | 9 1/4 | 9 1/2 |
| 38 | 38-11 13/16 | 9 1/2 | 9 3/4 |
| 39 | 40- 0 1/8 | 9 3/4 | 10 |
| 40 | 41- 0 7/16 | 10 | 10 1/4 |
| 41 | 42- 0 3/4 | 10 1/4 | 10 1/2 |
| 42 | 43- 1 1/16 | 10 1/2 | 10 3/4 |
| 43 | 44- 1 3/8 | 10 3/4 | 11 |
| 44 | 45- 1 11/16 | 11 | 11 5/16 |
| 45 | 46- 2 | 11 1/4 | 11 9/16 |
| 46 | 47- 2 5/16 | 11 1/2 | 11 13/16 |
| 47 | 48- 2 5/8 | 11 3/4 | 12 1/16 |
| 48 | 49- 2 15/16 | 12 | 12 5/16 |
| 49 | 50- 3 1/4 | | |
| 50 | 51- 3 9/16 | | |

### *Hip or Valley Rafters*
### 21 1/2 in 17 – 51 3/4°

| SPAN *Feet* | LENGTH *Feet Inches* | SPAN *Inches* | LENGTH *Inches* |
|---|---|---|---|
| 1 | 1- 1 11/16 | 1/4 | 5/16 |
| 2 | 2- 3 3/8 | 1/2 | 9/16 |
| 3 | 3- 5 1/16 | 3/4 | 7/8 |
| 4 | 4- 6 13/16 | 1 | 1 1/8 |
| 5 | 5- 8 1/2 | 1 1/4 | 1 7/16 |
| 6 | 6-10 3/16 | 1 1/2 | 1 11/16 |
| 7 | 7-11 7/8 | 1 3/4 | 2 |
| 8 | 9- 1 9/16 | 2 | 2 5/16 |
| 9 | 10- 3 1/4 | 2 1/4 | 2 9/16 |
| 10 | 11- 4 15/16 | 2 1/2 | 2 7/8 |
| 11 | 12- 6 5/8 | 2 3/4 | 3 1/8 |
| 12 | 13- 8 3/8 | 3 | 3 7/16 |
| 13 | 14-10 1/16 | 3 1/4 | 3 11/16 |
| 14 | 15-11 3/4 | 3 1/2 | 4 |
| 15 | 17- 1 7/16 | 3 3/4 | 4 1/4 |
| 16 | 18- 3 1/8 | 4 | 4 9/16 |
| 17 | 19- 4 13/16 | 4 1/4 | 4 7/8 |
| 18 | 20- 6 1/2 | 4 1/2 | 5 1/8 |
| 19 | 21- 8 3/16 | 4 3/4 | 5 7/16 |
| 20 | 22- 9 15/16 | 5 | 5 11/16 |
| 21 | 23-11 5/8 | 5 1/4 | 6 |
| 22 | 25- 1 5/16 | 5 1/2 | 6 1/4 |
| 23 | 26- 3 | 5 3/4 | 6 9/16 |
| 24 | 27- 4 11/16 | 6 | 6 7/8 |
| 25 | 28- 6 3/8 | 6 1/4 | 7 1/8 |
| 26 | 29- 8 1/16 | 6 1/2 | 7 7/16 |
| 27 | 30- 9 3/4 | 6 3/4 | 7 11/16 |
| 28 | 31-11 1/2 | 7 | 8 |
| 29 | 33- 1 3/16 | 7 1/4 | 8 1/4 |
| 30 | 34- 2 7/8 | 7 1/2 | 8 9/16 |
| 31 | 35- 4 9/16 | 7 3/4 | 8 7/8 |
| 32 | 36- 6 1/4 | 8 | 9 1/8 |
| 33 | 37- 7 15/16 | 8 1/4 | 9 7/16 |
| 34 | 38- 9 5/8 | 8 1/2 | 9 11/16 |
| 35 | 39-11 5/16 | 8 3/4 | 10 |
| 36 | 41- 1 1/16 | 9 | 10 1/4 |
| 37 | 42- 2 3/4 | 9 1/4 | 10 9/16 |
| 38 | 43- 4 7/16 | 9 1/2 | 10 13/16 |
| 39 | 44- 6 1/8 | 9 3/4 | 11 1/8 |
| 40 | 45- 7 13/16 | 10 | 11 7/16 |
| 41 | 46- 9 1/2 | 10 1/4 | 11 11/16 |
| 42 | 47-11 3/16 | 10 1/2 | 12 |
| 43 | 49- 0 7/8 | 10 3/4 | 12 1/4 |
| 44 | 50- 2 5/8 | 11 | 12 9/16 |
| 45 | 51- 4 5/16 | 11 1/4 | 12 13/16 |
| 46 | 52- 6 | 11 1/2 | 13 1/8 |
| 47 | 53- 7 11/16 | 11 3/4 | 13 7/16 |
| 48 | 54- 9 3/8 | 12 | 13 11/16 |
| 49 | 55-11 1/16 | | |
| 50 | 57- 0 3/4 | | |

## 22 in 12 – 61 1/2°

### *Cuts and Saw Settings*

**Tail Cut, Hip or Valley Rafters,** with Square Cut Common Rafters:
Framing Square: **7 1/8 & 24**; set saw at **19 3/4°**
(Compound Miter or Radial Arm Saws: set bevel at 19 3/4°, miter at 16 1/2°)
Explanation: Section 3(d), pages 12–13.

**Chamfer Cut, Hip or Valley Rafters:**
Set saw at **38 1/4°**. Explanation: Section 3(d), page 10.

**Fascia Board Miter Cut at Hip or Valley:**
Framing Square: **10 9/16 & 12**; set saw at **19 3/4°**
(Compound Miter or Radial Arm Saws: set bevel at 19 3/4°, miter at 41 1/4°)
Explanation: Section 5, pages 24–25.

**Frieze Block against Hip or Valley Rafter:**
Framing Square: **10 9/16 & 12**; set saw at **19 3/4°**
(Compound Miter or Radial Arm Saws: set bevel at 19 3/4°, miter at 41 1/4°)
Explanation: Section 6, page 26.

**Roof Sheathing at Hip or Valley:**
Framing Square: **6 & 12 1/2**; set saw at **38 1/4°**
(Compound Miter or Radial Arm Saws: set bevel at 38 1/4°, miter at 25 1/2°)
Distance "d" = **23"**. Explanation: Section 7, pages 26–27.

### *Calculating the Length of a Rafter*

**General Rule:** Span of building
less Thickness of ridge board (if any)
gives Length of rafter to plumb line of seat cut

*and* Span of building
less Thickness of ridge board (if any)
plus 2 × projection of overhang
gives Total length of rafter to tail cut

Explanation: Section 4, pages 13–19.

| *Jack Rafters* | | | | | |
|---|---|---|---|---|---|
| SPACED *Inches* | DIFFERENCE *Inches* | SPACED *Inches* | DIFFERENCE *Inches* | SPACED *Inches* | DIFFERENCE *Inches* |
| 1/4 | 1/2 | 15 | 31 5/16 | **32** | **66 13/16** |
| 1/2 | 1 1/16 | **16** | **33 7/16** | 33 | 68 15/16 |
| 3/4 | 1 9/16 | 17 | 35 1/2 | 34 | 71 |
| 1 | 2 1/16 | 18 | 37 9/16 | 35 | 73 1/16 |
| 2 | 4 3/16 | 19 | 39 11/16 | **36** | **75 3/16** |
| 3 | 6 1/4 | 20 | 41 3/4 | 37 | 77 1/4 |
| 4 | 8 3/8 | 21 | 43 7/8 | 38 | 79 3/8 |
| 5 | 10 7/16 | 22 | 45 15/16 | 39 | 81 7/16 |
| 6 | 12 1/2 | 23 | 48 1/16 | 40 | 83 9/16 |
| 7 | 14 5/8 | **24** | **50 1/8** | 41 | 85 5/8 |
| 8 | 16 11/16 | 25 | 52 3/16 | 42 | 87 11/16 |
| 9 | 18 13/16 | 26 | 54 5/16 | 43 | 89 13/16 |
| 10 | 20 7/8 | 27 | 56 3/8 | 44 | 91 7/8 |
| 11 | 23 | 28 | 58 1/2 | 45 | 94 |
| **12** | **25 1/16** | 29 | 60 9/16 | 46 | 96 1/16 |
| 13 | 27 1/8 | 30 | 62 5/8 | 47 | 98 1/8 |
| 14 | 29 1/4 | 31 | 64 3/4 | **48** | **100 1/4** |

**Common Rafters**
**22 in 12 – 61 1/2°**

| SPAN Feet | LENGTH Feet Inches | SPAN Inches | LENGTH Inches |
|---|---|---|---|
| 1 | 1- 0 1/2 | 1/4 | 1/4 |
| 2 | 2- 1 1/16 | 1/2 | 1/2 |
| 3 | 3- 1 9/16 | 3/4 | 13/16 |
| 4 | 4- 2 1/8 | 1 | 1 1/16 |
| 5 | 5- 2 5/8 | 1 1/4 | 1 5/16 |
| 6 | 6- 3 3/16 | 1 1/2 | 1 9/16 |
| 7 | 7- 3 11/16 | 1 3/4 | 1 13/16 |
| 8 | 8- 4 1/4 | 2 | 2 1/16 |
| 9 | 9- 4 3/4 | 2 1/4 | 2 3/8 |
| 10 | 10- 5 5/16 | 2 1/2 | 2 5/8 |
| 11 | 11- 5 13/16 | 2 3/4 | 2 7/8 |
| 12 | 12- 6 3/8 | 3 | 3 1/8 |
| 13 | 13- 6 7/8 | 3 1/4 | 3 3/8 |
| 14 | 14- 7 7/16 | 3 1/2 | 3 5/8 |
| 15 | 15- 7 15/16 | 3 3/4 | 3 15/16 |
| 16 | 16- 8 1/2 | 4 | 4 3/16 |
| 17 | 17- 9 | 4 1/4 | 4 7/16 |
| 18 | 18- 9 9/16 | 4 1/2 | 4 11/16 |
| 19 | 19-10 1/16 | 4 3/4 | 4 15/16 |
| 20 | 20-10 5/8 | 5 | 5 1/4 |
| 21 | 21-11 1/8 | 5 1/4 | 5 1/2 |
| 22 | 22-11 11/16 | 5 1/2 | 5 3/4 |
| 23 | 24- 0 3/16 | 5 3/4 | 6 |
| 24 | 25- 0 3/4 | 6 | 6 1/4 |
| 25 | 26- 1 1/4 | 6 1/4 | 6 1/2 |
| 26 | 27- 1 3/4 | 6 1/2 | 6 13/16 |
| 27 | 28- 2 5/16 | 6 3/4 | 7 1/16 |
| 28 | 29- 2 13/16 | 7 | 7 5/16 |
| 29 | 30- 3 3/8 | 7 1/4 | 7 9/16 |
| 30 | 31- 3 7/8 | 7 1/2 | 7 13/16 |
| 31 | 32- 4 7/16 | 7 3/4 | 8 1/16 |
| 32 | 33- 4 15/16 | 8 | 8 3/8 |
| 33 | 34- 5 1/2 | 8 1/4 | 8 5/8 |
| 34 | 35- 6 | 8 1/2 | 8 7/8 |
| 35 | 36- 6 9/16 | 8 3/4 | 9 1/8 |
| 36 | 37- 7 1/16 | 9 | 9 3/8 |
| 37 | 38- 7 5/8 | 9 1/4 | 9 11/16 |
| 38 | 39- 8 1/8 | 9 1/2 | 9 15/16 |
| 39 | 40- 8 11/16 | 9 3/4 | 10 3/16 |
| 40 | 41- 9 3/16 | 10 | 10 7/16 |
| 41 | 42- 9 3/4 | 10 1/4 | 10 11/16 |
| 42 | 43-10 1/4 | 10 1/2 | 10 15/16 |
| 43 | 44-10 13/16 | 10 3/4 | 11 1/4 |
| 44 | 45-11 5/16 | 11 | 11 1/2 |
| 45 | 46-11 7/8 | 11 1/4 | 11 3/4 |
| 46 | 48- 0 3/8 | 11 1/2 | 12 |
| 47 | 49- 0 15/16 | 11 3/4 | 12 1/4 |
| 48 | 50- 1 7/16 | 12 | 12 1/2 |
| 49 | 51- 1 15/16 | | |
| 50 | 52- 2 1/2 | | |

**Hip or Valley Rafters**
**22 in 17 – 52 1/4°**

| SPAN Feet | LENGTH Feet Inches | SPAN Inches | LENGTH Inches |
|---|---|---|---|
| 1 | 1- 1 7/8 | 1/4 | 5/16 |
| 2 | 2- 3 13/16 | 1/2 | 9/16 |
| 3 | 3- 5 11/16 | 3/4 | 7/8 |
| 4 | 4- 7 9/16 | 1 | 1 3/16 |
| 5 | 5- 9 7/16 | 1 1/4 | 1 7/16 |
| 6 | 6-11 3/8 | 1 1/2 | 1 3/4 |
| 7 | 8- 1 1/4 | 1 3/4 | 2 |
| 8 | 9- 3 1/8 | 2 | 2 5/16 |
| 9 | 10- 5 1/16 | 2 1/4 | 2 5/8 |
| 10 | 11- 6 15/16 | 2 1/2 | 2 7/8 |
| 11 | 12- 8 13/16 | 2 3/4 | 3 3/16 |
| 12 | 13-10 11/16 | 3 | 3 1/2 |
| 13 | 15- 0 5/8 | 3 1/4 | 3 3/4 |
| 14 | 16- 2 1/2 | 3 1/2 | 4 1/16 |
| 15 | 17- 4 3/8 | 3 3/4 | 4 5/16 |
| 16 | 18- 6 1/4 | 4 | 4 5/8 |
| 17 | 19- 8 3/16 | 4 1/4 | 4 15/16 |
| 18 | 20-10 1/16 | 4 1/2 | 5 3/16 |
| 19 | 21-11 15/16 | 4 3/4 | 5 1/2 |
| 20 | 23- 1 7/8 | 5 | 5 13/16 |
| 21 | 24- 3 3/4 | 5 1/4 | 6 1/16 |
| 22 | 25- 5 5/8 | 5 1/2 | 6 3/8 |
| 23 | 26- 7 1/2 | 5 3/4 | 6 11/16 |
| 24 | 27- 9 7/16 | 6 | 6 15/16 |
| 25 | 28-11 5/16 | 6 1/4 | 7 1/4 |
| 26 | 30- 1 3/16 | 6 1/2 | 7 1/2 |
| 27 | 31- 3 1/8 | 6 3/4 | 7 13/16 |
| 28 | 32- 5 | 7 | 8 1/8 |
| 29 | 33- 6 7/8 | 7 1/4 | 8 3/8 |
| 30 | 34- 8 3/4 | 7 1/2 | 8 11/16 |
| 31 | 35-10 11/16 | 7 3/4 | 9 |
| 32 | 37- 0 9/16 | 8 | 9 1/4 |
| 33 | 38- 2 7/16 | 8 1/4 | 9 9/16 |
| 34 | 39- 4 5/16 | 8 1/2 | 9 13/16 |
| 35 | 40- 6 1/4 | 8 3/4 | 10 1/8 |
| 36 | 41- 8 1/8 | 9 | 10 7/16 |
| 37 | 42-10 | 9 1/4 | 10 11/16 |
| 38 | 43-11 15/16 | 9 1/2 | 11 |
| 39 | 45- 1 13/16 | 9 3/4 | 11 5/16 |
| 40 | 46- 3 11/16 | 10 | 11 9/16 |
| 41 | 47- 5 9/16 | 10 1/4 | 11 7/8 |
| 42 | 48- 7 1/2 | 10 1/2 | 12 1/8 |
| 43 | 49- 9 3/8 | 10 3/4 | 12 7/16 |
| 44 | 50-11 1/4 | 11 | 12 3/4 |
| 45 | 52- 1 3/16 | 11 1/4 | 13 |
| 46 | 53- 3 1/16 | 11 1/2 | 13 5/16 |
| 47 | 54- 4 15/16 | 11 3/4 | 13 5/8 |
| 48 | 55- 6 13/16 | 12 | 13 7/8 |
| 49 | 56- 8 3/4 | | |
| 50 | 57-10 5/8 | | |

# $22\frac{1}{2}$ in 12 – 62°

## *Cuts and Saw Settings*

**Tail Cut, Hip or Valley Rafters,** with Square Cut Common Rafters:
Framing Square: **7 1/16 & 24**; set saw at **19 1/2°**
(Compound Miter or Radial Arm Saws: set bevel at 19 1/2°, miter at 16 1/4°)
Explanation: Section 3(d), pages 12–13.

**Chamfer Cut, Hip or Valley Rafters:**
Set saw at **38 1/2°**. Explanation: Section 3(d), page 10.

**Fascia Board Miter Cut at Hip or Valley:**
Framing Square: **10 9/16 & 12**; set saw at **19 1/2°**
(Compound Miter or Radial Arm Saws: set bevel at 19 1/2°, miter at 41 1/2°)
Explanation: Section 5, pages 24–25.

**Frieze Block against Hip or Valley Rafter:**
Framing Square: **10 9/16 & 12**; set saw at **19 1/2°**
(Compound Miter or Radial Arm Saws: set bevel at 19 1/2°, miter at 41 1/2°)
Explanation: Section 6, page 26.

**Roof Sheathing at Hip or Valley:**
Framing Square: **6 & 12 3/4**; set saw at **38 1/2°**
(Compound Miter or Radial Arm Saws: set bevel at 38 1/2°, miter at 25 1/4°)
Distance "d" = **22 9/16"**. Explanation: Section 7, pages 26–27.

## *Calculating the Length of a Rafter*

**General Rule:** Span of building
less Thickness of ridge board (if any)
gives Length of rafter to plumb line of seat cut

*and* Span of building
less Thickness of ridge board (if any)
plus 2 × projection of overhang
gives Total length of rafter to tail cut

Explanation: Section 4, pages 13–19.

| *Jack Rafters* | | | | | |
|---|---|---|---|---|---|
| SPACED *Inches* | DIFFERENCE *Inches* | SPACED *Inches* | DIFFERENCE *Inches* | SPACED *Inches* | DIFFERENCE *Inches* |
| 1/4 | 9/16 | 15 | 31 7/8 | **32** | **68** |
| 1/2 | 1 1/16 | **16** | **34** | 33 | 70 1/8 |
| 3/4 | 1 5/8 | 17 | 36 1/8 | 34 | 72 1/4 |
| 1 | 2 1/8 | 18 | 38 1/4 | 35 | 74 3/8 |
| 2 | 4 1/4 | 19 | 40 3/8 | **36** | **76 1/2** |
| 3 | 6 3/8 | 20 | 42 1/2 | 37 | 78 5/8 |
| 4 | 8 1/2 | 21 | 44 5/8 | 38 | 80 3/4 |
| 5 | 10 5/8 | 22 | 46 3/4 | 39 | 82 7/8 |
| 6 | 12 3/4 | 23 | 48 7/8 | 40 | 85 |
| 7 | 14 7/8 | **24** | **51** | 41 | 87 1/8 |
| 8 | 17 | 25 | 53 1/8 | 42 | 89 1/4 |
| 9 | 19 1/8 | 26 | 55 1/4 | 43 | 91 3/8 |
| 10 | 21 1/4 | 27 | 57 3/8 | 44 | 93 1/2 |
| 11 | 23 3/8 | 28 | 59 1/2 | 45 | 95 5/8 |
| **12** | **25 1/2** | 29 | 61 5/8 | 46 | 97 3/4 |
| 13 | 27 5/8 | 30 | 63 3/4 | 47 | 99 7/8 |
| 14 | 29 3/4 | 31 | 65 7/8 | **48** | **102** |

| ***Common Rafters* 22 1/2 in 12 – 62°** | | | |
|---|---|---|---|
| **SPAN** *Feet* | **LENGTH** *Feet Inches* | **SPAN** *Inches* | **LENGTH** *Inches* |
| 1 | 1- 0 3/4 | 1/4 | 1/4 |
| 2 | 2- 1 1/2 | 1/2 | 9/16 |
| 3 | 3- 2 1/4 | 3/4 | 13/16 |
| 4 | 4- 3 | 1 | 1 1/16 |
| 5 | 5- 3 3/4 | 1 1/4 | 1 5/16 |
| 6 | 6- 4 1/2 | 1 1/2 | 1 5/8 |
| 7 | 7- 5 1/4 | 1 3/4 | 1 7/8 |
| 8 | 8- 6 | 2 | 2 1/8 |
| 9 | 9- 6 3/4 | 2 1/4 | 2 3/8 |
| 10 | 10- 7 1/2 | 2 1/2 | 2 11/16 |
| 11 | 11- 8 1/4 | 2 3/4 | 2 15/16 |
| 12 | 12- 9 | 3 | 3 3/16 |
| 13 | 13- 9 3/4 | 3 1/4 | 3 7/16 |
| 14 | 14-10 1/2 | 3 1/2 | 3 3/4 |
| 15 | 15-11 1/4 | 3 3/4 | 4 |
| 16 | 17- 0 | 4 | 4 1/4 |
| 17 | 18- 0 3/4 | 4 1/4 | 4 1/2 |
| 18 | 19- 1 1/2 | 4 1/2 | 4 13/16 |
| 19 | 20- 2 1/4 | 4 3/4 | 5 1/16 |
| 20 | 21- 3 | 5 | 5 5/16 |
| 21 | 22- 3 3/4 | 5 1/4 | 5 9/16 |
| 22 | 23- 4 1/2 | 5 1/2 | 5 7/8 |
| 23 | 24- 5 1/4 | 5 3/4 | 6 1/8 |
| 24 | 25- 6 | 6 | 6 3/8 |
| 25 | 26- 6 3/4 | 6 1/4 | 6 5/8 |
| 26 | 27- 7 1/2 | 6 1/2 | 6 15/16 |
| 27 | 28- 8 1/4 | 6 3/4 | 7 3/16 |
| 28 | 29- 9 | 7 | 7 7/16 |
| 29 | 30- 9 3/4 | 7 1/4 | 7 11/16 |
| 30 | 31-10 1/2 | 7 1/2 | 8 |
| 31 | 32-11 1/4 | 7 3/4 | 8 1/4 |
| 32 | 34- 0 | 8 | 8 1/2 |
| 33 | 35- 0 3/4 | 8 1/4 | 8 3/4 |
| 34 | 36- 1 1/2 | 8 1/2 | 9 1/16 |
| 35 | 37- 2 1/4 | 8 3/4 | 9 5/16 |
| 36 | 38- 3 | 9 | 9 9/16 |
| 37 | 39- 3 3/4 | 9 1/4 | 9 13/16 |
| 38 | 40- 4 1/2 | 9 1/2 | 10 1/8 |
| 39 | 41- 5 1/4 | 9 3/4 | 10 3/8 |
| 40 | 42- 6 | 10 | 10 5/8 |
| 41 | 43- 6 3/4 | 10 1/4 | 10 7/8 |
| 42 | 44- 7 1/2 | 10 1/2 | 11 3/16 |
| 43 | 45- 8 1/4 | 10 3/4 | 11 7/16 |
| 44 | 46- 9 | 11 | 11 11/16 |
| 45 | 47- 9 3/4 | 11 1/4 | 11 15/16 |
| 46 | 48-10 1/2 | 11 1/2 | 12 1/4 |
| 47 | 49-11 1/4 | 11 3/4 | 12 1/2 |
| 48 | 51- 0 | 12 | 12 3/4 |
| 49 | 52- 0 3/4 | | |
| 50 | 53- 1 1/2 | | |

| ***Hip or Valley Rafters* 22 1/2 in 17 – 53°** | | | |
|---|---|---|---|
| **SPAN** *Feet* | **LENGTH** *Feet Inches* | **SPAN** *Inches* | **LENGTH** *Inches* |
| 1 | 1- 2 1/16 | 1/4 | 5/16 |
| 2 | 2- 4 3/16 | 1/2 | 9/16 |
| 3 | 3- 6 1/4 | 3/4 | 7/8 |
| 4 | 4- 8 3/8 | 1 | 1 3/16 |
| 5 | 5-10 7/16 | 1 1/4 | 1 7/16 |
| 6 | 7- 0 9/16 | 1 1/2 | 1 3/4 |
| 7 | 8- 2 5/8 | 1 3/4 | 2 1/16 |
| 8 | 9- 4 3/4 | 2 | 2 3/8 |
| 9 | 10- 6 13/16 | 2 1/4 | 2 5/8 |
| 10 | 11- 8 15/16 | 2 1/2 | 2 15/16 |
| 11 | 12-11 | 2 3/4 | 3 1/4 |
| 12 | 14- 1 1/8 | 3 | 3 1/2 |
| 13 | 15- 3 3/16 | 3 1/4 | 3 13/16 |
| 14 | 16- 5 1/4 | 3 1/2 | 4 1/8 |
| 15 | 17- 7 3/8 | 3 3/4 | 4 3/8 |
| 16 | 18- 9 7/16 | 4 | 4 11/16 |
| 17 | 19-11 9/16 | 4 1/4 | 5 |
| 18 | 21- 1 5/8 | 4 1/2 | 5 5/16 |
| 19 | 22- 3 3/4 | 4 3/4 | 5 9/16 |
| 20 | 23- 5 13/16 | 5 | 5 7/8 |
| 21 | 24- 7 15/16 | 5 1/4 | 6 3/16 |
| 22 | 25-10 | 5 1/2 | 6 7/16 |
| 23 | 27- 0 1/8 | 5 3/4 | 6 3/4 |
| 24 | 28- 2 3/16 | 6 | 7 1/16 |
| 25 | 29- 4 1/4 | 6 1/4 | 7 5/16 |
| 26 | 30- 6 3/8 | 6 1/2 | 7 5/8 |
| 27 | 31- 8 7/16 | 6 3/4 | 7 15/16 |
| 28 | 32-10 9/16 | 7 | 8 1/4 |
| 29 | 34- 0 5/8 | 7 1/4 | 8 1/2 |
| 30 | 35- 2 3/4 | 7 1/2 | 8 13/16 |
| 31 | 36- 4 13/16 | 7 3/4 | 9 1/8 |
| 32 | 37- 6 15/16 | 8 | 9 3/8 |
| 33 | 38- 9 | 8 1/4 | 9 11/16 |
| 34 | 39-11 1/8 | 8 1/2 | 10 |
| 35 | 41- 1 3/16 | 8 3/4 | 10 1/4 |
| 36 | 42- 3 5/16 | 9 | 10 9/16 |
| 37 | 43- 5 3/8 | 9 1/4 | 10 7/8 |
| 38 | 44- 7 7/16 | 9 1/2 | 11 1/8 |
| 39 | 45- 9 9/16 | 9 3/4 | 11 7/16 |
| 40 | 46-11 5/8 | 10 | 11 3/4 |
| 41 | 48- 1 3/4 | 10 1/4 | 12 1/16 |
| 42 | 49- 3 13/16 | 10 1/2 | 12 5/16 |
| 43 | 50- 5 15/16 | 10 3/4 | 12 5/8 |
| 44 | 51- 8 | 11 | 12 15/16 |
| 45 | 52-10 1/8 | 11 1/4 | 13 3/16 |
| 46 | 54- 0 3/16 | 11 1/2 | 13 1/2 |
| 47 | 55- 2 5/16 | 11 3/4 | 13 13/16 |
| 48 | 56- 4 3/8 | 12 | 14 1/16 |
| 49 | 57- 6 1/2 | | |
| 50 | 58- 8 9/16 | | |

# 23 in 12 – 62 1/2°

## *Cuts and Saw Settings*

**Tail Cut, Hip or Valley Rafters,** with Square Cut Common Rafters:
Framing Square: **6 15/16 & 24**; set saw at **19°**
(Compound Miter or Radial Arm Saws: set bevel at 19°, miter at 16 1/4°)
Explanation: Section 3(d), pages 12–13.

**Chamfer Cut, Hip or Valley Rafters:**
Set saw at **38 3/4°**. Explanation: Section 3(d), page 10.

**Fascia Board Miter Cut at Hip or Valley:**
Framing Square: **10 5/8 & 12**; set saw at **19°**
(Compound Miter or Radial Arm Saws: set bevel at 19°, miter at 41 1/2°)
Explanation: Section 5, pages 24–25.

**Frieze Block against Hip or Valley Rafter:**
Framing Square: **10 5/8 & 12**; set saw at **19°**
(Compound Miter or Radial Arm Saws: set bevel at 19°, miter at 41 1/2°)
Explanation: Section 6, page 26.

**Roof Sheathing at Hip or Valley:**
Framing Square: **6 & 13**; set saw at **38 3/4°**
(Compound Miter or Radial Arm Saws: set bevel at 38 3/4°, miter at 24 3/4°)
Distance "d" = **22 3/16"**. Explanation: Section 7, pages 26–27.

## *Calculating the Length of a Rafter*

**General Rule:** Span of building
less Thickness of ridge board (if any)
gives Length of rafter to plumb line of seat cut

*and* Span of building
less Thickness of ridge board (if any)
plus 2 × projection of overhang
gives Total length of rafter to tail cut

Explanation: Section 4, pages 13–19.

| *Jack Rafters* | | | | | |
|---|---|---|---|---|---|
| SPACED *Inches* | DIFFERENCE *Inches* | SPACED *Inches* | DIFFERENCE *Inches* | SPACED *Inches* | DIFFERENCE *Inches* |
| 1/4 | 9/16 | 15 | 32 7/16 | **32** | **69 3/16** |
| 1/2 | 1 1/16 | **16** | **34 9/16** | 33 | 71 5/16 |
| 3/4 | 1 5/8 | 17 | 36 3/4 | 34 | 73 1/2 |
| 1 | 2 3/16 | 18 | 38 15/16 | 35 | 75 11/16 |
| 2 | 4 5/16 | 19 | 41 1/16 | **36** | **77 13/16** |
| 3 | 6 1/2 | 20 | 43 1/4 | 37 | 80 |
| 4 | 8 5/8 | 21 | 45 3/8 | 38 | 82 1/8 |
| 5 | 10 13/16 | 22 | 47 9/16 | 39 | 84 5/16 |
| 6 | 13 | 23 | 49 3/4 | 40 | 86 1/2 |
| 7 | 15 1/8 | **24** | **51 7/8** | 41 | 88 5/8 |
| 8 | 17 5/16 | 25 | 54 1/16 | 42 | 90 13/16 |
| 9 | 19 7/16 | 26 | 56 3/16 | 43 | 92 15/16 |
| 10 | 21 5/8 | 27 | 58 3/8 | 44 | 95 1/8 |
| 11 | 23 3/4 | 28 | 60 9/16 | 45 | 97 5/16 |
| **12** | **25 15/16** | 29 | 62 11/16 | 46 | 99 7/16 |
| 13 | 28 1/8 | 30 | 64 7/8 | 47 | 101 5/8 |
| 14 | 30 1/4 | 31 | 67 | **48** | **103 3/4** |

| ***Common Rafters*** **23 in 12 – 62 1/2°** | | | |
|---|---|---|---|
| **SPAN** *Feet* | **LENGTH** *Feet Inches* | **SPAN** *Inches* | **LENGTH** *Inches* |
| 1 | 1- 1 | 1/4 | 1/4 |
| 2 | 2- 1 15/16 | 1/2 | 9/16 |
| 3 | 3- 2 15/16 | 3/4 | 13/16 |
| 4 | 4- 3 7/8 | 1 | 1 1/16 |
| 5 | 5- 4 7/8 | 1 1/4 | 1 3/8 |
| 6 | 6- 5 13/16 | 1 1/2 | 1 5/8 |
| 7 | 7- 6 13/16 | 1 3/4 | 1 7/8 |
| 8 | 8- 7 3/4 | 2 | 2 3/16 |
| 9 | 9- 8 3/4 | 2 1/4 | 2 7/16 |
| 10 | 10- 9 11/16 | 2 1/2 | 2 11/16 |
| 11 | 11-10 11/16 | 2 3/4 | 3 |
| 12 | 12-11 5/8 | 3 | 3 1/4 |
| 13 | 14- 0 5/8 | 3 1/4 | 3 1/2 |
| 14 | 15- 1 5/8 | 3 1/2 | 3 13/16 |
| 15 | 16- 2 9/16 | 3 3/4 | 4 1/16 |
| 16 | 17- 3 9/16 | 4 | 4 5/16 |
| 17 | 18- 4 1/2 | 4 1/4 | 4 5/8 |
| 18 | 19- 5 1/2 | 4 1/2 | 4 7/8 |
| 19 | 20- 6 7/16 | 4 3/4 | 5 1/8 |
| 20 | 21- 7 7/16 | 5 | 5 3/8 |
| 21 | 22- 8 3/8 | 5 1/4 | 5 11/16 |
| 22 | 23- 9 3/8 | 5 1/2 | 5 15/16 |
| 23 | 24-10 5/16 | 5 3/4 | 6 3/16 |
| 24 | 25-11 5/16 | 6 | 6 1/2 |
| 25 | 27- 0 1/4 | 6 1/4 | 6 3/4 |
| 26 | 28- 1 1/4 | 6 1/2 | 7 |
| 27 | 29- 2 1/4 | 6 3/4 | 7 5/16 |
| 28 | 30- 3 3/16 | 7 | 7 9/16 |
| 29 | 31- 4 3/16 | 7 1/4 | 7 13/16 |
| 30 | 32- 5 1/8 | 7 1/2 | 8 1/8 |
| 31 | 33- 6 1/8 | 7 3/4 | 8 3/8 |
| 32 | 34- 7 1/16 | 8 | 8 5/8 |
| 33 | 35- 8 1/16 | 8 1/4 | 8 15/16 |
| 34 | 36- 9 | 8 1/2 | 9 3/16 |
| 35 | 37-10 | 8 3/4 | 9 7/16 |
| 36 | 38-10 15/16 | 9 | 9 3/4 |
| 37 | 39-11 15/16 | 9 1/4 | 10 |
| 38 | 41- 0 7/8 | 9 1/2 | 10 1/4 |
| 39 | 42- 1 7/8 | 9 3/4 | 10 9/16 |
| 40 | 43- 2 7/8 | 10 | 10 13/16 |
| 41 | 44- 3 13/16 | 10 1/4 | 11 1/16 |
| 42 | 45- 4 13/16 | 10 1/2 | 11 3/8 |
| 43 | 46- 5 3/4 | 10 3/4 | 11 5/8 |
| 44 | 47- 6 3/4 | 11 | 11 7/8 |
| 45 | 48- 7 11/16 | 11 1/4 | 12 3/16 |
| 46 | 49- 8 11/16 | 11 1/2 | 12 7/16 |
| 47 | 50- 9 5/8 | 11 3/4 | 12 11/16 |
| 48 | 51-10 5/8 | 12 | 13 |
| 49 | 52-11 9/16 | | |
| 50 | 54- 0 9/16 | | |

| ***Hip or Valley Rafters*** **23 in 17 – 53 1/2°** | | | |
|---|---|---|---|
| **SPAN** *Feet* | **LENGTH** *Feet Inches* | **SPAN** *Inches* | **LENGTH** *Inches* |
| 1 | 1- 2 5/16 | 1/4 | 5/16 |
| 2 | 2- 4 9/16 | 1/2 | 5/8 |
| 3 | 3- 6 7/8 | 3/4 | 7/8 |
| 4 | 4- 9 3/16 | 1 | 1 3/16 |
| 5 | 5-11 7/16 | 1 1/4 | 1 1/2 |
| 6 | 7- 1 3/4 | 1 1/2 | 1 13/16 |
| 7 | 8- 4 1/16 | 1 3/4 | 2 1/16 |
| 8 | 9- 6 5/16 | 2 | 2 3/8 |
| 9 | 10- 8 5/8 | 2 1/4 | 2 11/16 |
| 10 | 11-10 15/16 | 2 1/2 | 3 |
| 11 | 13- 1 3/16 | 2 3/4 | 3 1/4 |
| 12 | 14- 3 1/2 | 3 | 3 9/16 |
| 13 | 15- 5 13/16 | 3 1/4 | 3 7/8 |
| 14 | 16- 8 1/16 | 3 1/2 | 4 3/16 |
| 15 | 17-10 3/8 | 3 3/4 | 4 7/16 |
| 16 | 19- 0 11/16 | 4 | 4 3/4 |
| 17 | 20- 2 15/16 | 4 1/4 | 5 1/16 |
| 18 | 21- 5 1/4 | 4 1/2 | 5 3/8 |
| 19 | 22- 7 9/16 | 4 3/4 | 5 11/16 |
| 20 | 23- 9 13/16 | 5 | 5 15/16 |
| 21 | 25- 0 1/8 | 5 1/4 | 6 1/4 |
| 22 | 26- 2 7/16 | 5 1/2 | 6 9/16 |
| 23 | 27- 4 11/16 | 5 3/4 | 6 7/8 |
| 24 | 28- 7 | 6 | 7 1/8 |
| 25 | 29- 9 5/16 | 6 1/4 | 7 7/16 |
| 26 | 30-11 9/16 | 6 1/2 | 7 3/4 |
| 27 | 32- 1 7/8 | 6 3/4 | 8 1/16 |
| 28 | 33- 4 3/16 | 7 | 8 5/16 |
| 29 | 34- 6 7/16 | 7 1/4 | 8 5/8 |
| 30 | 35- 8 3/4 | 7 1/2 | 8 15/16 |
| 31 | 36-11 1/16 | 7 3/4 | 9 1/4 |
| 32 | 38- 1 5/16 | 8 | 9 1/2 |
| 33 | 39- 3 5/8 | 8 1/4 | 9 13/16 |
| 34 | 40- 5 15/16 | 8 1/2 | 10 1/8 |
| 35 | 41- 8 3/16 | 8 3/4 | 10 7/16 |
| 36 | 42-10 1/2 | 9 | 10 11/16 |
| 37 | 44- 0 13/16 | 9 1/4 | 11 |
| 38 | 45- 3 1/16 | 9 1/2 | 11 5/16 |
| 39 | 46- 5 3/8 | 9 3/4 | 11 5/8 |
| 40 | 47- 7 11/16 | 10 | 11 15/16 |
| 41 | 48- 9 15/16 | 10 1/4 | 12 3/16 |
| 42 | 50- 0 1/4 | 10 1/2 | 12 1/2 |
| 43 | 51- 2 9/16 | 10 3/4 | 12 13/16 |
| 44 | 52- 4 13/16 | 11 | 13 1/8 |
| 45 | 53- 7 1/8 | 11 1/4 | 13 3/8 |
| 46 | 54- 9 7/16 | 11 1/2 | 13 11/16 |
| 47 | 55-11 11/16 | 11 3/4 | 14 |
| 48 | 57- 2 | 12 | 14 5/16 |
| 49 | 58- 4 5/16 | | |
| 50 | 59- 6 9/16 | | |

# 23 1/2 in 12 – 63°

## *Cuts and Saw Settings*

**Tail Cut, Hip or Valley Rafters,** with Square Cut Common Rafters:
Framing Square: **6 7/8 & 24**; set saw at **18 3/4°**
(Compound Miter or Radial Arm Saws: set bevel at 18 3/4°, miter at 16°)
Explanation: Section 3(d), pages 12–13.

**Chamfer Cut, Hip or Valley Rafters:**
Set saw at **39°**. Explanation: Section 3(d), page 10.

**Fascia Board Miter Cut at Hip or Valley:**
Framing Square: **10 11/16 & 12**; set saw at **18 3/4°**
(Compound Miter or Radial Arm Saws: set bevel at 18 3/4°, miter at 41 3/4°)
Explanation: Section 5, pages 24–25.

**Frieze Block against Hip or Valley Rafter:**
Framing Square: **10 11/16 & 12**; set saw at **18 3/4°**
(Compound Miter or Radial Arm Saws: set bevel at 18 3/4°, miter at 41 3/4°)
Explanation: Section 6, page 26.

**Roof Sheathing at Hip or Valley:**
Framing Square: **6 & 13 3/16**; set saw at **39°**
(Compound Miter or Radial Arm Saws: set bevel at 39°, miter at 24 1/2°)
Distance "d" = **21 13/16"**. Explanation: Section 7, pages 26–27.

## *Calculating the Length of a Rafter*

**General Rule:** Span of building
less Thickness of ridge board (if any)
gives Length of rafter to plumb line of seat cut

*and* Span of building
less Thickness of ridge board (if any)
plus 2 × projection of overhang
gives Total length of rafter to tail cut

Explanation: Section 4, pages 13–19.

*Jack Rafters*

| SPACED *Inches* | DIFFERENCE *Inches* | SPACED *Inches* | DIFFERENCE *Inches* | SPACED *Inches* | DIFFERENCE *Inches* |
|---|---|---|---|---|---|
| 1/4 | 9/16 | 15 | 33 | **32** | **70 3/8** |
| 1/2 | 1 1/8 | **16** | **35 3/16** | 33 | 72 9/16 |
| 3/4 | 1 5/8 | 17 | 37 3/8 | 34 | 74 3/4 |
| 1 | 2 3/16 | 18 | 39 9/16 | 35 | 76 15/16 |
| 2 | 4 3/8 | 19 | 41 3/4 | **36** | **79 3/16** |
| 3 | 6 5/8 | 20 | 44 | 37 | 81 3/8 |
| 4 | 8 13/16 | 21 | 46 3/16 | 38 | 83 9/16 |
| 5 | 11 | 22 | 48 3/8 | 39 | 85 3/4 |
| 6 | 13 3/16 | 23 | 50 9/16 | 40 | 87 15/16 |
| 7 | 15 3/8 | **24** | **52 3/4** | 41 | 90 1/8 |
| 8 | 17 9/16 | 25 | 55 | 42 | 92 3/8 |
| 9 | 19 13/16 | 26 | 57 3/16 | 43 | 94 9/16 |
| 10 | 22 | 27 | 59 3/8 | 44 | 96 3/4 |
| 11 | 24 3/16 | 28 | 61 9/16 | 45 | 98 15/16 |
| **12** | **26 3/8** | 29 | 63 3/4 | 46 | 101 1/8 |
| 13 | 28 9/16 | 30 | 65 15/16 | 47 | 103 3/8 |
| 14 | 30 13/16 | 31 | 68 3/16 | **48** | **105 9/16** |

## *Common Rafters*
## 23 1/2 in 12 – 63°

| SPAN *Feet* | LENGTH *Feet Inches* | SPAN *Inches* | LENGTH *Inches* |
|---|---|---|---|
| 1 | 1- 1 3/16 | 1/4 | 1/4 |
| 2 | 2- 2 3/8 | 1/2 | 9/16 |
| 3 | 3- 3 9/16 | 3/4 | 13/16 |
| 4 | 4- 4 3/4 | 1 | 1 1/8 |
| 5 | 5- 5 15/16 | 1 1/4 | 1 3/8 |
| 6 | 6- 7 3/16 | 1 1/2 | 1 5/8 |
| 7 | 7- 8 3/8 | 1 3/4 | 1 15/16 |
| 8 | 8- 9 9/16 | 2 | 2 3/16 |
| 9 | 9-10 3/4 | 2 1/4 | 2 1/2 |
| 10 | 10-11 15/16 | 2 1/2 | 2 3/4 |
| 11 | 12- 1 1/8 | 2 3/4 | 3 |
| 12 | 13- 2 5/16 | 3 | 3 5/16 |
| 13 | 14- 3 1/2 | 3 1/4 | 3 9/16 |
| 14 | 15- 4 11/16 | 3 1/2 | 3 7/8 |
| 15 | 16- 5 7/8 | 3 3/4 | 4 1/8 |
| 16 | 17- 7 1/16 | 4 | 4 3/8 |
| 17 | 18- 8 5/16 | 4 1/4 | 4 11/16 |
| 18 | 19- 9 1/2 | 4 1/2 | 4 15/16 |
| 19 | 20-10 11/16 | 4 3/4 | 5 1/4 |
| 20 | 21-11 7/8 | 5 | 5 1/2 |
| 21 | 23- 1 1/16 | 5 1/4 | 5 3/4 |
| 22 | 24- 2 1/4 | 5 1/2 | 6 1/16 |
| 23 | 25- 3 7/16 | 5 3/4 | 6 5/16 |
| 24 | 26- 4 5/8 | 6 | 6 5/8 |
| 25 | 27- 5 13/16 | 6 1/4 | 6 7/8 |
| 26 | 28- 7 | 6 1/2 | 7 1/8 |
| 27 | 29- 8 3/16 | 6 3/4 | 7 7/16 |
| 28 | 30- 9 7/16 | 7 | 7 11/16 |
| 29 | 31-10 5/8 | 7 1/4 | 8 |
| 30 | 32-11 13/16 | 7 1/2 | 8 1/4 |
| 31 | 34- 1 | 7 3/4 | 8 1/2 |
| 32 | 35- 2 3/16 | 8 | 8 13/16 |
| 33 | 36- 3 3/8 | 8 1/4 | 9 1/16 |
| 34 | 37- 4 9/16 | 8 1/2 | 9 3/8 |
| 35 | 38- 5 3/4 | 8 3/4 | 9 5/8 |
| 36 | 39- 6 15/16 | 9 | 9 7/8 |
| 37 | 40- 8 1/8 | 9 1/4 | 10 3/16 |
| 38 | 41- 9 3/8 | 9 1/2 | 10 7/16 |
| 39 | 42-10 9/16 | 9 3/4 | 10 3/4 |
| 40 | 43-11 3/4 | 10 | 11 |
| 41 | 45- 0 15/16 | 10 1/4 | 11 1/4 |
| 42 | 46- 2 1/8 | 10 1/2 | 11 9/16 |
| 43 | 47- 3 5/16 | 10 3/4 | 11 13/16 |
| 44 | 48- 4 1/2 | 11 | 12 1/8 |
| 45 | 49- 5 11/16 | 11 1/4 | 12 3/8 |
| 46 | 50- 6 7/8 | 11 1/2 | 12 5/8 |
| 47 | 51- 8 1/16 | 11 3/4 | 12 15/16 |
| 48 | 52- 9 1/4 | 12 | 13 3/16 |
| 49 | 53-10 1/2 | | |
| 50 | 54-11 11/16 | | |

## *Hip or Valley Rafters*
## 23 1/2 in 17 – 54 1/4°

| SPAN *Feet* | LENGTH *Feet Inches* | SPAN *Inches* | LENGTH *Inches* |
|---|---|---|---|
| 1 | 1- 2 1/2 | 1/4 | 5/16 |
| 2 | 2- 5 | 1/2 | 5/8 |
| 3 | 3- 7 1/2 | 3/4 | 7/8 |
| 4 | 4-10 | 1 | 1 3/16 |
| 5 | 6- 0 7/16 | 1 1/4 | 1 1/2 |
| 6 | 7- 2 15/16 | 1 1/2 | 1 13/16 |
| 7 | 8- 5 7/16 | 1 3/4 | 2 1/8 |
| 8 | 9- 7 15/16 | 2 | 2 7/16 |
| 9 | 10-10 7/16 | 2 1/4 | 2 11/16 |
| 10 | 12- 0 15/16 | 2 1/2 | 3 |
| 11 | 13- 3 7/16 | 2 3/4 | 3 5/16 |
| 12 | 14- 5 15/16 | 3 | 3 5/8 |
| 13 | 15- 8 7/16 | 3 1/4 | 3 15/16 |
| 14 | 16-10 15/16 | 3 1/2 | 4 1/4 |
| 15 | 18- 1 3/8 | 3 3/4 | 4 1/2 |
| 16 | 19- 3 7/8 | 4 | 4 13/16 |
| 17 | 20- 6 3/8 | 4 1/4 | 5 1/8 |
| 18 | 21- 8 7/8 | 4 1/2 | 5 7/16 |
| 19 | 22-11 3/8 | 4 3/4 | 5 3/4 |
| 20 | 24- 1 7/8 | 5 | 6 1/16 |
| 21 | 25- 4 3/8 | 5 1/4 | 6 5/16 |
| 22 | 26- 6 7/8 | 5 1/2 | 6 5/8 |
| 23 | 27- 9 3/8 | 5 3/4 | 6 15/16 |
| 24 | 28-11 7/8 | 6 | 7 1/4 |
| 25 | 30- 2 5/16 | 6 1/4 | 7 9/16 |
| 26 | 31- 4 13/16 | 6 1/2 | 7 7/8 |
| 27 | 32- 7 5/16 | 6 3/4 | 8 1/8 |
| 28 | 33- 9 13/16 | 7 | 8 7/16 |
| 29 | 35- 0 5/16 | 7 1/4 | 8 3/4 |
| 30 | 36- 2 13/16 | 7 1/2 | 9 1/16 |
| 31 | 37- 5 5/16 | 7 3/4 | 9 3/8 |
| 32 | 38- 7 13/16 | 8 | 9 11/16 |
| 33 | 39-10 5/16 | 8 1/4 | 9 15/16 |
| 34 | 41- 0 3/4 | 8 1/2 | 10 1/4 |
| 35 | 42- 3 1/4 | 8 3/4 | 10 9/16 |
| 36 | 43- 5 3/4 | 9 | 10 7/8 |
| 37 | 44- 8 1/4 | 9 1/4 | 11 3/16 |
| 38 | 45-10 3/4 | 9 1/2 | 11 1/2 |
| 39 | 47- 1 1/4 | 9 3/4 | 11 3/4 |
| 40 | 48- 3 3/4 | 10 | 12 1/16 |
| 41 | 49- 6 1/4 | 10 1/4 | 12 3/8 |
| 42 | 50- 8 3/4 | 10 1/2 | 12 11/16 |
| 43 | 51-11 1/4 | 10 3/4 | 13 |
| 44 | 53- 1 11/16 | 11 | 13 5/16 |
| 45 | 54- 4 3/16 | 11 1/4 | 13 9/16 |
| 46 | 55- 6 11/16 | 11 1/2 | 13 7/8 |
| 47 | 56- 9 3/16 | 11 3/4 | 14 3/16 |
| 48 | 57-11 11/16 | 12 | 14 1/2 |
| 49 | 59- 2 3/16 | | |
| 50 | 60- 4 11/16 | | |

# 24 in 12 – 63 1/2°

### *Cuts and Saw Settings*

**Tail Cut, Hip or Valley Rafters,** with Square Cut Common Rafters:
Framing Square: **6 13/16 & 24**; set saw at **18 1/2°**
(Compound Miter or Radial Arm Saws: set bevel at 18 1/2°, miter at 15 3/4°)
Explanation: Section 3(d), pages 12–13.

**Chamfer Cut, Hip or Valley Rafters:**
Set saw at **39 1/4°**. Explanation: Section 3(d), page 10.

**Fascia Board Miter Cut at Hip or Valley:**
Framing Square: **10 3/4 & 12**; set saw at **18 1/2°**
(Compound Miter or Radial Arm Saws: set bevel at 18 1/2°, miter at 41 3/4°)
Explanation: Section 5, pages 24–25.

**Frieze Block against Hip or Valley Rafter:**
Framing Square: **10 3/4 & 12**; set saw at **18 1/2°**
(Compound Miter or Radial Arm Saws: set bevel at 18 1/2°, miter at 41 3/4°)
Explanation: Section 6, page 26.

**Roof Sheathing at Hip or Valley:**
Framing Square: **6 & 13 7/16**; set saw at **39 1/4°**
(Compound Miter or Radial Arm Saws: set bevel at 39 1/4°, miter at 24°)
Distance "d" = **21 7/16"**. Explanation: Section 7, pages 26–27.

### *Calculating the Length of a Rafter*

**General Rule:** Span of building
less Thickness of ridge board (if any)
gives Length of rafter to plumb line of seat cut

*and* Span of building
less Thickness of ridge board (if any)
plus 2 × projection of overhang
gives Total length of rafter to tail cut

Explanation: Section 4, pages 13–19.

| *Jack Rafters* | | | | | |
|---|---|---|---|---|---|
| SPACED *Inches* | DIFFERENCE *Inches* | SPACED *Inches* | DIFFERENCE *Inches* | SPACED *Inches* | DIFFERENCE *Inches* |
| 1/4 | 9/16 | 15 | 33 9/16 | **32** | **71 9/16** |
| 1/2 | 1 1/8 | **16** | **35 3/4** | 33 | 73 13/16 |
| 3/4 | 1 11/16 | 17 | 38 | 34 | 76 |
| 1 | 2 1/4 | 18 | 40 1/4 | 35 | 78 1/4 |
| 2 | 4 1/2 | 19 | 42 1/2 | **36** | **80 1/2** |
| 3 | 6 11/16 | 20 | 44 3/4 | 37 | 82 3/4 |
| 4 | 8 15/16 | 21 | 46 15/16 | 38 | 85 |
| 5 | 11 3/16 | 22 | 49 3/16 | 39 | 87 3/16 |
| 6 | 13 7/16 | 23 | 51 7/16 | 40 | 89 7/16 |
| 7 | 15 5/8 | **24** | **53 11/16** | 41 | 91 11/16 |
| 8 | 17 7/8 | 25 | 55 7/8 | 42 | 93 15/16 |
| 9 | 20 1/8 | 26 | 58 1/8 | 43 | 96 1/8 |
| 10 | 22 3/8 | 27 | 60 3/8 | 44 | 98 3/8 |
| 11 | 24 5/8 | 28 | 62 5/8 | 45 | 100 5/8 |
| **12** | **26 13/16** | 29 | 64 7/8 | 46 | 102 7/8 |
| 13 | 29 1/16 | 30 | 67 1/16 | 47 | 105 1/8 |
| 14 | 31 5/16 | 31 | 69 5/16 | **48** | **107 5/16** |

| Common Rafters 24 in 12 – 63 1/2° | | | |
|---|---|---|---|
| SPAN *Feet* | LENGTH *Feet Inches* | SPAN *Inches* | LENGTH *Inches* |
| 1 | 1- 1 7/16 | 1/4 | 1/4 |
| 2 | 2- 2 13/16 | 1/2 | 9/16 |
| 3 | 3- 4 1/4 | 3/4 | 13/16 |
| 4 | 4- 5 11/16 | 1 | 1 1/8 |
| 5 | 5- 7 1/16 | 1 1/4 | 1 3/8 |
| 6 | 6- 8 1/2 | 1 1/2 | 1 11/16 |
| 7 | 7- 9 15/16 | 1 3/4 | 1 15/16 |
| 8 | 8-11 5/16 | 2 | 2 1/4 |
| 9 | 10- 0 3/4 | 2 1/4 | 2 1/2 |
| 10 | 11- 2 3/16 | 2 1/2 | 2 13/16 |
| 11 | 12- 3 9/16 | 2 3/4 | 3 1/16 |
| 12 | 13- 5 | 3 | 3 3/8 |
| 13 | 14- 6 7/16 | 3 1/4 | 3 5/8 |
| 14 | 15- 7 13/16 | 3 1/2 | 3 15/16 |
| 15 | 16- 9 1/4 | 3 3/4 | 4 3/16 |
| 16 | 17-10 11/16 | 4 | 4 1/2 |
| 17 | 19- 0 1/16 | 4 1/4 | 4 3/4 |
| 18 | 20- 1 1/2 | 4 1/2 | 5 |
| 19 | 21- 2 15/16 | 4 3/4 | 5 5/16 |
| 20 | 22- 4 5/16 | 5 | 5 9/16 |
| 21 | 23- 5 3/4 | 5 1/4 | 5 7/8 |
| 22 | 24- 7 3/16 | 5 1/2 | 6 1/8 |
| 23 | 25- 8 9/16 | 5 3/4 | 6 7/16 |
| 24 | 26-10 | 6 | 6 11/16 |
| 25 | 27-11 7/16 | 6 1/4 | 7 |
| 26 | 29- 0 13/16 | 6 1/2 | 7 1/4 |
| 27 | 30- 2 1/4 | 6 3/4 | 7 9/16 |
| 28 | 31- 3 11/16 | 7 | 7 13/16 |
| 29 | 32- 5 1/16 | 7 1/4 | 8 1/8 |
| 30 | 33- 6 1/2 | 7 1/2 | 8 3/8 |
| 31 | 34- 7 15/16 | 7 3/4 | 8 11/16 |
| 32 | 35- 9 5/16 | 8 | 8 15/16 |
| 33 | 36-10 3/4 | 8 1/4 | 9 1/4 |
| 34 | 38- 0 3/16 | 8 1/2 | 9 1/2 |
| 35 | 39- 1 9/16 | 8 3/4 | 9 13/16 |
| 36 | 40- 3 | 9 | 10 1/16 |
| 37 | 41- 4 7/16 | 9 1/4 | 10 5/16 |
| 38 | 42- 5 13/16 | 9 1/2 | 10 5/8 |
| 39 | 43- 7 1/4 | 9 3/4 | 10 7/8 |
| 40 | 44- 8 11/16 | 10 | 11 3/16 |
| 41 | 45-10 1/16 | 10 1/4 | 11 7/16 |
| 42 | 46-11 1/2 | 10 1/2 | 11 3/4 |
| 43 | 48- 0 7/8 | 10 3/4 | 12 |
| 44 | 49- 2 5/16 | 11 | 12 5/16 |
| 45 | 50- 3 3/4 | 11 1/4 | 12 9/16 |
| 46 | 51- 5 1/8 | 11 1/2 | 12 7/8 |
| 47 | 52- 6 9/16 | 11 3/4 | 13 1/8 |
| 48 | 53- 8 | 12 | 13 7/16 |
| 49 | 54- 9 3/8 | | |
| 50 | 55-10 13/16 | | |

| Hip or Valley Rafters 24 in 17 – 54 3/4° | | | |
|---|---|---|---|
| SPAN *Feet* | LENGTH *Feet Inches* | SPAN *Inches* | LENGTH *Inches* |
| 1 | 1- 2 11/16 | 1/4 | 5/16 |
| 2 | 2- 5 3/8 | 1/2 | 5/8 |
| 3 | 3- 8 1/16 | 3/4 | 15/16 |
| 4 | 4-10 13/16 | 1 | 1 1/4 |
| 5 | 6- 1 1/2 | 1 1/4 | 1 1/2 |
| 6 | 7- 4 3/16 | 1 1/2 | 1 13/16 |
| 7 | 8- 6 7/8 | 1 3/4 | 2 1/8 |
| 8 | 9- 9 9/16 | 2 | 2 7/16 |
| 9 | 11- 0 1/4 | 2 1/4 | 2 3/4 |
| 10 | 12- 3 | 2 1/2 | 3 1/16 |
| 11 | 13- 5 11/16 | 2 3/4 | 3 3/8 |
| 12 | 14- 8 3/8 | 3 | 3 11/16 |
| 13 | 15-11 1/16 | 3 1/4 | 4 |
| 14 | 17- 1 3/4 | 3 1/2 | 4 5/16 |
| 15 | 18- 4 7/16 | 3 3/4 | 4 9/16 |
| 16 | 19- 7 1/8 | 4 | 4 7/8 |
| 17 | 20- 9 7/8 | 4 1/4 | 5 3/16 |
| 18 | 22- 0 9/16 | 4 1/2 | 5 1/2 |
| 19 | 23- 3 1/4 | 4 3/4 | 5 13/16 |
| 20 | 24- 5 15/16 | 5 | 6 1/8 |
| 21 | 25- 8 5/8 | 5 1/4 | 6 7/16 |
| 22 | 26-11 5/16 | 5 1/2 | 6 3/4 |
| 23 | 28- 2 | 5 3/4 | 7 1/16 |
| 24 | 29- 4 3/4 | 6 | 7 3/8 |
| 25 | 30- 7 7/16 | 6 1/4 | 7 5/8 |
| 26 | 31-10 1/8 | 6 1/2 | 7 15/16 |
| 27 | 33- 0 13/16 | 6 3/4 | 8 1/4 |
| 28 | 34- 3 1/2 | 7 | 8 9/16 |
| 29 | 35- 6 3/16 | 7 1/4 | 8 7/8 |
| 30 | 36- 8 15/16 | 7 1/2 | 9 3/16 |
| 31 | 37-11 5/8 | 7 3/4 | 9 1/2 |
| 32 | 39- 2 5/16 | 8 | 9 13/16 |
| 33 | 40- 5 | 8 1/4 | 10 1/8 |
| 34 | 41- 7 11/16 | 8 1/2 | 10 7/16 |
| 35 | 42-10 3/8 | 8 3/4 | 10 11/16 |
| 36 | 44- 1 1/16 | 9 | 11 |
| 37 | 45- 3 13/16 | 9 1/4 | 11 5/16 |
| 38 | 46- 6 1/2 | 9 1/2 | 11 5/8 |
| 39 | 47- 9 3/16 | 9 3/4 | 11 15/16 |
| 40 | 48-11 7/8 | 10 | 12 1/4 |
| 41 | 50- 2 9/16 | 10 1/4 | 12 9/16 |
| 42 | 51- 5 1/4 | 10 1/2 | 12 7/8 |
| 43 | 52- 7 15/16 | 10 3/4 | 13 3/16 |
| 44 | 53-10 11/16 | 11 | 13 1/2 |
| 45 | 55- 1 3/8 | 11 1/4 | 13 3/4 |
| 46 | 56- 4 1/16 | 11 1/2 | 14 1/16 |
| 47 | 57- 6 3/4 | 11 3/4 | 14 3/8 |
| 48 | 58- 9 7/16 | 12 | 14 11/16 |
| 49 | 60- 0 1/8 | | |
| 50 | 61- 2 7/8 | | |

# 24½ in 12 – 64°

## *Cuts and Saw Settings*

**Tail Cut, Hip or Valley Rafters,** with Square Cut Common Rafters:
Framing Square: **6 11/16 & 24**; set saw at **18°**
(Compound Miter or Radial Arm Saws: set bevel at 18°, miter at 15 1/2°)
Explanation: Section 3(d), pages 12–13.

**Chamfer Cut, Hip or Valley Rafters:**
Set saw at **39 1/2°**. Explanation: Section 3(d), page 10.

**Fascia Board Miter Cut at Hip or Valley:**
Framing Square: **10 3/4 & 12**; set saw at **18°**
(Compound Miter or Radial Arm Saws: set bevel at 18°, miter at 42°)
Explanation: Section 5, pages 24–25.

**Frieze Block against Hip or Valley Rafter:**
Framing Square: **10 3/4 & 12**; set saw at **18°**
(Compound Miter or Radial Arm Saws: set bevel at 18°, miter at 42°)
Explanation: Section 6, page 26.

**Roof Sheathing at Hip or Valley:**
Framing Square: **6 & 13 5/8**; set saw at **39 1/2°**
(Compound Miter or Radial Arm Saws: set bevel at 39 1/2°, miter at 23 3/4°)
Distance "d" = **21 1/8"**. Explanation: Section 7, pages 26–27.

## *Calculating the Length of a Rafter*

**General Rule:** Span of building
less Thickness of ridge board (if any)
gives Length of rafter to plumb line of seat cut

*and* Span of building
less Thickness of ridge board (if any)
plus 2 × projection of overhang
gives Total length of rafter to tail cut

Explanation: Section 4, pages 13–19.

| *Jack Rafters* | | | | | |
|---|---|---|---|---|---|
| SPACED *Inches* | DIFFERENCE *Inches* | SPACED *Inches* | DIFFERENCE *Inches* | SPACED *Inches* | DIFFERENCE *Inches* |
| 1/4 | 9/16 | 15 | 34 1/8 | **32** | **72 3/4** |
| 1/2 | 1 1/8 | **16** | **36 3/8** | 33 | 75 |
| 3/4 | 1 11/16 | 17 | 38 5/8 | 34 | 77 5/16 |
| 1 | 2 1/4 | 18 | 40 15/16 | 35 | 79 9/16 |
| 2 | 4 9/16 | 19 | 43 3/16 | **36** | **81 13/16** |
| 3 | 6 13/16 | 20 | 45 7/16 | 37 | 84 1/8 |
| 4 | 9 1/16 | 21 | 47 3/4 | 38 | 86 3/8 |
| 5 | 11 3/8 | 22 | 50 | 39 | 88 11/16 |
| 6 | 13 5/8 | 23 | 52 5/16 | 40 | 90 15/16 |
| 7 | 15 15/16 | **24** | **54 9/16** | 41 | 93 3/16 |
| 8 | 18 3/16 | 25 | 56 13/16 | 42 | 95 1/2 |
| 9 | 20 7/16 | 26 | 59 1/8 | 43 | 97 3/4 |
| 10 | 22 3/4 | 27 | 61 3/8 | 44 | 100 |
| 11 | 25 | 28 | 63 5/8 | 45 | 102 5/16 |
| **12** | **27 1/4** | 29 | 65 15/16 | 46 | 104 9/16 |
| 13 | 29 9/16 | 30 | 68 3/16 | 47 | 106 7/8 |
| 14 | 31 13/16 | 31 | 70 1/2 | **48** | **109 1/8** |

## Common Rafters
## 24 1/2 in 12 – 64°

| SPAN Feet | LENGTH Feet Inches | SPAN Inches | LENGTH Inches |
|---|---|---|---|
| 1 | 1- 1 5/8 | 1/4 | 5/16 |
| 2 | 2- 3 1/4 | 1/2 | 9/16 |
| 3 | 3- 4 15/16 | 3/4 | 7/8 |
| 4 | 4- 6 9/16 | 1 | 1 1/8 |
| 5 | 5- 8 3/16 | 1 1/4 | 1 7/16 |
| 6 | 6- 9 13/16 | 1 1/2 | 1 11/16 |
| 7 | 7-11 1/2 | 1 3/4 | 2 |
| 8 | 9- 1 1/8 | 2 | 2 1/4 |
| 9 | 10- 2 3/4 | 2 1/4 | 2 9/16 |
| 10 | 11- 4 3/8 | 2 1/2 | 2 13/16 |
| 11 | 12- 6 1/16 | 2 3/4 | 3 1/8 |
| 12 | 13- 7 11/16 | 3 | 3 7/16 |
| 13 | 14- 9 5/16 | 3 1/4 | 3 11/16 |
| 14 | 15-10 15/16 | 3 1/2 | 4 |
| 15 | 17- 0 5/8 | 3 3/4 | 4 1/4 |
| 16 | 18- 2 1/4 | 4 | 4 9/16 |
| 17 | 19- 3 7/8 | 4 1/4 | 4 13/16 |
| 18 | 20- 5 1/2 | 4 1/2 | 5 1/8 |
| 19 | 21- 7 3/16 | 4 3/4 | 5 3/8 |
| 20 | 22- 8 13/16 | 5 | 5 11/16 |
| 21 | 23-10 7/16 | 5 1/4 | 5 15/16 |
| 22 | 25- 0 1/16 | 5 1/2 | 6 1/4 |
| 23 | 26- 1 3/4 | 5 3/4 | 6 9/16 |
| 24 | 27- 3 3/8 | 6 | 6 13/16 |
| 25 | 28- 5 | 6 1/4 | 7 1/8 |
| 26 | 29- 6 5/8 | 6 1/2 | 7 3/8 |
| 27 | 30- 8 5/16 | 6 3/4 | 7 11/16 |
| 28 | 31- 9 15/16 | 7 | 7 15/16 |
| 29 | 32-11 9/16 | 7 1/4 | 8 1/4 |
| 30 | 34- 1 3/16 | 7 1/2 | 8 1/2 |
| 31 | 35- 2 7/8 | 7 3/4 | 8 13/16 |
| 32 | 36- 4 1/2 | 8 | 9 1/16 |
| 33 | 37- 6 1/8 | 8 1/4 | 9 3/8 |
| 34 | 38- 7 3/4 | 8 1/2 | 9 11/16 |
| 35 | 39- 9 7/16 | 8 3/4 | 9 15/16 |
| 36 | 40-11 1/16 | 9 | 10 1/4 |
| 37 | 42- 0 11/16 | 9 1/4 | 10 1/2 |
| 38 | 43- 2 5/16 | 9 1/2 | 10 13/16 |
| 39 | 44- 4 | 9 3/4 | 11 1/16 |
| 40 | 45- 5 5/8 | 10 | 11 3/8 |
| 41 | 46- 7 1/4 | 10 1/4 | 11 5/8 |
| 42 | 47- 8 7/8 | 10 1/2 | 11 15/16 |
| 43 | 48-10 9/16 | 10 3/4 | 12 1/4 |
| 44 | 50- 0 3/16 | 11 | 12 1/2 |
| 45 | 51- 1 13/16 | 11 1/4 | 12 13/16 |
| 46 | 52- 3 7/16 | 11 1/2 | 13 1/16 |
| 47 | 53- 5 1/8 | 11 3/4 | 13 3/8 |
| 48 | 54- 6 3/4 | 12 | 13 5/8 |
| 49 | 55- 8 3/8 | | |
| 50 | 56-10 | | |

## Hip or Valley Rafters
## 24 1/2 in 17 – 55 1/4°

| SPAN Feet | LENGTH Feet Inches | SPAN Inches | LENGTH Inches |
|---|---|---|---|
| 1 | 1- 2 7/8 | 1/4 | 5/16 |
| 2 | 2- 5 13/16 | 1/2 | 5/8 |
| 3 | 3- 8 11/16 | 3/4 | 15/16 |
| 4 | 4-11 5/8 | 1 | 1 1/4 |
| 5 | 6- 2 1/2 | 1 1/4 | 1 9/16 |
| 6 | 7- 5 7/16 | 1 1/2 | 1 7/8 |
| 7 | 8- 8 5/16 | 1 3/4 | 2 3/16 |
| 8 | 9-11 3/16 | 2 | 2 1/2 |
| 9 | 11- 2 1/8 | 2 1/4 | 2 13/16 |
| 10 | 12- 5 | 2 1/2 | 3 1/8 |
| 11 | 13- 7 15/16 | 2 3/4 | 3 7/16 |
| 12 | 14-10 13/16 | 3 | 3 3/4 |
| 13 | 16- 1 3/4 | 3 1/4 | 4 1/16 |
| 14 | 17- 4 5/8 | 3 1/2 | 4 3/8 |
| 15 | 18- 7 1/2 | 3 3/4 | 4 11/16 |
| 16 | 19-10 7/16 | 4 | 4 15/16 |
| 17 | 21- 1 5/16 | 4 1/4 | 5 1/4 |
| 18 | 22- 4 1/4 | 4 1/2 | 5 9/16 |
| 19 | 23- 7 1/8 | 4 3/4 | 5 7/8 |
| 20 | 24-10 1/16 | 5 | 6 3/16 |
| 21 | 26- 0 15/16 | 5 1/4 | 6 1/2 |
| 22 | 27- 3 13/16 | 5 1/2 | 6 13/16 |
| 23 | 28- 6 3/4 | 5 3/4 | 7 1/8 |
| 24 | 29- 9 5/8 | 6 | 7 7/16 |
| 25 | 31- 0 9/16 | 6 1/4 | 7 3/4 |
| 26 | 32- 3 7/16 | 6 1/2 | 8 1/16 |
| 27 | 33- 6 3/8 | 6 3/4 | 8 3/8 |
| 28 | 34- 9 1/4 | 7 | 8 11/16 |
| 29 | 36- 0 1/8 | 7 1/4 | 9 |
| 30 | 37- 3 1/16 | 7 1/2 | 9 5/16 |
| 31 | 38- 5 15/16 | 7 3/4 | 9 5/8 |
| 32 | 39- 8 7/8 | 8 | 9 15/16 |
| 33 | 40-11 3/4 | 8 1/4 | 10 1/4 |
| 34 | 42- 2 11/16 | 8 1/2 | 10 9/16 |
| 35 | 43- 5 9/16 | 8 3/4 | 10 7/8 |
| 36 | 44- 8 7/16 | 9 | 11 3/16 |
| 37 | 45-11 3/8 | 9 1/4 | 11 1/2 |
| 38 | 47- 2 1/4 | 9 1/2 | 11 13/16 |
| 39 | 48- 5 3/16 | 9 3/4 | 12 1/8 |
| 40 | 49- 8 1/16 | 10 | 12 7/16 |
| 41 | 50-11 | 10 1/4 | 12 3/4 |
| 42 | 52- 1 7/8 | 10 1/2 | 13 1/16 |
| 43 | 53- 4 3/4 | 10 3/4 | 13 3/8 |
| 44 | 54- 7 11/16 | 11 | 13 11/16 |
| 45 | 55-10 9/16 | 11 1/4 | 14 |
| 46 | 57- 1 1/2 | 11 1/2 | 14 1/4 |
| 47 | 58- 4 3/8 | 11 3/4 | 14 9/16 |
| 48 | 59- 7 5/16 | 12 | 14 7/8 |
| 49 | 60-10 3/16 | | |
| 50 | 62- 1 1/16 | | |

# 25 in 12 – 64 1/4°

## *Cuts and Saw Settings*

**Tail Cut, Hip or Valley Rafters,** with Square Cut Common Rafters:
Framing Square: **6 5/8 & 24**; set saw at **17 3/4°**
(Compound Miter or Radial Arm Saws: set bevel at 17 3/4°, miter at 15 1/2°)
Explanation: Section 3(d), pages 12–13.

**Chamfer Cut, Hip or Valley Rafters:**
Set saw at **39 1/2°**. Explanation: Section 3(d), page 10.

**Fascia Board Miter Cut at Hip or Valley:**
Framing Square: **10 13/16 & 12**; set saw at **17 3/4°**
(Compound Miter or Radial Arm Saws: set bevel at 17 3/4°, miter at 42°)
Explanation: Section 5, pages 24–25.

**Frieze Block against Hip or Valley Rafter:**
Framing Square: **10 13/16 & 12**; set saw at **17 3/4°**
(Compound Miter or Radial Arm Saws: set bevel at 17 3/4°, miter at 42°)
Explanation: Section 6, page 26.

**Roof Sheathing at Hip or Valley:**
Framing Square: **6 & 13 7/8**; set saw at **39 1/2°**
(Compound Miter or Radial Arm Saws: set bevel at 39 1/2°, miter at 23 1/2°)
Distance "d" = **20 3/4"**. Explanation: Section 7, pages 26–27.

## *Calculating the Length of a Rafter*

**General Rule:** Span of building
less Thickness of ridge board (if any)
gives Length of rafter to plumb line of seat cut

*and* Span of building
less Thickness of ridge board (if any)
plus 2 × projection of overhang
gives Total length of rafter to tail cut

Explanation: Section 4, pages 13–19.

| *Jack Rafters* | | | | | |
|---|---|---|---|---|---|
| SPACED *Inches* | DIFFERENCE *Inches* | SPACED *Inches* | DIFFERENCE *Inches* | SPACED *Inches* | DIFFERENCE *Inches* |
| 1/4 | 9/16 | 15 | 34 11/16 | **32** | **73 15/16** |
| 1/2 | 1 1/8 | **16** | **37** | 33 | 76 1/4 |
| 3/4 | 1 3/4 | 17 | 39 5/16 | 34 | 78 9/16 |
| 1 | 2 5/16 | 18 | 41 5/8 | 35 | 80 7/8 |
| 2 | 4 5/8 | 19 | 43 15/16 | **36** | **83 3/16** |
| 3 | 6 15/16 | 20 | 46 3/16 | 37 | 85 1/2 |
| 4 | 9 1/4 | 21 | 48 1/2 | 38 | 87 13/16 |
| 5 | 11 9/16 | 22 | 50 13/16 | 39 | 90 1/8 |
| 6 | 13 7/8 | 23 | 53 1/8 | 40 | 92 7/16 |
| 7 | 16 3/16 | **24** | **55 7/16** | 41 | 94 3/4 |
| 8 | 18 1/2 | 25 | 57 3/4 | 42 | 97 1/16 |
| 9 | 20 13/16 | 26 | 60 1/16 | 43 | 99 3/8 |
| 10 | 23 1/8 | 27 | 62 3/8 | 44 | 101 11/16 |
| 11 | 25 7/16 | 28 | 64 11/16 | 45 | 104 |
| **12** | **27 3/4** | 29 | 67 | 46 | 106 5/16 |
| 13 | 30 1/16 | 30 | 69 5/16 | 47 | 108 5/8 |
| 14 | 32 3/8 | 31 | 71 5/8 | **48** | **110 15/16** |

## Common Rafters
## 25 in 12 – 64 1/4°

| SPAN Feet | LENGTH Feet Inches | SPAN Inches | LENGTH Inches |
|---|---|---|---|
| 1 | 1- 1 7/8 | 1/4 | 5/16 |
| 2 | 2- 3 3/4 | 1/2 | 9/16 |
| 3 | 3- 5 5/8 | 3/4 | 7/8 |
| 4 | 4- 7 7/16 | 1 | 1 1/8 |
| 5 | 5- 9 5/16 | 1 1/4 | 1 7/16 |
| 6 | 6-11 3/16 | 1 1/2 | 1 3/4 |
| 7 | 8- 1 1/16 | 1 3/4 | 2 |
| 8 | 9- 2 15/16 | 2 | 2 5/16 |
| 9 | 10- 4 13/16 | 2 1/4 | 2 5/8 |
| 10 | 11- 6 5/8 | 2 1/2 | 2 7/8 |
| 11 | 12- 8 1/2 | 2 3/4 | 3 3/16 |
| 12 | 13-10 3/8 | 3 | 3 7/16 |
| 13 | 15- 0 1/4 | 3 1/4 | 3 3/4 |
| 14 | 16- 2 1/8 | 3 1/2 | 4 1/16 |
| 15 | 17- 4 | 3 3/4 | 4 5/16 |
| 16 | 18- 5 7/8 | 4 | 4 5/8 |
| 17 | 19- 7 11/16 | 4 1/4 | 4 15/16 |
| 18 | 20- 9 9/16 | 4 1/2 | 5 3/16 |
| 19 | 21-11 7/16 | 4 3/4 | 5 1/2 |
| 20 | 23- 1 5/16 | 5 | 5 3/4 |
| 21 | 24- 3 3/16 | 5 1/4 | 6 1/16 |
| 22 | 25- 5 1/16 | 5 1/2 | 6 3/8 |
| 23 | 26- 6 7/8 | 5 3/4 | 6 5/8 |
| 24 | 27- 8 3/4 | 6 | 6 15/16 |
| 25 | 28-10 5/8 | 6 1/4 | 7 1/4 |
| 26 | 30- 0 1/2 | 6 1/2 | 7 1/2 |
| 27 | 31- 2 3/8 | 6 3/4 | 7 13/16 |
| 28 | 32- 4 1/4 | 7 | 8 1/16 |
| 29 | 33- 6 1/8 | 7 1/4 | 8 3/8 |
| 30 | 34- 7 15/16 | 7 1/2 | 8 11/16 |
| 31 | 35- 9 13/16 | 7 3/4 | 8 15/16 |
| 32 | 36-11 11/16 | 8 | 9 1/4 |
| 33 | 38- 1 9/16 | 8 1/4 | 9 9/16 |
| 34 | 39- 3 7/16 | 8 1/2 | 9 13/16 |
| 35 | 40- 5 5/16 | 8 3/4 | 10 1/8 |
| 36 | 41- 7 1/8 | 9 | 10 3/8 |
| 37 | 42- 9 | 9 1/4 | 10 11/16 |
| 38 | 43-10 7/8 | 9 1/2 | 11 |
| 39 | 45- 0 3/4 | 9 3/4 | 11 1/4 |
| 40 | 46- 2 5/8 | 10 | 11 9/16 |
| 41 | 47- 4 1/2 | 10 1/4 | 11 13/16 |
| 42 | 48- 6 3/8 | 10 1/2 | 12 1/8 |
| 43 | 49- 8 3/16 | 10 3/4 | 12 7/16 |
| 44 | 50-10 1/16 | 11 | 12 11/16 |
| 45 | 51-11 15/16 | 11 1/4 | 13 |
| 46 | 53- 1 13/16 | 11 1/2 | 13 5/16 |
| 47 | 54- 3 11/16 | 11 3/4 | 13 9/16 |
| 48 | 55- 5 9/16 | 12 | 13 7/8 |
| 49 | 56- 7 3/8 | | |
| 50 | 57- 9 1/4 | | |

## Hip or Valley Rafters
## 25 in 17 – 55 3/4°

| SPAN Feet | LENGTH Feet Inches | SPAN Inches | LENGTH Inches |
|---|---|---|---|
| 1 | 1- 3 1/8 | 1/4 | 5/16 |
| 2 | 2- 6 3/16 | 1/2 | 5/8 |
| 3 | 3- 9 5/16 | 3/4 | 15/16 |
| 4 | 5- 0 7/16 | 1 | 1 1/4 |
| 5 | 6- 3 9/16 | 1 1/4 | 1 9/16 |
| 6 | 7- 6 5/8 | 1 1/2 | 1 7/8 |
| 7 | 8- 9 3/4 | 1 3/4 | 2 3/16 |
| 8 | 10- 0 7/8 | 2 | 2 1/2 |
| 9 | 11- 4 | 2 1/4 | 2 13/16 |
| 10 | 12- 7 1/16 | 2 1/2 | 3 1/8 |
| 11 | 13-10 3/16 | 2 3/4 | 3 7/16 |
| 12 | 15- 1 5/16 | 3 | 3 3/4 |
| 13 | 16- 4 3/8 | 3 1/4 | 4 1/16 |
| 14 | 17- 7 1/2 | 3 1/2 | 4 7/16 |
| 15 | 18-10 5/8 | 3 3/4 | 4 3/4 |
| 16 | 20- 1 3/4 | 4 | 5 1/16 |
| 17 | 21- 4 13/16 | 4 1/4 | 5 3/8 |
| 18 | 22- 7 15/16 | 4 1/2 | 5 11/16 |
| 19 | 23-11 1/16 | 4 3/4 | 6 |
| 20 | 25- 2 3/16 | 5 | 6 5/16 |
| 21 | 26- 5 1/4 | 5 1/4 | 6 5/8 |
| 22 | 27- 8 3/8 | 5 1/2 | 6 15/16 |
| 23 | 28-11 1/2 | 5 3/4 | 7 1/4 |
| 24 | 30- 2 9/16 | 6 | 7 9/16 |
| 25 | 31- 5 11/16 | 6 1/4 | 7 7/8 |
| 26 | 32- 8 13/16 | 6 1/2 | 8 3/16 |
| 27 | 33-11 15/16 | 6 3/4 | 8 1/2 |
| 28 | 35- 3 | 7 | 8 13/16 |
| 29 | 36- 6 1/8 | 7 1/4 | 9 1/8 |
| 30 | 37- 9 1/4 | 7 1/2 | 9 7/16 |
| 31 | 39- 0 3/8 | 7 3/4 | 9 3/4 |
| 32 | 40- 3 7/16 | 8 | 10 1/16 |
| 33 | 41- 6 9/16 | 8 1/4 | 10 3/8 |
| 34 | 42- 9 11/16 | 8 1/2 | 10 11/16 |
| 35 | 44- 0 3/4 | 8 3/4 | 11 |
| 36 | 45- 3 7/8 | 9 | 11 5/16 |
| 37 | 46- 7 | 9 1/4 | 11 5/8 |
| 38 | 47-10 1/8 | 9 1/2 | 11 15/16 |
| 39 | 49- 1 3/16 | 9 3/4 | 12 1/4 |
| 40 | 50- 4 5/16 | 10 | 12 9/16 |
| 41 | 51- 7 7/16 | 10 1/4 | 12 7/8 |
| 42 | 52-10 9/16 | 10 1/2 | 13 1/4 |
| 43 | 54- 1 5/8 | 10 3/4 | 13 9/16 |
| 44 | 55- 4 3/4 | 11 | 13 7/8 |
| 45 | 56- 7 7/8 | 11 1/4 | 14 3/16 |
| 46 | 57-10 15/16 | 11 1/2 | 14 1/2 |
| 47 | 59- 2 1/16 | 11 3/4 | 14 13/16 |
| 48 | 60- 5 3/16 | 12 | 15 1/8 |
| 49 | 61- 8 5/16 | | |
| 50 | 62-11 3/8 | | |